Advances in Wood Based Composites

Advances in Wood Based Composites

Guest Editors

Ľuboš Krišťák
Roman Reh
Marius Cătălin Barbu
Eugenia Mariana Tudor

MDPI

Basel • Beijing • Wuhan • Barcelona • Belgrade • Novi Sad • Cluj • Manchester

Guest Editors
Ľuboš Krišťák
Department of Physics, Electrical Engineering and Applied Mechanics
Technical University in Zvolen
Zvolen
Slovakia

Roman Reh
Department of Wood Technology
Technical University in Zvolen
Zvolen
Slovakia

Marius Cătălin Barbu
Design and Green Engineering Department
Salzburg University of Applied Sciences
Kuchl
Austria

Eugenia Mariana Tudor
Design and Green Engineering Department
Salzburg University of Applied Sciences
Kuchl
Austria

Editorial Office
MDPI AG
Grosspeteranlage 5
4052 Basel, Switzerland

This is a reprint of the Special Issue, published open access by the journal *Polymers* (ISSN 2073-4360), freely accessible at: www.mdpi.com/journal/polymers/special_issues/S54YU995MS.

For citation purposes, cite each article independently as indicated on the article page online and using the guide below:

Lastname, A.A.; Lastname, B.B. Article Title. *Journal Name* **Year**, *Volume Number*, Page Range.

ISBN 978-3-7258-4012-0 (Hbk)
ISBN 978-3-7258-4011-3 (PDF)
https://doi.org/10.3390/books978-3-7258-4011-3

Contents

polymers

MDPI

Editorial

Advances in Wood-Based Composites

Lubos Kristak [1,*], Roman Reh [1], Marius Catalin Barbu [2,3] and Eugenia Mariana Tudor [2,3]

1 Faculty of Wood Sciences and Technology, Technical University in Zvolen, T. G. Masaryka 24, SK-960 01 Zvolen, Slovakia; reh@tuzvo.sk
2 Forest Products Technology and Timber Construction Department, Salzburg University of Applied Sciences, Markt 136 a, 5431 Kuchl, Austria; marius.barbu@fh-salzburg.ac.at (M.C.B.); eugenia.tudor@fh-salzburg.ac.at (E.M.T.)
3 Faculty of Furniture Design and Wood Engineering, Transylvania University of Brasov, B-dul. Eroilor nr. 29, 500036 Brasov, Romania
* Correspondence: kristak@tuzvo.sk

1. Introduction

The significance of wood-based composites has grown substantially in recent years due to their enhanced material efficiency, sustainability, and versatile applications. These composites are engineered by integrating lignocellulosic wood elements, such as fibers, particles, or veneers, with thermoplastic or thermosetting polymeric adhesives through various manufacturing processes, including hot pressing, extrusion, or injection molding. Recent advancements in wood-based composites have focused on multiple facets, including processing techniques, product innovations, surface and color treatments, and the examination of key properties. Furthermore, bio-based layered hybrid composites are being researched to provide sustainable alternatives in structural applications, alongside a focus on diversifying the sourcing of raw materials. To optimize performance characteristics, specific additives such as hardeners, coupling agents, plasticizers, and fillers are incorporated into the composite formulation. These additives play a critical role in improving interfacial adhesion, mechanical strength, dimensional stability, and resistance to environmental degradation, thereby enhancing the overall durability and functionality of the material. Additionally, energy efficiency is critical; therefore, improved manufacturing effectiveness and sustainability are targeted. Finally, significant insights have been gained into the acoustic properties of wood through the analysis of sound propagation velocity, combined with impacting design choices. In addition, the evaluation of cellulose macromolecular properties after chemical treatment helps us understand the changes that affect the performance of wood composites, indicating a comprehensive approach to enhancing wood-based materials.

To optimize the structural integrity and performance of wood-based composites, specific additives are incorporated into their formulation. These additives include hardeners, coupling agents, plasticizers, flame retardants, fungicides, and fillers, each serving unique roles in modifying the physical and chemical properties of the polymer matrix [1–4]. Hardeners improve curing efficiency, leading to superior bonding strength, while coupling agents enhance interfacial adhesion between hydrophilic wood components and hydrophobic polymer matrices. Plasticizers increase flexibility, reducing brittleness, whereas fillers such as calcium carbonate, talc, or nanocellulose enhance dimensional stability and mechanical resilience. Additionally, flame retardants and fungicides improve fire resistance and biological durability, respectively, extending the material's service life in demanding environments [5–9]. Wood-based composites have gained widespread acceptance in both structural and non-structural applications, making them integral

check for updates

Received: 1 April 2025
Accepted: 11 April 2025
Published: 18 April 2025

Citation: Kristak, L.; Reh, R.; Barbu, M.C.; Tudor, E.M. Advances in Wood-Based Composites. *Polymers* **2025**, *17*, 1104. https://doi.org/10.3390/polym17081104

to various industries. In the construction sector, they are used for load-bearing and non-load-bearing elements such as beams, wall panels, sheathing, and formwork due to their high strength-to-weight ratio and resistance to warping. In interior applications, they are commonly utilized in furniture manufacturing, cabinetry, flooring, decorative panels, and moldings, offering aesthetic appeal combined with functional benefits [10–14]. Exterior applications include decking, fencing, cladding, and doors, where enhanced weather resistance and UV stability are critical factors. The automotive and aerospace industries also incorporate wood–plastic composites in lightweight, durable components to improve fuel efficiency and reduce environmental impact [15]. Advancements in wood-based composite technology continue to drive improvements in mechanical performance, moisture resistance, biodegradability, and recyclability. Emerging trends focus on the development of bio-based and eco-friendly adhesives derived from renewable resources, reducing reliance on synthetic petrochemical-based binders. Additionally, nanotechnology and fiber modification techniques are being explored to further enhance composite strength, thermal stability, and fire resistance. With increasing emphasis on sustainable materials and circular economic principles, wood-based composites are expected to play a crucial role in reducing deforestation, minimizing waste, and promoting green construction practices in the coming decades [16–18]. One of the ways to reduce deforestation is to use alternative raw materials from which composites can be made [19–25]. In the case of particleboards, a promising method is the use of alternative, lesser-known wood species in particleboard production.

2. An Overview of Published Articles

Reh et al. (contribution 1) analyzed perspectives on using alder, larch, and birch wood species to maintain the increasing particleboard production flow. These three wood species represent an eco-friendly and sustainable wood alternative to the conventional wood raw materials used. This review confirms the diversity of the use of these three species in different fields and proves their suitability for particleboard production.

Wood-based composites have gained significant attention in the global market, necessitating a deeper understanding of their physical and mechanical behavior, particularly in relation to adhesive polymerization. Silva et al. (contribution 2) investigated the effects of incorporating aluminum oxide and aluminum oxide copper nanoparticles into a urea-formaldehyde-based polymeric adhesive using an environmentally friendly approach. The manufactured *Eucalyptus urophylla* var. grandis wood composites were analyzed for their resin properties such as viscosity, gel time, and pH, as well as panel characteristics including density, moisture content, thickness swelling, modulus of elasticity, modulus of rupture, and thermal conductivity. The results showed that nanoparticle addition influenced viscosity, while all treatments exhibited a basic pH. However, gel time could not be determined after 10 min. No significant improvement was observed in swelling resistance, density, or moisture content. An increase in pressing temperature from 150 °C to 180 °C positively affected all physical and mechanical properties, suggesting enhanced polymerization of the adhesive. Overall, the incorporation of 0.5% nanoparticles had a limited effect on improving the physical–mechanical performance of the particleboards.

The modification of urea-formaldehyde adhesives involves altering its chemical composition or incorporating additives to enhance its performance in terms of bonding strength, durability, moisture resistance, and environmental impact, the last of which was the goal of the research by Reh et al. (contribution 3). The potential of using ground beech (*Fagus sylvatica* L.) bark as an eco-friendly additive in UF adhesives for molded plywood manufacturing was investigated. Nine-layered, flat, and molded plywood was produced under industrial conditions from beech veneers bonded with a UF adhesive mixture. The positive effect of beech bark in the UF adhesive mixture on a reduction in formaldehyde emissions from the molded plywood was also confirmed. Beech bark, considered to be wood-processing industry waste or a byproduct, has significant potential to be used as a filler in UF resins for molded plywood production, providing an environmentally friendly, inexpensive solution for the industrial valorization of bark as a bio-based formaldehyde scavenger.

The ongoing advancements in science and technology are driving a growing demand for environmentally friendly products derived from natural sources, as well as the enhanced reutilization of forestry and agricultural byproducts, which are often regarded as waste. The use of polymer blends to bond fibers for wood-based products seems an attractive substituent because of the low cost and suitability for a wide range of applications. Natural fiber-reinforced polymers are increasingly replacing synthetic fiber-reinforced plastics across various industrial sectors, including automotive manufacturing, packaging, and furniture production. These bio-based composites offer reduced weight and enhanced thermal properties, contributing to improved sustainability and performance. Gumowska et al. (contribution 4) elaborated on layered composites produced with different biopolymer adhesive layers, including biopolymer polylactic acid, polycaprolactone, and biopolymer blends of PLA + polyhydroxybutyrate with the addition of microcrystalline cellulose and triethyl citrate for these blends, which acted as binders and co-created the five layers in the elaborated composites. The modulus of rupture, modulus of elasticity, internal bonding strength, and density profile were obtained, and differential scanning calorimetry, thermogravimetric analysis, and scanning electron microscopy analysis were performed. The results showed that among the composites in which two pure biopolymers were used, PLA obtained the best results. The results of this study demonstrate the feasibility of producing layered wood-based composites using various biopolymers and their blends as specialized property layers and binders. This approach enables the development of formaldehyde-free wood-based composites that enhance the inherent properties of both wood and biopolymers, offering improved performance and sustainability.

The physicochemical and mechanical properties of wood-based materials can be significantly improved through impregnation, a process in which functional substances penetrate the wood's cellular structure, modifying its intrinsic characteristics. Impregnation involves deep penetration of resins, chemicals, or nanoparticles into the wood structures. Rahayu et al. (contribution 5) used TiO_2 nanoparticle impregnation and analyzed their effect on the density and dimensional stability of mangium wood and the effectiveness of the presence of the impregnation in wood in degrading pollutants. The samples were analyzed for density, weight percent gain (WPG), and bulking effect (BE). The samples were also analyzed via X-ray diffraction (XRD) and Fourier-transform infrared spectroscopy (FTIR). TiO_2 nanoparticles resulted in an increase in density, WPG, and BE-treated mangium. Based on the XRD and FTIR results, the TiO_2 nanoparticle was successfully impregnated into mangium wood. A scanning electron microscopy–energy-dispersive X-ray spectroscopy analysis indicated that TiO_2 nanoparticles covered the surface of the wood cells. The TiO_2-impregnated mangium wood has a higher photocatalyst activity than the untreated wood, indicating better protection from UV radiation and pollutants.

Wood preservation involves various methods to extend its lifespan and protect it from decay, insects, mold, and weathering. The effect of using selected inorganic chemicals as the main components of waterborne wood preservative systems on the degradation of the cellulose constituent in wood from model samples was examined by Jurczykova et al. (contribution 6). Whatman papers, as pure cellulose model samples, were impregnated with 10 different 5 wt% solutions of inorganic salts and distilled water and consequently subjected to wet-thermal accelerated aging. The samples were then derivatized to cellulose tricarbanilates through two different procedures (by precipitation in a methanol–water mixture/by evaporation of pyridine from the reaction mixture) and finally analyzed using size exclusion chromatography. Chemically treated and aged cellulose samples showed different changes in their degree of polymerization and polydispersity in terms of untreated non-aged standards caused by different ongoing degradation reactions, such as dehydration, hydrolysis, oxidation, and cross-linking. This paper brings new insights regarding the complex evaluation of the polymeric properties of degraded cellulose by considering all important factors affecting the sample and the analysis itself through the use of statistics.

Thermal modification is another environmentally friendly process that enhances the durability, stability, and resistance of wood by exposing it to elevated temperatures, mostly in the absence of oxygen. This treatment alters the wood's chemical structure, reducing its hygroscopicity and improving its resistance to biological degradation. Jurczkyova et al. (contribution 7) examined the effect of thermal modification temperature on the selected optical properties of six tropical wood species—Sp. cedar (*Cedrala odorata*), iroko (*Chlorophora excelsa*), merbau (*Intsia* spp.), meranti (*Shorea* spp.), padouk (*Pterocarpus soyauxii*), and teak (*Tectona grandis*). The CIELAB color space parameters, yellowness, ISO brightness, and UV-Vis diffuse reflectance spectra were obtained. Subsequently, these wood samples were extracted into three individual solvents (acetone, ethanol, and ethanol-toluene). The yields of the extracted compounds; their absorption spectra; and again, their L*, a*, b*, DE*, and Yi parameters were determined. The results showed that thermal modification above 200 °C causes tropical wood to lose its characteristic color, which contributes to the unique aesthetic appeal of individual species. Although the process results in a more homogeneous appearance, it significantly reduces the wood's brightness and yellowness. At temperatures above 220 °C, color differences become negligible, as confirmed by a cluster analysis within the PCA (Principal Component Analysis). This uniformity may not always be desirable, as some consumers select wood primarily for its decorative properties. Ultimately, the suitability of thermally modified wood depends on its intended application and whether color consistency is a priority or if other advantages of ThermoWood, such as enhanced durability and stability, take precedence.

Fine art coatings are typically formulated by combining metal fillers with water-based coatings, which are then applied to the surfaces of wood structures, furniture, and crafts for decorative purposes. However, their durability is often constrained by limited mechanical strength. In contrast, the dispersion of metal fillers and the mechanical performance of the coating can be significantly enhanced through the use of coupling agents, which facilitate strong interfacial bonding between the resin matrix and the metal filler. Han and Yan (contribution 8) examined the effect of silane coupling agent modification and its influence on a brass powder–water-based acrylic coating. Three different silane coupling agents, 3-aminopropyltriethoxysilane, -(2,3-epoxypropoxy)propytrimethoxysilane, and -methacryloxypropyltrimethoxysilane, were used to modify the brass powder filler in orthogonal tests. The artistic effect and optical properties of the modified art coating induced by different proportions of brass powder, silane coupling agents, and pH were compared. The results demonstrated that the amount of brass powder and the kind of coupling agent used had a substantial impact on the coating's optical characteristics. These results provide a technological foundation for the development of artistic coatings for wood.

The influence of coating treatments on sound propagation speeds in thin boards made of spruce and maple resonance wood was examined by Faktorova et al. (contribution 9). Wood samples, both unvarnished and coated with oil-based or alcohol varnish, were exposed to UV radiation and saline fog. Lamb wave propagation was analyzed using a semicircular test model to assess acoustic responses in the longitudinal and radial directions. A statistical analysis revealed that oil-based varnish reduced propagation speed mainly in the radial direction, while alcohol varnish had a greater effect longitudinally in spruce wood. In maple wood, increasing varnish layers decreased the anisotropy ratio between the longitudinal and radial directions, regardless of varnish type. Research showed that surface treatments applied to wooden resonance plates affect sound propagation speeds in both the longitudinal and radial directions. The acoustic behavior of the wood is influenced by both the type of treatment and the wood's anatomical characteristics, with variations in propagation speeds observed even among samples of the same species due to differences in their anatomical structure.

In furniture production, efforts are being made to lower energy consumption while adopting green practices. These include using sustainable materials, optimizing manufacturing processes, and implementing energy-efficient technologies. By reducing energy use and minimizing environmental impact, the industry can produce eco-friendly furniture that meets both consumer demands and sustainability goals. Pakula et al. (contribution 10) analyzed the support of eco-design ideas and sustainable manufacturing techniques by examining the energy consumption related to drilling holes for various furniture connections such as eccentric joints, confirmat screws, and dowels. The energy consumption was measured using a portable power quality analyzer. The measurement process involved recording energy consumption at different stages of the machining process, allowing for an analysis of specific cutting work and total energy consumption for various joint types. The results indicate that connecting furniture with dowels consumes the least energy but is the least user-friendly, as it does not allow for disassembly, reassembly, or easy transport. While eccentric joints consume more energy than dowels, they offer greater convenience for transportation and easier assembly. However, each time the furniture is reassembled, it loses strength due to material tearing caused by the removal of the confirmat screw. The highest energy consumption for drilling holes is associated with the eccentric joint, which, despite requiring more energy and larger holes, provides the best durability, ease of assembly, and ability to disassemble and transport the furniture without compromising its strength. Ultimately, it can be concluded that enhancing the usability of furniture increases the use of technology and, consequently, energy consumption. This study offers valuable insights into energy consumption associated with different furniture joints, providing critical information for eco-design and sustainable manufacturing practices. The analysis highlights the varying energy demands of each joint type, which can inform the development of more energy-efficient and environmentally friendly methods for furniture production.

3. Conclusions

The collective contributions in this compilation reflect a comprehensive and forward-looking exploration of sustainable practices and innovative technologies in the field of wood-based materials and products. Several studies (Reh et al. and Silva et al.) focus on alternative raw materials and eco-friendly adhesive formulations, highlighting the potential of underutilized wood species and nanoparticle-enhanced resins to improve performance while reducing environmental impact. Complementary research (Reh et al., Gumowska et al.) advances the use of biopolymers, natural fillers, and bio-based additives, aiming to minimize formaldehyde emissions and develop fully biodegradable, formaldehyde-free composites.

The enhancement of material properties through physical and chemical modifications is another recurring theme. The contributions by Rahayu et al. and Jurczykova et al. demonstrate how impregnation and inorganic chemical treatments can improve wood's durability, dimensional stability, and pollutant resistance, while thermal modification (Jurczykova et al.) offers improved structural properties at the expense of aesthetic value.

Specialized applications, such as artistic coatings (Han and Yan) and acoustic performance (Faktorová et al.), reveal the nuanced influence of surface treatments and additives on the visual, tactile, and auditory properties of wood, expanding its functional and creative potential. Finally, Pakula et al. contribute practical insights into the energy demands of furniture manufacturing, balancing environmental responsibility with user convenience and mechanical durability.

The articles collectively emphasize a growing commitment to sustainability, innovation, and efficiency within the wood-based materials industry. Across all contributions, there is a clear shift toward eco-friendly practices, whether through the use of alternative wood species, bio-based adhesives, or advanced material modifications.

Researchers are increasingly exploring ways to reduce environmental impact, enhance performance, and utilize natural or waste-derived materials more effectively. This includes the integration of nanoparticles, biopolymers, and plant-based additives, as well as process improvements such as thermal treatments and impregnation.

Another common theme is the importance of balancing functionality, aesthetics, and sustainability. Whether optimizing acoustic performance, durability, or manufacturing energy use, the studies demonstrate that thoughtful material selection and processing techniques can lead to better, greener products.

Overall, these contributions reflect a dynamic and evolving field, where science and technology are being applied to meet modern demands for environmentally responsible, high-performing wood-based materials across various applications—from construction and furniture to fine arts and acoustics.

Author Contributions: All authors contributed equally to this work. All authors have read and agreed to the published version of the manuscript.

Funding: This research was supported by the Slovak Research and Development Agency under contracts No. SK-CZ-RD-21-0100 and APVV-19-0269, and by the Ministry of Education, Science, Research and Sport of the Slovak Republic under the project VEGA 1/0077/24.

Conflicts of Interest: The authors declare no conflicts of interest.

List of Contributions

1. Reh, R.; Kristak, L.; Kral, P.; Pipiska, T.; Jopek, M. Perspectives on Using Alder, Larch, and Birch Wood Species to Maintain the Increasing Particleboard Production Flow. *Polymers* **2024**, *16*, 1532. https://doi.org/10.3390/polym16111532.
2. Silva, L.C.L.; Lima, F.O.; De Araujo, V.A.; Dos Santos, H.F.; Lahr, F.A.R.; Christoforo, A.L.; Favarim, H.R.; de Campos, C.I. Influence of Pressing Temperatures on Physical–Mechanical Properties of Wood Particleboards Made with Urea-Formaldehyde Adhesive Containing Al_2O_3 and CuO Nanoparticles. *Polymers* **2024**, *16*, 1652. https://doi.org/10.3390/polym16121652.
3. Reh, R.; Kristak, L.; Sedliacik, J.; Bekhta, P.; Wronka, A.; Kowaluk, G. Molded Plywood with Proportions of Beech Bark in Adhesive Mixtures: Production on an Industrial Scale. *Polymers* **2024**, *16*, 966. https://doi.org/10.3390/polym16070966.
4. Gumowska, A.; Robles, E.; Bikoro, A.; Wronka, A.; Kowaluk, G. Selected Properties of Bio-Based Layered Hybrid Composites with Biopolymer Blends for Structural Applications. *Polymers* **2022**, *14*, 4393. https://doi.org/10.3390/polym14204393.

5. Rahayu, I.; Darmawan, W.; Nawawi, D.S.; Prihatini, E.; Ismail, R.; Laksono, G.D. Physical Properties of Fast-Growing Wood-Polymer Nano Composite Synthesized through TiO_2 Nanoparticle Impregnation. *Polymers* **2022**, *14*, 4463. https://doi.org/10.3390/polym14204463.
6. Jurczyková, T.; Kačík, F.; Čabalová, I.; Hájková, K. Evaluation of Selected Cellulose Macromolecular Properties after Its Chemical Treatment Using Size Exclusion Chromatography. *Polymers* **2023**, *15*, 573. https://doi.org/10.3390/polym15030573.
7. Jurczyková, T.; Šárovec, O.; Kačík, F.; Hájková, K.; Jurczyk, T.; Hrčka, R. Chromophores' Contribution to Color Changes of Thermally Modified Tropical Wood Species. *Polymers* **2023**, *15*, 4000. https://doi.org/10.3390/polym15194000.
8. Han, Y.; Yan, X. Effect of Silane Coupling Agent Modification on Properties of Brass Powder-Water-Based Acrylic Coating on Tilia europaea. *Polymers* **2023**, *15*, 1396. https://doi.org/10.3390/polym15061396.
9. Faktorová, D.; Stanciu, M.D.; Krbata, M.; Savin, A.; Kohutiar, M.; Chlada, M.; Năstac, S.M. Analysis of the Anisotropy of Sound Propagation Velocity in Thin Wooden Plates Using Lamb Waves. *Polymers* **2024**, *16*, 753. https://doi.org/10.3390/polym16060753.
10. Pakuła, W.; Prałat, B.; Potok, Z.; Wiaderek, K.; Rogoziński, T. Energy Consumption for Furniture Joints during Drilling in Birch Plywood. *Polymers* **2024**, *16*, 1045. https://doi.org/10.3390/polym16081045.

References

1. Kawalerczyk, J.; Dziurka, D.; Mirski, R.; Trociński, A. Flour fillers with urea-formaldehyde resin in plywood. *BioResources* **2019**, *14*, 6727–6735. [CrossRef]
2. Walkiewicz, J.; Kawalerczyk, J.; Mirski, R.; Dziurka, D.; Wieruszewski, M. The application of various bark species as a fillers for UF resin in plywood manufacturing. *Materials* **2022**, *15*, 7201. [CrossRef] [PubMed]
3. Mirski, R.; Derkowski, A.; Kawalerczyk, J.; Dziurka, D.; Walkiewicz, J. The possibility of using pine bark particles in the chipboard manufacturing process. *Materials* **2022**, *15*, 5731. [CrossRef]
4. Antov, P.; Savov, V.; Mantanis, G.I.; Neykov, N. Medium-density fibreboards bonded with phenol-formaldehyde resin and calcium lignosulfonate as an eco-friendly additive. *Wood Mater. Sci. Eng.* **2021**, *16*, 42–48. [CrossRef]
5. Zanuttini, R.; Negro, F. Wood-based composites: Innovation towards a sustainable future. *Forests* **2021**, *12*, 1717. [CrossRef]
6. Beims, R.F.; Arredondo, R.; Carrero, D.J.S.; Yuan, Z.; Li, H.; Shui, H.; Zhang, Y.; Leitch, M.; Xu, C.C. Functionalized wood as bio-based advanced materials: Properties, applications, and challenges. *Renew. Sustain. Energy Rev.* **2022**, *157*, 112074. [CrossRef]
7. Nguyen, D.L.; Luedtke, J.; Nopens, M.; Krause, A. Production of wood-based panel from recycled wood resource: A literature review. *Eur. J. Wood Wood Prod.* **2023**, *81*, 557–570. [CrossRef]
8. Madyaratri, E.W.; Ridho, M.R.; Aristri, M.A.; Lubis, M.A.R.; Iswanto, A.H.; Nawawi, D.S.; Antov, P.; Kristak, L.; Majlingová, A.; Fatriasari, W. Recent advances in the development of fire-resistant biocomposites—A review. *Polymers* **2022**, *14*, 362. [CrossRef]
9. Lee, S.H.; Lum, W.C.; Boon, J.G.; Kristak, L.; Antov, P.; Pędzik, M.; Rogoziński, T.; Taghiyari, H.R.; Lubis, M.A.R.; Fatriasari, W. Particleboard from agricultural biomass and recycled wood waste: A review. *J. Mater. Res. Technol.* **2022**, *20*, 4630–4658. [CrossRef]
10. Wei, P.X.; Chen, J.X.; Zhang, Y.; Pu, L.J. Wood-based sandwich panels: A review. *Wood Res.* **2021**, *66*, 875–890. [CrossRef]
11. Sutkowska, M.; Stefańska, A.; Vaverkova, M.D.; Dixit, S.; Thakur, A. Recent advances in prefabrication techniques for biobased materials towards a low-carbon future: From modules to sustainability. *J. Build. Eng.* **2024**, *91*, 109558. [CrossRef]
12. Ramesh, M.; Rajeshkumar, L.; Sasikala, G.; Balaji, D.; Saravanakumar, A.; Bhuvaneswari, V.; Bhoopathi, R. A critical review on wood-based polymer composites: Processing, properties, and prospects. *Polymers* **2022**, *14*, 589. [CrossRef] [PubMed]
13. Krapež Tomec, D.; Kariž, M. Use of wood in additive manufacturing: Review and future prospects. *Polymers* **2022**, *14*, 1174. [CrossRef] [PubMed]
14. Ratnasingam, J. *Furniture Manufacturing*; Springer: Singapore, 2022.
15. Hussin, M.H.; Abd Latif, N.H.; Hamidon, T.S.; Idris, N.N.; Hashim, R.; Appaturi, J.N.; Brosse, N.; Ziegler-Devin, I.; Chrusiel, L.; Fatriasari, W. Latest advancements in high-performance bio-based wood adhesives: A critical review. *J. Mater. Res. Technol.* **2022**, *21*, 3909–3946. [CrossRef]
16. Dunky, M. Wood Adhesives Based on Natural Resources: A Critical Review: Part I. Protein-Based Adhesives. *Prog. Adhes. Adhes.* **2021**, *6*, 203–336.
17. Kristak, L.; Antov, P.; Bekhta, P.; Lubis, M.A.R.; Iswanto, A.H.; Reh, R.; Sedliacik, J.; Savov, V.; Taghiyari, H.R.; Papadopoulos, A.N. Recent progress in ultra-low formaldehyde emitting adhesive systems and formaldehyde scavengers in wood-based panels: A review. *Wood Mater. Sci. Eng.* **2023**, *18*, 763–782. [CrossRef]
18. Calvano, S.; Negro, F.; Blanc, S.; Bruzzese, S.; Brun, F.; Zanuttini, R. Adding Value to Wood-Based Products: A Systematic Literature Review on Drivers. *Forests* **2025**, *16*, 590. [CrossRef]

19. Mirski, R.; Kawalerczyk, J.; Dziurka, D.; Wieruszewski, M.; Trociński, A. Effects of using bark particles with various dimensions as a filler for urea-formaldehyde resin in plywood. *BioResources* **2020**, *15*, 1692. [CrossRef]
20. Kidalova, L.; Stevulova, N.; Terpakova, E.; Sicakova, A. Utilization of alternative materials in lightweight composites. *J. Clean. Prod.* **2012**, *34*, 116–119. [CrossRef]
21. Podlena, M.; Böhm, M.; Saloni, D.; Velarde, G.; Salas, C. Tuning the adhesive properties of soy protein wood adhesives with different coadjutant polymers, nanocellulose and lignin. *Polymers* **2021**, *13*, 1972. [CrossRef]
22. Pędzik, M.; Janiszewska, D.; Rogoziński, T. Alternative lignocellulosic raw materials in particleboard production: A review. *Ind. Crops Prod.* **2021**, *174*, 114162. [CrossRef]
23. Hýsek, Š.; Wimmer, R.; Böhm, M. Optimal processing of flax and hemp fibre nonwovens. *BioResources* **2016**, *11*, 8522–8534. [CrossRef]
24. Böhm, M.; Salem, M.Z.; Srba, J. Formaldehyde emission monitoring from a variety of solid wood, plywood, blockboard and flooring products manufactured for building and furnishing materials. *J. Hazard. Mater.* **2012**, *221*, 68–79. [CrossRef] [PubMed]
25. Dukarska, D.; Pędzik, M.; Rogozińska, W.; Rogoziński, T.; Czarnecki, R. Characteristics of straw particles of selected grain species purposed for the production of lignocellulose particleboards. *Part. Sci. Technol.* **2021**, *39*, 213–222. [CrossRef]

polymers MDPI

Review

Perspectives on Using Alder, Larch, and Birch Wood Species to Maintain the Increasing Particleboard Production Flow

Roman Reh [1], Lubos Kristak [1,*], Pavel Kral [2], Tomas Pipiska [2] and Miroslav Jopek [3]

1 Faculty of Wood Science and Technology, Technical University in Zvolen, T.G. Masaryka 24, 960 01 Zvolen, Slovakia; reh@tuzvo.sk
2 Faculty of Forestry and Wood Technology, Mendel University in Brno, Zemedelska 1665/1, 613 00 Brno, Czech Republic; kral@mendelu.cz (P.K.); tpipiska@gmail.com (T.P.)
3 Faculty of Mechanical Engineering, Brno University of Technology, Technicka 2, 616 69 Brno, Czech Republic; jopek@fme.vutbr.cz
* Correspondence: kristak@tuzvo.sk

Abstract: Particleboard, engineered wood products as part of a large family of wood composite materials, developed in use mainly in the 1950s and 1960s to utilize inferior wood and wood waste when good-quality wood was in short supply; the annual production capacity worldwide is over 100 million m^3. It is also necessary to have a lot of wood raw material for its production, although raw material resources are limited on our planet. In addition to the main wood species, it is therefore possible to think about the wider use of alternative, lesser-known European species of alder, larch, and birch in particleboard production. These three wood species represent an eco-friendly and sustainable wood alternative to the conventional wood raw materials used. This review confirms the diversity of the use of these three species in different fields and proves their suitability in relation to particleboard production. Fundamental research is ongoing in certain universities to determine the proportional shares of use of these tree species in particleboard (in a certain weight proportion in their core layers) for the purpose of formulating the correct technology shares and rules for their application in the wood-based panel industry.

Keywords: wood-based panels; raw materials; lesser-known European species; alder; larch; birch; particleboard

check for updates

Citation: Reh, R.; Kristak, L.; Kral, P.; Pipiska, T.; Jopek, M. Perspectives on Using Alder, Larch, and Birch Wood Species to Maintain the Increasing Particleboard Production Flow. *Polymers* **2024**, *16*, 1532. https://doi.org/10.3390/polym16111532

Academic Editor: Hiroshi Yoshihara

Received: 11 March 2024
Revised: 23 May 2024
Accepted: 27 May 2024
Published: 29 May 2024

1. Introduction

Particleboard, engineered wood products as part of a large family of wood composite materials, dates from the early 20th century and developed in use mainly in the 1950s and 1960s to utilize inferior wood and wood waste when good-quality wood was in short supply. The annual particleboard production capacity worldwide is over 100 million m^3, and it is expected to reach 122 million m^3 by 2029. In Europe, almost 30.0 million m^3 of particleboard are produced each year, mainly for furniture applications (another particleboard volume is manufactured for construction purposes, but these waterproof particleboard are not the subject of this paper) [1–5].

Particleboard type P2 is suitable for dry areas. The triple-layer structure of the standard product consists of a robust core layer and two fine surface layers in order to allow clean usage. Particleboard type P2 is mainly manufactured for surface coatings. The particleboard production and its use are progressing. Due to the impact of COVID-19, several restrictions have been imposed by the world's governments, and thus, construction and furniture activities experienced a modest downturn in 2020–2021. The conditions gradually improved after 2021, restoring the particleboard market's growth trajectory. We are certainly aware that the significant growth in the construction of homes on account of rapid urbanization and the rising global population should represent one of the key factors fueling the particleboard market growth in following years unless there are unexpected changes, such as COVID-19 was [1–5].

Since particleboard is produced in a large volume worldwide, it is also necessary to have a lot of wood raw material from which the particleboard is made. Raw material resources are limited on our planet. Although wood raw material is renewable, its reserves are still not as high as would be desirable and there is also competition from other sectors (pulp and paper industry, combustion, and energy production) [6–9].

The question of what to produce such a large number of particleboards from is not underestimated, and worldwide research and manufacturing plants have been exploring it for several decades. Lignocellulose raw materials that do not come from trees are used, recycling of used wood has progressed significantly, and non-wood raw materials are tested. The density of particleboard is purposefully reduced by manufacturers. The lower density of particleboard means less wood raw material is needed. The lower density of particleboard may be partly to the detriment of its properties, so a fundamental reduction in density is out of the question [10–15].

Therefore, a question arises as to whether it is preferable to make more use of real wood for the production of particleboard, namely wood species that are not used at all or only used to a very limited extent and only locally for the production of particleboard. It would entail wood species of less or not at all used wood species to make a particleboard. It is the main reason for compiling this review paper [1,12].

Global research is trying to solve the possibilities of processing lesser-known and lesser-used forest trees for their use. In particleboard production, these attempts to process and correctly use lesser-known trees are limited, unfinished, and often completely absent [1,15].

It may be a matter of debate as to what are lesser-known or lesser-used trees in terms of a particular location or a certain period in terms of the available stocks of such trees in particular forests [10,16]. We will not address the representation of tree species in the forests of the world in this paper, that is, let us say, a proposal for some other paper for the authors' collective, including forestry experts. However, some facts about the stocks of these trees in the region of Central Europe, which is mainly of concern, are obvious.

Alders are represented in the forest stands of Slovakia in the range of 1.5–2% of the total number of trees [17]. In the forests of the Czech Republic, black alder (*Alnus glutinosa* (L.) Gaertn.) is mainly represented, and overall, it is estimated that all alders can be up to 2% of the total area of forest stands [18]. The representation of alders in Austria is in the range of 2.5–3% of the total number of trees in forest stands. Even in Germany, the most widespread species is black alder and the representation of alder in the forest stands of Germany is around 3% of the total share of trees. In Poland, alders have a more significant presence, not only black alder but also gray alder (*Alnus incana* (L.) Moench). In Polish forests, alders represent about 5% of the total forest area, which makes them an important deciduous species in this country [19]. In all the mentioned countries, the presence of alder is important for biodiversity and soil stabilization in humid areas.

Data on the processed volumes of alder wood vary depending on the annual production, demand, and forestry policy of individual countries. In Slovakia, alders are not considered economically important trees, and their annual processing is around 18,000–20,000 m^3. They are mainly used for the production of furniture and small handicraft products. In the Czech Republic, the annual processing of alder is around 35,000–38,000 m^3. Their wood is mainly used for the production of plywood, furniture, and cladding. The annual processing of alder in Austria is around 14,000–15,000 m^3. They are used for smaller wood products and cladding. In Germany, the annual processing of alder is estimated at approximately 40,000–50,000 m^3. They are mainly used for the production of furniture, veneers, and smaller wooden products. In Poland, alder processing is higher, ranging between 60,000 and 70,000 m^3 per year. They are popular for the production of furniture, plywood, and decorative elements [20–22].

In the forests of the Slovak Republic, birch trees represent around 2% of the total share of trees [17]. In the Czech Republic, birch makes up around 2.5–3% of the total share of trees in forests [18]. In Austria, the representation of birch is relatively small, mostly below 1% of the total number of trees in forests. In Germany, birches are mainly found in the northern parts of the country and reach a share of around 2% in forest stands. White birch (*Betula pubescens* Ehrh.) and other birch species are relatively abundant in Poland, with the representation of birches reaching approximately 7–8% of the total share of woody plants in the country [19].

The annual volume of birch processing in Slovakia is relatively low, around 14,000–15,000 m^3. Birches are used for the production of furniture and small art products. In the Czech Republic, birch processing is around 25,000–28,000 m^3 per year. The wood is mainly used for the production of plywood, furniture, and smaller wood products. Austria processes approximately 14,000–15,000 m^3 of birch wood per year, primarily for the production of furniture and smaller decorative items. In Germany, approximately 35,000–40,000 m^3 of birch wood is processed annually. Its use is similar to that in other countries: production of furniture, veneers, and decorative products. In Poland, birch processing is somewhat higher, approximately 45,000–55,000 m^3 per year. In addition to furniture, birches are used for the production of plywood [20,22,23].

European larch (*Larix decidua* Mill.) makes up about 3.5% of the total share of woody plants in Slovakia, and its share is slightly increasing over the years [17]. In the Czech Republic, larch has a representation of around 3–4% [18]. In Austria, larch is common in mountainous areas, its share reaches approximately 5% of the total forest area, and it is an important tree for Austria. In Germany, larch has a representation of around 2–3%. In Poland, larch occurs with a share of about 3–4% within the total forest area [19].

Approximately 60,000–65,000 m^3 of larch wood is processed in Slovakia annually. It is mainly used in the construction industry for the production of durable structural elements and cladding. In the Czech Republic, larch processing is around 80,000–85,000 m^3 per year. Its wood is mainly used in construction, for the production of floors, cladding and durable external structures. In Austria, the annual volume of larch wood processing is around 120,000–130,000 m^3. Austria has a strong tradition of using larch in construction due to its durability and resistance to weather conditions. In Germany, the processing of larch is approximately 130,000–140,000 m^3 per year. It is popular in construction and furniture production due to its strength and durability. In Poland, larch is processed in a volume of approximately 95,000–100,000 m^3 per year. Its wood is mainly used for the production of durable building structures, furniture, and cladding. These values are also qualified estimates and depend on the annual extraction and demand for larch wood [20–22,24].

We will assume that some wood species have greater prerequisites for processing them into particleboard, while others do not have these prerequisites [8]. In recent years, few studies have been carried out on various trees growing around the world and their possibilities for use in particleboard [25,26]. The basic question can be whether these studied trees are represented by their volume in forests in certain regions of the world in such a way that their research in terms of particleboard production is meaningful. Of course, there is no point in conducting research about a tree species that occurs, say, in a park with a representation of several individuals. Such wood species will not provide the necessary volumes of wood raw material to produce the usual large number of particleboard produced, and even if the parameters of such wood species are advantageous, for production plants such information is completely negligible. Therefore, it is necessary to know how much wood raw material from a particular tree species is available in the catchment area of a certain particleboard plant [11].

In this paper, we focused on the region of Central Europe, and we selected three types of forest tree species that are not used in Central Europe for particleboard production. These are the alder—*Alnus* spp., birch—*Betula* spp., and larch—*Larix* spp. tree species. We aimed to point out the suitability of these three trees for particleboard production from various points of view because research has already processed and published quite a diverse range of information about the possibilities of utilization and incorporation of these tree species for various purposes and some selected information from them is instructive [7,13,15].

Each of these three tree species is reviewed theoretically from the point of view of their potential for mechanical processing into chips and particles in two subsections: (1) the potential of the tree species in terms of the suitability of chipping and fragmentation to fine particles and (2) a literature overview of the previous experience of other authors in processing these three tree species into particleboard.

All this information will aid in designing new compositions of particleboard structures. We are deeply aware that the examined woods are not present in forest stands in such a way that they can be processed to 100% proportions in particleboard. Since particleboard consists of three layers, we will not plan and we do not anticipate influencing the composition of the particleboard surface layers composed of fine particles; the surface layers will remain composed exclusively of spruce species. We plan to influence the composition of the particleboard core layer and replace the part of the spruce particles with a certain proportion of one of the three investigated species. The particleboard will have an identical appearance to ordinary boards, and even a less experienced customer will not recognize the difference between 100% particleboard from spruce and our proposed particleboard with a slightly modified core layer of particleboard.

It will be a proposal for a possible and acceptable modification of the composition of the particleboard core layer, replacing part of the wood in it but replacing it with wood, real wood grown in the forest, not non-wood materials, not strange cheap non-wood materials that reduce the properties and value of the particleboard.

Our goal is to create a wood mixture for the particleboard core layer with the incorporation certain proportion of particles from these three wood species. We choose this approach from the point of view that these three forest trees are available in Central Europe, as well as in North Europe, and they are not industrially used for particleboard purposes, given their lower quantity of available volumes. However, as suitable raw materials in a certain proportion in the core particleboard layer, they can be a recordable source of wood raw material and they can enable smooth particleboard production with a steady increase in their production volume and with a relative shortage of common wood raw materials for their production [1,2,5,27].

This review paper was prepared on the basis of a competitive and approved international research project (2022–2025) with the name Analysis of the Properties of the Less-Known European Wood Species in Composite Materials, which is financed from the research funds of two states (Slovak Republic, Czech Republic), with the participation of three research universities. This project is focused on the analysis of the properties of lesser-known wood species, primarily birch, larch, and alder, and their comparison with spruce for use in wood-based composites, especially oriented strand boards and particleboard and in the subsequent processing. We understand this project as a starting point for other international projects in broader contexts with the involvement of several countries because the lack of the majority wood species (spruce) for particleboard production is acute and, even if a minor source of wood raw material, every type is in practice, alongside other alternative raw material sources (recycled), very welcome. This research is aimed toward particleboard type P2 development with the triple-layer structure with the core layer modification and two fine surface layers conservation. However, the idea of incorporating minor sources of wood raw material into the particleboard core layer must be research-verified and sufficiently proven so that the physical and mechanical properties of the particleboard should be of the required level in terms of international technical standards and customer expectations.

2. Alder Species

2.1. Potential of Alder in Terms of Its Suitability for Chipping and Fragmentation into Fine Particles

In Section 2.1, we will try to indicate the potential of alder in terms of its suitability for chipping and fragmentation into fine particles. Our point is that the possible use and processing of alder for the purpose of particleboard should be theoretically elaborated in this review paper in such a way that the research project on our part is meaningful and that our work of a research nature is based on theoretically acquired knowledge and tested procedures published by other authors.

A genus of trees, alder belongs to the group of soft deciduous trees, and its importance in industrial processing is limited by the proportion of occurrence. Alder wood has a remarkable structure. It is a very high-demand raw material, although its mechanical processing requires more attention. Each object made of alder is unique due to the richness of the natural structures of the annual rings, bumps, and irregular course of the core in a composition with a variety of colors—from light golden, through deep, honey orange, to dark brown [28–33].

The wood of the alder is scattered porous, without a noticeable darker core. The vessels on the transverse section are not visible, while on longitudinal sections they appear as only small cracks. Fine string rays are grouped into wide bands (combined rays), on a transverse section in irregular distances and on the tangential section noticeable as long darker bands. Individual string rays are not visible, not even with a magnifying glass. The vessels are narrow and very numerous. The radial diameter of the vessels is 0.065 mm, tangential 0.05 mm, and on average, there are 100 vessels per 1 mm^2. The length of the wood fibers is 0.30–1.01–1.65 mm, and the proportion of wood fibers is 46–58–74%, indicating good possibilities for splitting and chipping the alder wood [28,34–38].

An overview of the basic physical properties of alder wood is provided in Table 1. An overview of the basic mechanical properties of alder wood at a moisture content of 12% is provided in Table 2.

Table 1. Overview of the basic physical properties of alder wood [29,30,32,35].

Density (kg/m^3)		**The Moisture Content of Freshly Cut Wood (%)**		**Fiber Saturation Point (FSP) (%)**
In Absolute Dry State ρ_o	**Moisture Content of 30% ρ_{rf}**	**Sapwood**	**Mature Wood**	
450–510–600	530	80–110		31–34
Shrinkage (%)	Longitudinal	Radial	Tangential	Volumetric
	0.5	4.6	8.5	13.8

Table 2. Overview of the basic mechanical properties of alder wood at a moisture content of 12% [29,32,35,36].

Property	**Parallel to the Fibers w = 12%**	**Perpendicular to the Fibers w = 12%**
Tensile strength (MPa)	94	7.3
Compressive strength (MPa)	55	
Shear strength (MPa)	4.5	
Bending strength (MPa)	97	
Modulus of elasticity in bending (MPa)	11,700	
Brinell hardness (MPa)	36	14

Published collected data on the physical and mechanical properties of alder wood, although only partially reviewed in the literature, already indicate that alder wood can be processed for the purpose of particleboard production. The individual values of the physical and mechanical properties of alder wood are basically not very different from other more commonly processed softwoods or hardwoods with a lower density similar to that of alder for the purpose of particleboard production. Therefore, alder wood seems to be applicable for our intended purpose. In Section 2.2, we will try to summarize the previous experiences of various authors in the use of alder for particleboard production and the context of alder processing.

2.2. Alder in Particleboard and the Context of Its Processing

The industrial production of particleboard made with the help of alder is unknown, but several research workers have published partial results of alder wood processing, either from the premise of researching its properties in chipping or gluing or directly in pressing and laboratory production of particleboard.

2.2.1. Previous Experience of Particleboard Production from Alder Wood

Nemlį, G. (2003) was dealing with particleboard laboratory manufacturing from alder (*Alnus glutinosa* subsp. *Barbata*) wood. He aimed to evaluate the circumstances of the moisture content of the particles, the shelling ratio, and the effect of wood dust on the particleboard mechanical properties (static bending, modulus of elasticity, and internal bond) as well as selected physical properties (thickness swelling). His research results demonstrate that alder particles, if used for particleboard production, exceeded the EN standards for internal bond, modulus of elasticity, and static bending, although the thickness swelling values were poorer than the requirements [39].

While Nemlį, G. (2003) [39] was producing a pure 100% particleboard from alder and this is probably not entirely justified considering the not very large stocks of alder wood in forests in Europe, another research study has focused on a more sober proportion of alder in particleboards. This idea is justified, and we also prefer it. There were investigations of the wood particles (with the sieve fraction 0.4–2 mm) as a mixture of gray alder (*Alnus incana* (L.) Moench) and the lignin powder with a concentration of 15 or 25%. Their thickness swelling, density, water absorption, internal bond strength, modulus of rupture, and modulus of elasticity in bending were determined. It was detected that the amount and the source of the added lignin clearly affected all the tested particleboard properties [40]. It seems to us from this that alder wood is acceptable for the purpose of particleboard production.

Kumaş et al. [41] conducted research on the effects of the drying temperature of the particles and the pressing temperature on particleboard's technological properties, which were produced from black alder wood. Wood particles with different drying temperatures (120 and 150 °C) and pressing temperatures (150 and 200 °C) were used. If the particle-drying temperature was higher, the pressing temperature had an absolute effect on the particleboard's dimensional stability. An increased pressing temperature improved the board's mechanical properties. This research needs to be appreciated and taken into account, as the board's mechanical properties are important and the tried and tested path to achieve them must be followed.

In their research, other scientific workers dealt with the assumptions of alder processing from various points of view, and they came to conclusions and findings that are helpful in the investigation of this wood as a suitable raw material for partially covering the needs of wood raw material to produce particleboard.

2.2.2. Analysis of the Chemical Composition of Alder Wood as a Factor in Particleboard Production

It is very important to know the chemical composition of alder wood in detail because the individual components can partially affect its processing and the future particleboard properties from different aspects.

Geffert at al. [42] evaluated the chemical components of alder wood (*Alnus glutinosa* L.) and also investigated the changes in the chemical components caused by steaming with saturated steam at three temperatures—105 °C, 125 °C and 135 °C. They reported that cellulose is present in regular alder wood at 44.1%, polysaccharides at 77.2%, lignin at 22.0% and extractives at 6.6%. The greatest changes in the alder wood after steaming were observed in the polysaccharides content, with the greatest decrease at a temperature of 135 °C. The cellulose content was relatively stable and small changes was caused by the degradation of the more unstable wood components were apparent. Alder contains a relatively wide range of different substances, including various polyphenols, diarylheptanoids, flavonoids, triterpenoids and steroids. Black alder and gray alder contain the triterpenoids alnusenone, taraxerol and taraxerone. The steroids brassinolide and castasterone have been isolated from sticky alder. Among the biologically active components of alder, diarylheptanoids are the most important. Phenolic compounds and flavonoids are carriers of the antioxidant effect. The yellow dye kaempferol is contained in the bark of *Alnus nitida*.

Alder wood contains extractive substances affecting its gluing process. Janceva et al. [43] studied the possibility of using condensed tannins-rich extracts from gray alder (*Alnus incana*) as well as from black alder (*Alnus glutinosa*) bark in the production of particleboard and plywood adhesives. The extracted microparticles from alder bark residues used into the composition of the adhesive system showed positive potential for application as a filler. They certainly play a significant role in particleboard production.

A seemingly less related study was published by Öztürk et al. [44], but the results of the studied alder in it, among other tree species, are very important. The waste-expanded polystyrene particles of five different densities were used in the production of composite particleboard. Half of the alder or other species chips were dried in a drying oven and the other half just naturally conditioned at room temperature, and then a three-layer composite particleboard 18 mm thick was produced. The thermal insulation performance of the composite particleboard panels increased with the use of waste-expanded polystyrene foams with a density of 30 kg/m^3, according to the analysis findings. The lowest thermal conductivity values were acquired from the expanded polystyrene waste foams with the low density of 13, 18, and 22 kg/m^3 for the panels produced with alder, pine, and poplar in the natural-drying condition. Technical drying showed better thermal performance than natural drying. The accompanying substances in alder wood have been identified, as in previous research [43], and need to be considered.

In their previous study, in comparison with [44], Öztürk et al. [45] dealt with the thermal conductivity of wood material, which is superior to other building materials because of its substantial porous structure. Their efforts aimed to produce a relatively new wood composite material with insulating properties by using insulating material polystyrene instead of formaldehyde-based adhesives as the bonding material. Five wood species (alder, beech, poplar, pine, spruce) were used and three-layer particleboard with the thickness of 18 mm was produced in this study. The thermal conductivity values obtained from natural drying were found to be higher than from technical drying, according to the results of the study. Alder wood was fully operational in this trial, and it was recommended for further research.

2.2.3. Investigations of Alder Wood Particle Storage Methods

For our needs, it seems that the research by Szadkowska et al. [46] is very stimulating. Their study was focused on investigating the effect of the storing methods for wood particles from two species, alder (*Alnus* Mill.), and pine (*Pinus sylvestris* L.), on the basic chemical composition. In the study, two storing methods for woody biomass were used: an open pile and a covered pile. The wood was cut down at the end of November and it was freely laid as industrial rough chips for 8 months from December onwards. After this time had passed, the material was collected again for its chemical composition analyses as well as the enzymatic hydrolysis. The influence of the type of storage on the composition of the individual structural components of the wood was demonstrated as the results of the chemical composition analysis of the wood for both studied species.

On the basis of the results of the chemical determinations carried out on the wood biomass of alder wood stored in an open pile and covered pile for 8 months, it can be concluded that the wood storage method plays a significant role in the enzymatic hydrolysis efficiency, which has been statistically confirmed. High yields of enzymatic hydrolysis were obtained for alder wood biomass being stored in an open pile; moreover, the alder biomass had a higher yield of enzymatic hydrolysis when using dyadic enzymes than biomass from softwoods. A higher cellulose content (+3.6%) was observed when the alder raw material was stored in a covered pile. The holocellulose content had a similar level for alder raw material storage, being 70.1% and 71.9% for an open and covered pile, respectively. The hydrolysis effectiveness increased with the time as well, independent of the material type that was hydrolyzed. Thus, it is obvious that raw alder wood (like other types of wood) stored for too long is not suitable for the production of quality particleboard.

2.2.4. Consequences of the Thermal Treatment of Alder Wood for Particleboard Production

The effect of elevated temperatures on alder wood quality as studied in the past by other authors is important in terms of determining the correct pressing temperature for particleboard using alder wood. Thermal treatment of alder wood has been investigated by Lee et al. [47]. They wanted to offer several advantages to the treated wood, such as exclusion of oxygen during the treatment process and even more and the rapid transfer of heat into the wood samples. Alder wood treated in soya oil at 180 °C and 200 °C for 6 h and 10 h exhibited superior resistance against *Postia placenta*. Urea formaldehyde-bonded particleboard was soaked in palm oil for 24 h before being thermally treated in the oven at 180 °C, 200 °C, and 220 °C. Such temperatures did not destroy the alder wood and pressing it into a particleboard at the level of 230–240 °C, or even slightly higher, is feasible.

The color changes and decay resistance against white rot fungus (*Pycnoporus sanguineus*) of the samples after thermal treatment were investigated. After thermal treatment, the lightness of the samples decreased, and the extent of the darkening increased along with an increasing treatment temperature. With the expected lamination of the particleboard surface, this argument will not matter, but when using raw particleboards without surface treatments, it is necessary to draw attention to this.

2.2.5. Comparisons of Selected Alder Wood Resources from Various Stem Parts for Particleboard Production

The selection of a specific stem part of alder wood resources for particleboard production is not possible during production. Nevertheless, it is good to know that the use of an insufficient general wood mixture in particleboard production can affect the properties, as further research has pointed out. Alder was also the subject of the project on binderless bark particleboard made from bark waste obtained from gelam (*Melaleuca viridiflora* Sol. ex Gaertn.) [48]. The particleboard's thermal insulation properties were determined as well. Up to four different temperatures (140 °C, 160 °C, 180 °C, and 200 °C) were set up to make single-layer binderless bark particleboard with a density of less than or equal to 590 kg/m^3. The results showed that the pressing temperature affected the physical and mechanical properties of the manufactured boards. As the pressing temperature was increased from 140 °C to 200 °C, the average values of the mechanical properties also increased, even though the increase was not significant for the MoR and MoE values at pressing temperatures of 180 °C and 200 °C. The best physical and mechanical properties were obtained from the particleboard pressed at a temperature of 200 °C, with an MoR value of 3.97 N/mm^2, MoE value of 758.05 N/mm^2, internal bonding (IB) value of 0.6 MPa, TS24h value of 4.94%, and WA24h value of 16.3%. For us, this implies that higher pressing temperatures for the particleboard are fine even if the pressed chip mixture is diverse.

The characteristics of particles made of straw of selected grain species, including comparisons with alder particleboard purposed to produce lignocellulose particleboard, were tested by Dukarska et al. [49]. Their research offered the characteristics of three different species of grain straw particles applied in lignocellulose particleboard production. The research aimed to determine the physical properties of particles of less traditionally input raw materials and to evaluate their usefulness in the process of manufacturing lignocellulose particleboards purposed for building constructions and to measure the particles' usefulness in the process of the lignocellulose particleboards manufacturing oriented for the building constructions. Every investigated material was characterized by the different tapped bulk density, varied angle of repose, geometry, poured, and the slippery angle of repose. The material produced was sorted using screens with a mesh diameter of 4 mm and 2 mm in order to select the material for the surface layers and the core layer. The desired fraction for the core layer consisted of particles retained on a sieve with a mesh size of 2 mm. Particles larger than 4 mm were reground and sorted, and particles smaller than 2 mm were used for the surface layers. These values need to be known and evaluated, and this is also the case with our intended exploration of less traditional trees, which has been confirmed by this research [49]. Alder wood has not been primarily researched, but references to it and its connections are also applicable to the processing of alder wood into particles. The paper's conclusion, and for us a lesson, is that each material is characterized by different geometry, poured, and tapped bulk density, varied angles of repose, and slippery angles of repose. Differences in the sizes and shapes of particles of various grain species and in their moisture content strongly affect the properties of particleboard produced with their use. We must be aware that difficulties in storage, transport, and particleboard production may happen.

Salca [50] examined black alder species (*Alnus glutinosa* L.) from various points of view as a resource for value-added products. She assumed that due to its workability, properties, and appearance, black alder can be considered a suitable particleboard raw material. Her review offers a contour of the potential of black alder as a wood raw material. She emphasized the application and alder wood properties in terms of light exposure and the color changes under air, the surface quality of processed particleboard during sanding and milling, coating properties, and the process optimization. The findings of her study help with the practical applications so that this woody species can be processed and utilized more efficiently for value-added products. The average bending strength values found for alder wood were in the range of 44–110 MPa. Black alder as a less-utilized woody species has been shown to have considerable potential for industry so that its use can successfully reduce the import activities of other woody species.

2.2.6. Possibilities of Chemical Modifications of Alder Wood for the Purpose of Improving the Quality of Particleboard Production

As in Section 2.2.4, this subsection summarizes the chemical properties of alder wood and their potential impact on particleboard production. The synergistic influence of the intumescent coating that contains titanium dioxide and antimony trioxide onto alder and spruce woody species was brought to the attention of a team of researchers led by Torun et al. [51]. It is good to know the reactions of alder wood with various chemicals. The results confirmed that the wood surfaces coated by different percentages of ammonium trioxide and titanium dioxide added to paint had a positive influence on the wood's physical properties as a promising raw material for particleboard production with a certain share of alder particles.

Lovaglio et al. [52] tested the wetting behavior of alder wood surface (*Alnus cordata* (Loisel) Duby) from the point of view of the influence of alkyl ketene dimer and the thermo-treatment before and after a current artificial weathering test. The FTIR analysis carried out by them promoted their hypothesis that the alkyl ketene dimer could make the bonds relatively chemically stable, even when using a rather small concentration. Their findings provide important information for any future research and for alkyl ketene dimer utilization on lesser-used wood species.

2.2.7. Use of Alder Wood for the Production of Other Types of Wood Composites and Other Context

Alder wood plywood has been produced for many years and has been studied in detail. The context of gluing alder veneers is fully applicable to gluing alder chips, since in the vast majority of cases it will involve the use of an identical type of glue. Bekhta et al. [53] dealt with the properties of plywood composed of thermally densified and non-densified birch and alder veneers. This, in turn, is another view of the possible behavior of alder wood, applicable to its intended processing into particleboard. Their observation aimed to develop plywood with two wood species (black alder and birch) and two veneer treatments (thermally densified and non-densified) to evaluate the influence of various lay-up schemes on the plywood's properties. The results demonstrated that the wood species, type of construction, and thermal densification of the veneer applied influenced the examined mechanical and physical properties. Black alder veneers can be placed into the inner layers of plywood without the danger of reducing their shear strength. An increase in the share of the thermally densified veneer in one panel leads to the higher density, shear strength, and bending strength of the plywood. It was also demonstrated that non-treated alder veneer, despite exhibiting lower strength properties than birch veneer, could be used with a proper lay-up scheme in the plywood-based product industry. Therefore, mixed-species plywood enables the increased utilization of low-grade, low-density, and low-cost alder wood veneers as inner veneers in panels to reduce the production costs and increase the mechanical properties of predominately low-density alder wood plywood. This is a good insight when further reflecting on the processing of alder wood.

In a broader context, these issues are also dealt with in a book compiled by the editors Bekhta and Krystofiak [54].

Alder plywood is also of interest to other researchers. Öncel et al. [55] aimed to determine both the tensile–shear strengths of plywood produced from Uludağ fir (*Abies nordmanniana* subsp. *bornmülleriana* Mattf.), alder (*Alnus glutinosa* L.), Scots pine (*Pinus sylvestris* L.) and Samsun poplar (77/51 *Populus deltoides* Bartr.) by using phenol formaldehyde resin. They tested the rotary cut veneers in combinations of poplar–fir, poplar–pine, poplar–alder, and poplar wood, along with the effect of wood types on the adhesion glue quality. As a result, the tensile–shear strength values of alder–poplar plywood were found to be average. It was found as well that there was not a significant difference between the alder–poplar and fir–poplar plywood types regarding the tensile–shear strength. All the plywood types tested in this study, including alder plywood, are of suitable quality for outdoor use. Another research project went even further. The authors continued the research on laminated veneer lumber (LVL) from alder wood, the properties of which also passed the standard values. Since alder wood has passed these tests, it will certainly behave appropriately, while observing all the necessary technological conditions, including in the context of its gluing in combination with spruce in the core layers of particleboard.

An extension of the idea of this research was reported by İlkay and Mengeloğ [56]. They verified the layer combination effect on the equilibrium moisture content, oven-dry specific gravity, water absorption and thickness swelling, modulus of rupture, modulus of elasticity tensile–shear strength, and compression strength parallel to the grain of LVL manufactured from alder and poplar. Urea-formaldehyde resin was used in LVL production. LVL was produced in 10 different combinations. The highest compression strength, MOR, MOE, and oven-dry specific gravity values were obtained with alder veneers used in overall plies. But, it is interesting that the highest tensile–shear strength values were obtained with poplar veneers used in overall plies. With the increased use of poplar veneer in LVL production, an increase in the amount of water absorption and thickness swelling was observed. It was noticed as well that as the contribution rate of the alder veneer in lamination increases, the modulus of rupture, modulus of elasticity, oven dry specific gravity, and compression strength values were raised. Also, based on this study, alder is a suitable wood for its mechanical processing and use in composite materials.

Suvorova and Romanenko [57] tried to prove that it was possible to obtain wood with high-performance properties based on mechanical and chemical action because of the optimization of technological processes and use of temperature exposure. The initial raw material was hardwood (alder, aspen), which is not too commonly used in construction and in the production of finished materials. The condition for obtaining wood with high operating properties (increasing strength, density, reduction of water saturation, ensuring the samples dimensional stability for a long time) is the ability of wood as a natural polymer to change properties under the combined effect of pressure and temperature.

Romaneko himself supported these activities in another published paper [57].

It is a very correct aspect when we also evaluate the possibility of dust formation during alder processing. This view of alder was dealt with by Pędzik et al. [58]. Wood dust poses the risk of explosion and fire and a threat to the health of employees, worsens the quality of processing, accelerates the wear of machines, and requires large financial outlays for its removal. This study is focused on research on the extent to which the grit size of the sanding paper affects the size of the wood dust elements and the share of the finest elements that, when dispersed in room air, may form the respirable fraction. Six species of hardwood (alder, beech, ash, oak, walnut, and hornbeam) and three species of softwood (larch, spruce, and pine) were used in this research. Based on the results obtained, they found that the dust elements' sizes were also obtained in the case of alder, which means that we must be careful when processing it.

The dust generated during the processing of alder was also investigated by Sydor et al., and they achieved similar results, with an emphasis on dust arising from alder wood [59].

If we look at the issue of alder processing from the perspective of the future, we must also be aware of the issue of disposal of alder products at the end of their useful life. Qiu et al. [60] were active in this regard. The paper deals with the thermal decomposition and the behavior in combustion of two kinds of plywood used in composite floorings. The research study was prepared on the use of a fire propagation apparatus. The combustion parameters, including the heat release rate, ignition time, effective heat of combustion, and mass loss rate, were analyzed. The ignition risk and heat release of alder plywood exist. It follows that the disposal of alder products by incineration is not an operationally suitable activity, as it will be more appropriate to extend their life through recycling and circular economy.

The paper by Sedliacikova et al. [61] deals with the issue of the color tones of wooden and furniture products. This issue can be interesting from the point of view of particle color changes in the composition of particleboard, not so much from the point of view of the real color but from the point of view of chemical changes in wood that must be perfectly bonded together in the form of particles within the particle-pressing process. The main aim of their study was to identify the supply of the color tones of wooden and furniture products and to map the interest in these wood color tones for potential customers in Slovakia. It is necessary to increase the supply of wooden and furniture products with the natural color of the wood and at the same time in the color tones of white, gray, and brown. The current demand for modified alder wood and thermowood is significantly higher than the supply of such products on the market. The results of this study represent an opportunity for woodworking and furniture factories to adapt their product range according to the needs of potential customers, which will bring them a higher turnover and help to overcome the possible problems. Also, based on this study, we can conclude that alder wood can also be processed into the particleboard core layer from the point of view of the supply of wood products with the natural color of the wood.

2.2.8. Overall Evaluation of the Context of Alder Wood's Possibilities in Particleboard Production

It is clear from the mentioned and other unmentioned literature sources that alder has suitable prerequisites for its processing into particleboard. The sticky alder (*Alnus glutinosa* (L.) Gaertn.) is a deciduous tree, but it is one of the softest woods, which is not very common in the category of deciduous trees. This could be interesting for the purpose of our project, as the low density of particleboard and OSB boards is currently in high demand all over the world. Another characteristic of alder wood is its resistance to impacts, but also its low stiffness. Alder wood has a high resistance to exposure in outdoor conditions in humid or even drier places; there is a high probability of creating few, if any, cracks.

The durability of alder is mediocre in the range of hardwoods. It is impossible to expect the quality of alder wood to be comparable to oak or maple. Alder wood can absorb moisture, which can affect the quality of particleboard and OSB boards. Alder wood is very suitable for any manual processing, and it also polishes well and accepts all kinds of colors. It is suitable for making models of various products. Winter harvesting of alder and its quick cutting is recommended, and the protection of the ends of the logs is also suitable. Lumber must be stored in airy cages; thin lintels should be used. The use of alder wood for furniture has been known for a long time. The appearance and, usually, the lower price of alder wood are chosen by many people. In addition, its durability also makes alder home furniture long-lasting. Alder is used to make chairs, tables, cabinets, kitchen furniture, and other types of furniture. Alder wood is known for the production of interior doors thanks to its uniform structure with prominent knots. Alder is a relatively soft wood, which makes it easier to repair those places where damage has occurred. It is not suitable for the production of exterior doors; it may not be sufficiently resistant to the fluctuations of harsher weather.

The quality of alder wood may not be sufficient for high-quality windowsill boards or shaped wooden products. Alder wood is extremely suitable for decorative purposes in the form of covering decorative veneers (or even micro-veneers) for large-area panels for wall coverings. Its distinctive pattern and regular structure lend the slatted panels a beautiful appearance. Stabilization of alder veneers with a durable transparent coating system is necessary. Due to its softness, alder wood is suitable for turning, milling, and the production of various noble objects. It is also suitable for carving purposes; alder products are attractive. Alder is one of the most suitable types of wood for the production of kitchen utensils.

Alder plywood is well-known and of high quality. Alder is easily processed by centric peeling technology; the manufactured structural veneers are suitable for the surface or middle layers of plywood. The insufficient strength or softness of alder for some applications can be eliminated by making plywood from it. Alder plywood is used in the construction of doors, furniture, and the like. Some musical instruments, such as guitars, are made from alder. The guitar body is characterized by a robust and clean sound, a balanced entire sound spectrum, and other suitable acoustic properties. Alder wood is only used to a limited extent for pulp and paper production [28].

It can be concluded, based on this information from the published literature, that alder wood has the potential to be pressed with spruce particle mixes into the core layer of particleboard.

3. Birch Species

3.1. Potential of Birch in Terms of Its Suitability for Chipping and Fragmentation into Fine Particles

Even in the case of birch wood, there is a collection of available data on the potential of birch in terms of its suitability for chipping and fragmentation into fine particles. Birch is an appropriately more analyzed tree in the literature context, because its use in the veneer-based and plywood industry is so decisive that birch wood research is more extensive and long-term. The literature also contains data on the possible uses and processing of birch wood for the purpose of particleboard. Therefore, in this subsection of the review paper, birch wood is theoretically elaborated, which includes utilizable meaningful works of a research nature. These laboratory works must include theoretically acquired knowledge and birch wood-processing procedures published by other authors.

Birch (*Betula verrucosa* Ehrh.) and other birch species belong to the group of hardwoods, and its importance in industrial processing is limited by the proportion of occurrence. It is a very important tree species for the north of Europe and the experience of using it to produce particleboard is greater than in the case of alder [28,33,53]. The color of birch wood is white with a hint of red. The boundaries of the annual rings are indistinct. It creates a very nice texture on a tangential cut, especially wood with sleeping eyes or flaming. Birch is a sapwood tree species; the sapwood is not visible to the naked eye. It is a scattered porous wood. The width of the annual rings ranges from 1.2 to 2.0 mm. We can only see the shaft rays in birch on a radial section [28,33].

Of the total volume of wood, wood fibers make up 65.8–75.7%, vessels 10.6–21.4%, stem rays 10.8–11.7%, and wood parenchyma 2%. This is scattered (diffuse) in birch wood, i.e., the cells are scattered evenly over the annual rings. Spindle-shaped parenchyma cells occur in birch wood. The diameter of the vessels is 0.02–0.1 mm. The thickness of the wall is very small at 1 μm. The perforation of birch vessels is always stepped. The colons do not have a torus. Birch has a vascular tracheid, the length of which does not exceed

0.5 mm. The width of the libriform fiber in birch is 0.219 mm, and the thickness of the cell wall of the libriform fiber is approximately 5 μm. All the microscopic parameters indicate good possibilities for splitting and chipping the birch wood. Birch wood belongs to the medium–heavy woods, and it is slightly hard. It impregnates and stains well [28,34–38].

An overview of the basic physical properties of birch wood is provided in Table 3. An overview of the basic mechanical properties of birch wood at a moisture content of 12% is provided in Table 4.

Table 3. Overview of the basic physical properties of birch wood [30,62–64].

Density (kg/m^3)		The Moisture Content of Freshly Cut Wood (%)		Fiber Saturation Point (FSP) (%)
In Absolute Dry State ρ_o	Moisture Content of 30% ρ_{rf}	Sapwood	Mature Wood	
460–800	520	80–90		30–35
Shrinkage (%)	Longitudinal	Radial	Tangential	Volumetric
	0.6	5.3	7.8	14.2

Table 4. Overview of the basic mechanical properties of birch wood at a moisture content of 12% and a moisture content higher than 30% [30,62–64].

Property	Parallel to the Fibers		Perpendicular to the Fibers	
	w = 12%	w > 30%	w = 12%	w > 30%
Tensile strength (MPa)	137		7.0	4.0
Compressive strength (MPa)	59.9	26.3		
Shear strength (MPa)	16.2	7.7		
Bending strength (MPa)	123	63		
Modulus of elasticity in tension (MPa)	18,100		600	
Modulus of rigidity (MPa)	15,800		600	
Shear elastic modulus (MPa)	1450		800	
Modulus of elasticity in bending (MPa)	13,300	9900	800	
Toughness (J·cm^{-2})			9.32	7.85
Brinell hardness (MPa)	48		20	
Janka hardness (MPa)			40	

Birch, in terms of the air durability, can be classified among the low-lasting wood species. In terms of durability in contact with the ground, where there is a risk of damage by all types of rot, it is classified as a non-perennial woody species (Table 5).

Table 5. The durability of birch wood in years [28].

Exposure	Unprotected and Unimpregnated	Under the Roof	Under the Water	Always Dry
Durability (years)	3–8–15	5–20–30	20–40–60	300–500

The values of the physical and mechanical properties of birch wood collected from the literature are much more comprehensive than the information collected for alder wood, as listed in Section 2.1. It follows that the processing of birch wood into particleboard will be more refined and less experimental. Birch wood is well-studied in many, especially Nordic, countries, and this literature review is proof of that. The values of the physical and mechanical properties of birch wood are very favorable for the production of structural and decorative veneers, and the literature research shows that they are favorable for chipping wood and the production of various chips. Therefore, it is assumed that birch wood will also be suitable for the production of specific chips needed for the purpose of particleboard production. Section 3.2 provides a summary of the previous experience of various authors in the use of birch wood for particleboard purposes and in the context of birch processing and its possible impact on the spruce and alder particles mixture.

3.2. *Birch in Particleboard and the Context of Its Processing*

Kollmann et al. [65] and Maloney [66] drew attention to the great importance of the quality of wood as an input raw material for the production of particleboard in the 1970s. In their well-known publications, they emphasized the importance of the technical parameters of the particles used for particleboard production (dimensions, thickness, moisture content), and they also emphasized the suitability of certain types of wood species in terms of their structure. They stated that it is impossible to produce high-quality boards from inferior raw materials. From the range of available woods, in addition to conifers, they also recommend hardwoods with a lower density, and birch (as well as alder) meets their requirements.

3.2.1. Previous Experience of Particleboard Production from Birch Wood

Varis [67] focused on Nordic wood processing and particleboard production in Finland, mainly from spruce but also from birch, which is a reality. The main raw material in Finland is small-diameter spruce timber from thinning, sawdust generated in sawmilling, and particles and veneer trash created in the plywood industry. It is trash from the plywood industry that is mainly the source of birch particles, in addition to spruce particles, in the production of particleboard. Birch particles are a little heavier, but they are suitable for the quality of Finnish particleboard.

Varis [67] noted that in Europe, it is more difficult with raw materials for the production of particleboard, because in addition to a larger assortment of tree species, including birch, recycled wood and wood from demolitions are used, but the research also coped with such raw materials, as long as they do not exceed a certain reasonable share in particleboard.

The importance of suitable input raw materials for the production of particleboard is emphasized in the publication by Irle et al. [12] and, with an emphasis on Central and Eastern Europe, in a compiled publication by Reh [68].

Varis provided up-to-date information usable in pressing birch particles mixed with spruce, but also older research was performed by Liiri [69] is impressive and incorporating. Three-layer interior particleboards, prepared from birch, alder, pine, spruce, and aspen, were tested for their strength properties and also their water absorption and dimensional stability during soaking or variations in relative humidity. The wood density proved to be the main factor in determining the suitability of the raw material, affecting both the strength and dimensional stability. The bark in the particles for the middle layer did not adversely affect the board's properties. All the tree species appeared to be suitable for particleboards. Birch provides the boards with a density of more than 800 kg/m^3, which is already an unacceptably high density from today's point of view, but it tells us about the suitability of birch (and alder) for making particleboard. However, it could be satisfactorily mixed with lighter wood species.

Good news for processing birch into particleboard also comes from the research by Baharoğlu et al. [70]. They did not deal directly with specific tree species but compiled a very decent overview of the technological matters necessary to produce high-quality particleboard from alternative raw materials. They analyzed a variety of literature sources, and they summarized that the main factors that determine the potential of alternative raw materials to be used in composites production are their physical and chemical properties, the possibility of seasonal storage, local availability, the supply volume, and their carbon neutrality. Their study aimed to review the current state of knowledge on the possibilities of the use of various raw materials in the production of particleboard. Based on this review, an experiment was performed to identify the raw materials with the highest potential, considering their quality and the availability of boards made from them. Suggestions were made as well regarding the search for new raw materials, based on their production potential. Their summaries clearly show that some raw materials, even non-wood, may be acceptable for particleboard production, but real wood is the most suitable.

Similarly, Krug et al. [71] considered and accumulated technical information on particleboard that is presented as an important product. They also stated that particleboard, if it is to be of good quality and meet all customer requirements, must be made from the right raw materials and with the right manufacturing process. Real wood is an irreplaceable raw material.

Particles made from birch were researched by Çamlıbel and Aydın [72]. They closely watched the effects of a continuous pressing speed (580 or 600 mm/s) and the time of the conditioning (without and 72 h) on selected physical and mechanical properties and the formaldehyde content in particleboard. They based their research on the use of birch elements in MDF boards and particleboard using UF glue mixtures, where birch has proven itself well.

Skurydin and Skurydina [73] summarized the results of the digital differential spectrometry application in the processing of the results of the study of the structure of wood and wood-based materials by dynamic mechanical analysis (DMA). The results of the application of the DMA method and digital data-processing are presented in the study of the dynamic mechanical characteristics of wood of birch and apple, a composite made of hydrolyzed wood, as well as particleboard. It is proven that the use of digital processing of experimental data provides accurate results, simplifies their interpretation, as well as expands the understanding of the processes occurring in wood-based materials at the level of the intermolecular interaction. According to this method, birch turned out to be a suitable wood for particleboard manufacturing.

Singh et al., in their study, discussed birch and aspen used for the development of modern engineered wood [74]. Wooden elements used for the development of engineered wood have various factors in terms of the shape, size, surface conditions, density, and other dimensional parameters that affect the mechanical as well as physical properties of the final product. Also, the permeability, pH, element shape, and aspect ratio play a vital role in the manufacturing process for engineered wood. A wide range of compositions and percentages of raw elements and resins could be used to obtain different results in terms of the physical and mechanical properties as per the application of particleboard. The addition of selected suitable resins in small quantities along with the resin is also recommended as it upholds the mixing process by assisting in achieving homogeneity. Birch stands out as a suitable material comparable to well-known raw materials used to produce particleboard.

Razinkov and Ishchenko [75] looked at the issue from the perspective of the shortcoming of particleboard for the production of cabinet furniture. This is related to the properties, which must meet the standards even though particleboard-produced volumes are increased. Important disadvantages of particleboard include the toxicity associated with the release of formaldehyde, rather low mechanical (strength) properties, and limited areas of application of particleboard (mainly indoor). They watched the dynamics of the technical requirements for the properties of particleboard according to the standards and they looked for the reason for the reduction in the requirements concerning the strength of particleboard in bending and, in this regard, the constraint of the application areas of particleboard. Their research included the type of wood species used for particleboard production (mainly birch, alder, pine, aspen, and some others), and birch comes out positively from this research.

3.2.2. Possibilities of Particleboard Production from Birch Wood and Alternative Adhesives

Since birch is a very important wood species for the production of plywood for Northern Europe, quite a lot of scientific works deal with the processing and use of birch wood or birch bark in terms of processing it with cutting tools. This is interesting for any cutting tool operation, which can also be chipping machines. Tupciauskas et al. [76] investigated residual birch veneer shorts and the birch outer bark that are among the main residues from birch plywood production. They have a strong potential to be converted into value-added products. Birch (*Betula pendula*) outer bark contains up to 45% (on the dry basis) suberin and it has been recognized as a promising resource of suberinic acids; their bonding potential has been proven. The study evaluated the particleboard production made of birch wood particles obtained from residual veneer short pieces and the adhesive enriched with birch bark extracts, and it indicates to us that a mixture of birch wood, even in a partially inhomogeneous mixture, with the possible presence of glue, is acceptable in a reasonably small amount for particleboard core layer production.

Jeżo and Wronka [77] focused on post-extraction birch bark residues as a potential binder in particleboards. Their paper also concerned the possibility of using post-extraction residues obtained during the extraction of suberinic acid as a formaldehyde-free and ecological binder in the production of particleboards. Their research aimed to address the following issue: since suberinic acid itself is a good binder in the production of particle boards, it is worth checking whether the post-extraction residues also have similarly good properties in joining particles in particleboards, depending on the size of the wood particles. In addition, the use of post-extraction residues of bark, and thus the elimination of synthetic adhesives in the wood-based composites production process, allows the reuse of wood raw material, which fits perfectly with the idea of upcycling. The tests showed that using post-extraction residues of birch bark, using 10% and 20% resination, the requirements of the EN 312: 2010 standard [78] were met in the case of the modulus of elasticity for boards made of the largest wood particles used in the tests. The resination and the size of wood particles contributed to the improvement of the properties of the tested boards.

This topic also caught the attention of Makars et al. [79], and they dealt with the investigation of the furfural formation and mechanical properties of suberinic acid-bonded particleboard depending on the parameter preparation. The adhesive included the silver birch (*Betula pendula*) outer bark suberinic acidic, which can catalyze furfural formation from xylans in wood particles that are used for particleboard preparation. As part of their study, the impacts on the technological parameters (wet adhesive pH) of glycerol as an additive to the adhesive and the hot-pressing temperature were investigated in particleboard. The mechanical properties (modulus of elasticity, bending strength, and thickness swelling) of particleboard were studied.

Makars et al. [80], in their subsequent publication, expanded on this idea and the research on birch (*Betula* spp.). They also discussed the outer bark from the point of view of betulin as a valuable product. They realized that after the removal of betulin extractives, suberin-containing tissues were left.

3.2.3. Use of Birch Wood for the Production of Other Types of Wood Composites

It has already been stated several times that birch plywood and other birch products are essential for the Nordic countries. Hence, a substantial part of the published scientific information is applicable to the theory of birch particle production and their pressing. Tupciauskas et al. [81] tried to confine, replace, or maybe even eliminate the synthetic resins from wood-based composites because expanding the range of raw lignocellulosic is still important and attractive. Many pretreatments of lignocellulosic have been carried out, among which steam explosion resulted in the superior physical and mechanical properties of the obtained binder-less particleboard. Emissions of volatile organic compounds were investigated in the framework of the study from binderless particleboard obtained from different raw lignocellulosic and birch suberinic acid-bonded particleboard. The results showed that the number of detected volatile organic compounds and their chromatographic peak area varied significantly depending on the raw lignocellulosic, board density, and post-treatment (overlayering), decreasing over time.

The surface properties, like the roughness, color changes, hardness, wetting, and density, of heat-treated silver birch (*Betula pendula*) veneers were studied by Dudik et al. [82]. None of the parameters studied has proved the negative changes due to the temperature adjustment. Both in terms of the properties and in terms of the valuation, there is considerable potential for birch to replace, for example, beech, especially in the furniture industry, by application in the form of heat-treated veneers after the required heat treatment.

Kowaluk and Jeżo [83] investigated the modulus of elasticity and the contractual compression strength when compressing (MOEC) three different wood-based composites with several structure types with birch wood elements (veneers, particles). The density profiles and density shares of the samples were analyzed as well, and all of the examined wood composites exhibited a U-shaped density profile. The study results pointed out that there is no significant correlation between the density and the obtained parameters under compression. Observation of density share led the authors to conclude conversely than the results suggested. Therefore, the key factor effecting the compression performance of the samples was the adhesive area and solid glue content within the wood composites. It can be assumed that the bigger the total contact wood particles surface coated with adhesive resin (and thus the sum of the effective surfaces of the adhesive joint), the better the mechanical properties of the wood composites.

3.2.4. Other Birch Biomass for Particleboard Production

Dumitrascu et al. [84] considered three different fast-growing wood species, namely birch, poplar, and willow, in their research. They drew attention to the fact that the referred to species may have certain potential to replace the softwoods or wood species mixtures frequently used in the production of OSB (oriented strand boards). Their study described selected mechanical and physical properties of the wood species used that influence the OSBs' performance when made using 100% strands from each wood species. The strands of wood were first cut, then dried, screened, and sorted to form the middle and surface layers of the OSBs. The strands were blended with pMDI adhesive (polymeric diphenyl methane diisocyanate) and compressed with a heated hydraulic press. OSBs made of birch presented excellent elastic properties. The results of the experimental study can have industrial applications for the efficient utilization of less used and less known raw material or, as in our case, for the use of birch particles mixed with spruce particles in the particleboard core layer.

Xu [85] thought in 1994 that a huge number of material and processing variables influence the properties of particleboard. He thought that not too much is known about the particleboard's internal structure, and an essential principle or theory interrelating the structure, processing, and properties has yet to be developed. Basic knowledge of the particleboard structure is necessary to fully understand the present particleboard technology, but it is also important for the future upgrading and development of wood-based composites. The study was elaborated to develop some of this knowledge base. The two major objectives of this study were (1) to understand the influence of raw material characteristics on the horizontal density distribution, and (2) to specify the effect of particleboard nonuniformity defined by the horizontal density distribution on selected properties. He used birch particles in the research, and with the help of them, he determined 26 particleboard panels. All of them were made with precisely cut particles to study the first objective, while 30 particleboards involving different element sizes and distributions, different wood species and combinations were prepared to study the second objective. With the use of relatively larger specimen sizes, particleboard made with larger particles exhibited more substantial density variation, while particleboard made with smaller particles showed larger variations at smaller specimen sizes. Two aspects of lacks, namely the size and number, were identified as factors contributing to the relation between the particle size and horizontal the density distribution.

Carre [86] analyzed forest biomass for the manufacture of particleboard. He used Belgian samples of the proportion of aerial biomass greater than 7 cm in diameter, 1–7 cm, and less than 1 cm, for birch, beech, sessile oak, and hornbeam. Particleboards were made from the first thinnings of different types using the proportions of various grades of chips. Similar results were obtained from all the chips made from material greater than 1 cm in diameter. Incorporation of material less than 1 cm diameter greatly reduced the strength and stability of the boards, but acceptable boards could be made if the proportion of this material was reduced to about 5% (from 15–20% in the total biomass), e.g., by using it only in the core layer of three-layer boards. The tree species used turned out to be suitable, as the problem was only with first thinning diameters.

Pedieu et al. [87] proposed to substitute the traditional wood raw material in the surface layers of particleboard with the water-resistant white birch (*Betula papyrifera*) outer bark particles, which helps to improve the dimensional stability of manufactured mixed particleboard and thus reduces the shortages of raw material in a cost-efficient manner. Mixed particleboard was manufactured under laboratory conditions using untreated or alkali-treated white birch outer bark particles as a substitute material. These particles were successively resinated with 3% phenol-formaldehyde resin. Overall, the results of the study demonstrate that the wood-based panels could be manufactured using up to 45% of the proposed substitute material and still maintain the required physical and mechanical properties of particleboard.

3.2.5. Overall Evaluation of the Context of Birch Wood's Possibilities in Particleboard Production

From the listed and other unmentioned literature sources, birch has excellent prerequisites for its processing into particleboard. Quality assortments of all classes are used for the mechanical processing of warty birch (*Betula verrucosa* Ehrh.) raw wood. To produce birch lumber and blanks, no sawing machine is dominant, and therefore, neither is the cutting technology. Birchwood is well-processed by cutting, which is why we can produce quality lumber with frame saws, log band saws, and log circular saws. Birch dries more slowly and shrivels. Wood tends to vaporize, so sufficient air flow must be ensured during natural drying. Hot-air dryers are mainly used for artificial drying. This wood must be dried more slowly, and the drying temperatures are lower than for conifers. Compared to beech, birch's drying times are shorter. Birch does not significantly change color during drying. Cuts from manufactured lumber can be found in furniture products, floors, parquets, toy products, and products with useful properties in the home and in restaurant facilities. Birchwood is excellently processed by peeling and cutting. This feature is used in the production of tangential decorative veneers and the structural veneers for the elaboration of all types of plywood panels, including molded plywood and laminated wood. In the Nordic countries, birch is considered the most important wood for plywood. Birchwood in a lower quality class is used for the pulp and paper industry, and also as a welcome raw material in the production of agglomerated materials. The good machinability of birch wood is used in artistic carving. An inherent commercial use of birch wood is its sale as fireplace fuel [28].

Birch wood has proven itself in so many cases in terms of its processing that there are almost no doubts about its incorporation into the core layer of particleboard. It will be necessary to follow the technical principles and reveal the minor imperfections of birch wood by other authors in the past, and the processing of birch wood in some proportion in particleboard will probably be possible.

4. Larch Species

4.1. Potential of Larch in Terms of its Suitability for Chipping and Fragmentation into Fine Particles

Unlike the two previous trees (alder, birch), European larch and other larch species are softwood (coniferous) species. In our considerations, we also consider it a lesser-known and lesser-used wood species for the purpose of particleboard production, and we also assume that in a certain amount it could be added to the mixture of particles in the middle layer of particleboard, similarly to alder and birch woods, as discussed in Sections 2.1 and 3.1. Available data on the potential of larch in terms of its suitability for chipping and fragmentation to fine particles were collected. Due to the specifics of this softwood species, some knowledge gained is different from hardwood (alder, birch) species, but it is usable and valuable for us. This knowledge will also be used in the laboratory works and in the preparations for them in such a way that they serve the final goal—to correctly lay out and think about the composition of the middle layers of the particleboard with the possible incorporation of certain proportions of larch wood.

European larch (*Larix decidua* MILL.) has a colored core in the wood. According to the size of the coloration of the core, two basic forms are distinguished. The sapwood can be narrow and yellowish, the core reddish-brown to deep red; at other times, the sapwood is wider and the core lighter. The annual rings are delimited by darker summer wood. Resin channels are scattered in the spring—lighter wood. The wood is valuable for its color and durability. It should be applicable to the particleboard production as a whole, including all the connections resulting from its detailed composition.

The predominant element of the anatomical structure are tracheids—coils that occupy up to 89–93% of the total volume of the wood. They fulfill a conductive and mechanical function. In summer wood, they are thick-walled and tangentially thickened. The cell wall thickness of spring wood is 3.4–5.9–8.2 μm, and of summer wood it is 6.7–8.9–11.4 μm. The length of the tracheids is 2.3–3.4–4.3 mm. The lumen dimensions of spring wood are in the interval of 35.5–43.0–55.0 μm, and summer wood of 4.6–13.0–21.4 μm. The microscopic parameters of larch wood are different than in the case of the hardwoods investigated and presented; in any case, they indicate good possibilities for splitting and chipping the wood.

Larch wood is light, flexible, easy to work with, splitable, and durable even under water.

An overview of the basic physical properties of larch wood is provided in Table 6. An overview of the basic mechanical properties of larch wood at a moisture content of 12% and a moisture content higher than 30% is provided in Table 7. The durability of larch wood in years is listed in Table 8.

Table 6. Overview of the basic physical properties of larch wood [30,88–90].

Density (kg/m^3)		The Moisture Content of Freshly Cut Wood (%)		Fiber Saturation Point (FSP) (%)
In Absolute Dry State ρ_o	**Moisture Content of 30% ρ_{rf}**	**Sapwood**	**Mature Wood**	
400–550–820	460	100	30–40	23–28
Shrinkage (%)	Longitudinal	Radial	Tangential	Volumetric
	0.3	4.3	10.4	15.0

Table 7. Overview of the basic mechanical properties of larch wood at a moisture content of 12% and a moisture content higher than 30% [29,32,35,36].

Property	Parallel to the Fibers		Perpendicular to the Fibers	
	w = 12%	**w > 30%**	**w = 12%**	**w > 30%**
Tensile strength (MPa)	107			
Compressive strength (MPa)	55.0	41.0	9.7	
Shear strength (MPa)	12.4	6.9		
Bending strength (MPa)	92	64		
Modulus of elasticity in tension (MPa)	14,500			
Modulus of rigidity (MPa)	14,000		600	
Modulus of elasticity in bending (MPa)	13,800	6300		
Toughness (J·cm^{-2})			6.0	4.0
Brinell hardness (MPa)	53		19	
Janka hardness (MPa)	38		36.5	24.5

Larch can be classified as a very durable tree species in terms of its durability in the air. In terms of its durability in contact with the ground, where there is a risk of damage by all types of rot, it is classified as a medium-durable tree species.

Table 8. The durability of larch wood in years [28].

Exposure	Unprotected, Unimpregnated, and Impregnated	Under the Roof	Under the Water	Always Dry
Durability (years)	20–60–80	100–120–150	300–500–700	800–1000

The chemical composition of larch wood is varied by the origin, sampling, age, and state of health. Larch wood contains approximately 50.0% carbon, approximately 42.12% cellulose, 31.95% hemicelluloses, and 26.40% lignin by mass in its absolutely dry state. Larch extractives are typically lipids, phenolic compounds, terpenoids, fatty acids, resin acids, sterols and sterol esters, and waxes. The amount of extractives is always low; in the case of larch, 3.32 [91].

Larch wood was examined in relative detail by many authors, as is evident from the collected values of its physical and mechanical properties from the literature cited. We can therefore assume that these data will be beneficial for the direction of our further research regarding this species. It seems to us that this wood can be very interesting for its use in the central zone of the particleboard, since there is a relationship with spruce, which is currently the main wood used for particleboard production. Section 4.2 provides a summary of the knowledge of various authors about the use of larch wood for particleboard purposes and about the context of larch processing.

4.2. Larch in Particleboard and the Context of Its Processing

This subsection also considers the well-known fact that particleboard is at present a necessary, high-quality, highly represented, and very popular engineered wood-based panel composite manufactured from wood elements and synthetic adhesives or other suitable binders. It consists of varying sizes and shapes of particles from lignocellulosic materials, known and tested raw material, but also lesser-known and even non-wood raw material, bonded together with an adhesive under a high temperature and pressure. Particleboard production has positive effects on the environment due to the utilization of various wood residues. The replacement of input raw materials, even if only in a minor volume, brings various technological changes and must be perfectly solved in terms of the quality of the board offered to customers.

The gluing of larch wood is comprehensively addressed, e.g., in the related paper by Chauzov et al. [92]. Their paper presents a study concerning the influence of the density of larch wood on the strength of adhesive joints and its ability in relation to wet modified adhesive. The physical and mechanical properties of adhesive joints are investigated based on a modified melamine-urea-formaldehyde (MUF) adhesive. The application of MUF adhesives allows the creation of glue joints of larch wood with the required strength characteristics. In the formation of adhesive bonds, a sorting of larch by density is necessary. The use of the MUF adhesive composition may reduce the costs of the glue by 26%.

A clear trend in the production of particleboard today is to achieve quality by reducing the input costs associated with the lack of suitable input raw materials and reducing the density of the particleboards produced. Input raw material is not always an incentive for the quality of the particleboard and a meaningful proportion of different input components is decisive for the quality of the particleboard. From this point of view, we also investigate the possible proportion of larch in the central layer of the particleboard so that the necessary furniture quality of the boards is achieved and the unprocessed larch wood is used for an appropriate purpose.

4.2.1. Previous Experience of Particleboard Production from Larch Wood

Pazio and Boruszewski [93] performed an analysis of the influence of larch fibers and particles on selected properties of particleboard and fiberboard. Their paper presents the results of research on the effect of the addition of particles and fibers obtained from European larch wood (*Larix decidua* Mill.) from plantations on selected properties of particleboard and fiberboard in comparison to boards of the same structure based on typical industrial raw material uses by the wood-based panels industry. The differences were proven in the tests of the internal bond, modulus of rupture, modulus of elasticity in static bending, soaking in water, thickness swelling after 2 and 24 h and the density profile. Larch boards with a minimum 50% fiber share were characterized by the comparable values of the properties determined to standard boards regarding the modulus of rupture and modulus of elasticity tests; however, in the remaining variants, pine boards had better properties. The larch boards were evaluated as boards with significantly lower values of swelling by thickness than boards made of wood from forest cultivation. On the other hand, the density profile of the boards on the cross-section of the plantation raw material did not differ from the boards made of pine raw material.

Some authors have examined particleboard using larch in 100% representation, which is not a long-term solution in terms of practice, but it is necessary to become acquainted with the results of their research. Muhcu et al. [94] investigated the log position impact of the European Larch (*Larix decidua* Mill.) tree on the particleboard surface properties as well as the physical and mechanical properties. The logs used by them were split into five segments from the tree butt to the treetop. The wall thickness and fiber length of the wood decreased with the increasing tree height, while the lumen diameter reduced. Likewise, the amount of lignin and cellulose decreased with an increasing tree height, and on the other hand, the number of hemicelluloses increased. The highest values of the solubility (cold and hot water, alcohol-benzene, NaOH) and wood pH were detected in the butt log, continued by the middle log, and finally, the top log. The mechanical properties (internal bond, modulus of rupture, modulus of elasticity) and physical properties (water absorption, thickness swelling), as well as the particleboard surface quality (contact angle and surface roughness), were negatively affected by an increased tree height. The best particleboard properties were obtained for the boards produced from the wood elements of the butt log (0–3 m).

Bardak et al. [95] dealt with the effect of the utilization of sapwood and heartwood on the particleboard surface properties, mechanical and physical properties, and the formaldehyde emissions from particleboard. Trees of European Larch (*Larix decidua* Mill.) were chosen by them as raw materials. The logs were divided into only up to three segments: heartwood, sapwood, and total wood. The authors were interested in the proportions of cellulose. The amounts of it were 51.54%, and that was, of course, the highest number and hemicelluloses (22.24%) in the sapwood, followed by the total wood, and finally, the heartwood. On the other hand, the amount of lignin (30.54%) was found in the heartwood and that was the highest proportion. The extractive values were also surveyed; the highest were obtained from heartwood, followed by total wood, and finally, sapwood. The lowest pH value (3.03) was found in heartwood, while the sapwood samples provided the highest values (4.95). The test samples manufactured from pure sapwood have the smoothest surface and the lowest contact angles; on the other hand, the roughest surface and highest contact angle were acquired from the panel boards of pure heartwood. The formaldehyde emission values and the thickness swelling of the particleboard prepared from pure heartwood were significantly lower than the particleboard prepared from total wood and pure sapwood. The highest internal bond values, and the values of the modulus of rupture and modulus of elasticity, were identified in pure sapwood. The particleboard mechanical strength values are followed by the total wood and the pure heartwood. The particleboard wettability and the surface smoothness manufactured from pure sapwood are higher than those of total wood and pure heartwood.

Lee and Chung [96] assessed larch wood from the point of view of the effect of the temperature and press time on the physical properties of larch particleboard and it appeared to them that larch can be used for making particleboard.

Matsumoto et al. [97] looked at the problem of the effects of the particle size on the properties of particleboard made from larch. They determined the appropriate sizes of the elements, and they concluded that larch is a suitable tree species for making particleboard.

This is crucial information if we consider that this research took place at the end of the last century and its results are fully applicable for the intended pressing of larch particles in the mixture in particleboard. We are aware that adhesives and adhesive mixtures for particle pressing are different, more environmentally friendly and less harmful, but the published dimensions of the particles (and fibers) are clearly valuable and workable.

Also, Dix and Roffael [98] have been active in larch research since 40 years ago and they examined the pure larch particleboard from the point of view of the mechanical and technological properties and, in particular, the impact of larch sapwood and heartwood on the particleboard quality. This is all very stimulating information for us.

Shupe et al. [99] determined the effect of five different silvicultural strategies and wood types on the physical and mechanical properties of loblolly pine (*Pinus taeda* L.) particleboard as well as fiberboard. Their entire research is based on comparing their particleboard with particleboard made from larch (*Larix decidua* Mill.). The furnish was prepared unconventionally from veneer outerwood and innerwood for each stand. Regarding the particleboard, the modulus of rupture differences between the stands were insignificant. Certain significant moduli of elasticity differences existed between the stands for fiberboard and particleboard. The differences between the wood types were irrelevant for each stand. For most of the stands, innerwood yielded a higher mean of mechanical values than outerwood. The differences between the stand and wood types for the 2 and 24 h thickness swelling and water adsorption were very small. These trials have shown that innerwood can produce fiberboard as well as particleboard panels with very comparable physical and mechanical properties to outerwood at the level of particleboard made from larch. The effect of the silvicultural strategy (i.e., stand) was almost insignificant for most properties.

Simatupang et al. [100] examined larch, birch, poplar, and oak wood cut in spring when seasoned in the open air. Birch and poplar were processed by chipping after periods of 2, 4, 8, 16, and 32 weeks. One portion of the chips was artificially dried or extracted with water. The other portion was used immediately, and the laboratory boards were elaborated. The researchers used a rather unconventional binder, but their research is nonetheless remarkable. The binders used by them were pure cement or a mixture of cement with condensed silica fume. The boards' bending strength increased with seasoning time for all the investigated woods but decreased after 32 weeks of seasoning. Artificial drying of the elements and the addition of condensed silica fume led to higher bending strength values. Larch, birch, and poplar wood seasoned for two weeks and oakwood seasoned for two and twelve weeks were used to make composites with different amounts of condensed silica fume (optimum between 25 and 45%). For the usually unsuitable larch and oak wood, the addition of silicon provided boards with good strength values. In their research, it is summarized that nearly all lignocellulosic materials can be used to produce cement-bonded particleboard if condensed silica fume is present.

4.2.2. Inorganic Adhesives Used for the Particleboard Production from Larch

A well-known expert in joining wood with cement, Moslemi [101] considered a full range of particleboard treatments to determine the treatments' impact on wood–cement cohesion. This was another study of his with a cement binder; however, the species used was larch. His research treatments included hot water extraction of water-soluble components in wood or in chemical additives. The wood species involved included western larch and lodgepole pine. These species turned out to be highly inhibitory in a series of hydration tests. The obtained data presented in the paper show that substantial improvements in cement setting can be achieved by the removal of sugars and water-soluble extractives from western larch. In the case of lodgepole pine, such improvements did not take place. The data present that lodgepole pine has far less of an inhibitory effect than larch in the untreated state. The subsequent addition of chemical additives, especially calcium chloride, appears to enhance compatibility in cement–wood–water mixes. Thus, larch turned out to be a more suitable tree species than pine.

An important study was performed by Simatupang and Geimer [102] with larch wood particles to increase the compatibility of wood-based composites manufactured with inorganic binders of gypsum, Portland cement, or magnesia cement. The result of a series of chemical reactions is the setting of inorganic binders, causing a succession of crystallization stages. Crystallization is inhibited to a diverse degree by various wood kinds and their extractives. The inhibition effect can be reduced by wood aging, and it can be counteracted to certain degree with chemical additives. There have been developed new incorporating processes of solid, liquid, or gaseous additives with the aim of promoting the rapid curing of inorganic bonded boards based on the chemical composition of the wood used. We know from this paper as well as from other papers that larch is an acceptable wood species for these technological operations. The raw material for the binder was obtained as a by-product of other processes in some cases, thus lowering the costs as well. The expansion of inorganic bonded board markets worldwide depends on an increased need for building homes, the availability of materials, as well as on the regulatory codes and changes in established building practices.

Keegan et al. [103] have dealt in detail with larch processing from the point of view of its use in reconstituted board products such as particleboard and medium-density fiberboard. They stated that although there exists information on the use of larch as a raw material to produce pulp and paper, and this is also a good solution, larch as a raw material is a more suitable and economical solution in particleboard or fiberboard products.

4.2.3. Analysis of Chemical Composition of Larch Wood and Bark as a Factor in Particleboard Production

The chemical analyses performed of larch wood or larch bark published in scientific publications seem to be stimulating the research results for incorporating larch particles into particleboards, e.g., Gou et al. [104]. They characterized the nitrogen forms in larch wood particleboard. They used X-ray photoelectron spectroscopy (XPS) in the research to test the existing forms of nitrogen in larch wood, urea-formaldehyde (UF) adhesive, and finally, the particleboard made from larch wood elements and UF adhesive. Their results pointed out that the present nitrogen forms in larch wood are the pyrrolic structures and the amine family, and the nitrogen form in UF adhesive is the amine family only. As expected, the forms of nitrogen in larch wood particleboard and UF adhesive are the amine family and structures, and the percentage of amine nitrogen structures partially increased if compared to the percentage in larch wood. Their results contribute to the research on the release characterization of nitrogen wood species during heating and during wood-based panel pressing.

Liu et al. [105], in their study, investigate the volatile organic compound (VOC) emission of particleboard made of larch under diverse processing conditions. To analyze the VOC components and quantities, they used the equipment of a VOC collection chamber and gas chromatography–mass spectrometer. The emission rate and the concentration of the VOC were substantially affected by the hot-pressing time and temperature. With an increasing hot-pressing time and temperature, both the earlier emission concentration and the amount of total VOC increased. The composition of VOC was also influenced by the time and temperature, mostly the variety of benzene, terpene, and derivative. As for the esters, their existence and quantities were still the main components of VOC emissions.

Lo and Liew [106] divided their study of the Acacia hybrid into bark, sapwood, and heartwood, and they compared their results with other woody species, including larch. Each portion underwent Soxhlet extraction. During the particleboard production, a considerable number of wood elements was extracted using the ratio of extraction of wood particles to solvent extraction. The wood elements were bonded using urea-formaldehyde adhesive and the targeted particleboard density was 500 kg/m^3. Particleboard made from methanol-extracted wood elements for heartwood and sapwood had a provable difference in mechanical properties compared to the control sapwood particleboard. On the other hand, the heartwood extracted by methanol was higher as compared to the control heartwood. In general, it can be stated that the wood elements extracted by different extraction solvents could be used in particleboard production, but better particleboard properties were proven for methanol-extracted sapwood and heartwood and for hexane-extracted bark. The importance of other wood species for particleboard production is not disputed, and their proposals for achieving higher mechanical properties are probably more complex in terms of the instrumentation and more time-consuming.

Tudor et al. [107] analyzed the possible bark capacity of larch for formaldehyde removal in wood adhesives. Their research activities described the methods of wood-based composite production from larch bark. Five different types of adhesive systems were used (urea-formaldehyde, the mixture of 70% urea-formaldehyde + 30% polyvinyl acetate, polyvinyl acetate, polyurethane, and tannin-based adhesive) and the formaldehyde content of the composites based on larch bark (600 kg/m^3) was analyzed. A self-agglomerated board was analyzed as well. The findings of the research are that all the tested samples reached the E1 classification ($\leq$8 mg/100 oven dry), and especially in the case of particleboard bonded with tannin-based adhesive, the natural polymer acted as a significant formaldehyde scavenger.

Kain et al. [108], within their study, discussed thermal insulation wood-based panels made from larch, spruce, fir, pine, and oak tree bark with various adhesives (urea-formaldehyde, melamine-formaldehyde, Quebracho, Mimosa) as the wood element binder. The properties of wood-based panels made from larch bark mixed with industrial popcorn were investigated as well. The mechanical and physical properties of the panels were analyzed. They are significantly dependent on the particle size, panel density, bark species, adhesive type, and adhesive content. The bark species has an irrelevant influence on the mechanical properties of the wood-based panels, but the compression ratio is important for the panel strength and stiffness, and therefore, barks with lower bulk density are more preferable. Panels made with green tannin resins under laboratory conditions proved to have adequate properties for practical applications. The addition of a certain amount of popcorn is a relatively suitable means of decreasing the panel density, but the water absorption of such panels shall be comparatively increased. The bark type has a meaningless influence on the thermal conductivity of the wood-based panels; this parameter is predominantly affected by the panel density.

Further scientific research by Tudor et al. [109] aimed at the sound-absorption coefficient of larch bark-based insulation panel products and that is another look at the use of larch wood (and bark) and information for incorporating larch wood into particleboard. The main objective of this study was to investigate the sound absorption coefficient of bark-based insulation panel products made of softwood spruce bark (*Picea abies* (L.) H. Karst.) and larch bark (*Larix decidua* Mill.) with the help of the use of an impedance tube, with its frequency range set between 125 and 4000 Hz. The highest efficiency of sound absorption was noticed for spruce bark-based insulation panels bonded with urea-formaldehyde adhesive, at a level of 1000 and 2000 Hz. The same frequency interval covers the potential for noise reduction of larch bark-based panels glued with tannin-based adhesives. The experimental data show that softwood bark, which is generally considered an underrated material, can substitute for rather expensive materials that involve more gray energy in sound insulation applications. The sound absorption coefficient values therefore strengthen the application of insulation panels based on bark from the forest trees as structural elements for noise reduction in current residential buildings, and they concurrently open up new ways for deeper research in this field.

Kain et al. [110] performed an interesting qualitative investigation on VOC emissions from Norway spruce (*Picea abies*) and larch (*Larix decidua*) loose bark and bark-based panels. Spruce and larch barks were dried first using different methods. Then, the barks were prepared as loose wood elements and panel products. The VOC emissions were measured in a small chamber after 3 and 14 days using gas chromatography and coupled mass spectroscopy. This quantified the influence of the bark type, drying method, and time of hot pressing on the VOC emissions. There were higher total VOC emissions from spruce bark than those from larch bark. High pressing temperature treatments and pressing time significantly decrease emissions from the investigated bark types. In the emitting gases, terpenes, aldehydes, and acids were analyzed within the study.

Liu et al. [111] conducted research on the particleboard mechanical properties and particleboard formaldehyde emissions with nanomaterial-added melamine-impregnated papers, and larch particles were the input raw material for particleboard production. They investigated the influences of the different surface hot-pressing parameters, including the time, temperature, and pressing pressure, on the particleboard properties with nanomaterial-added melamine-impregnated papers. An orthogonal experimental design was employed to investigate the impact of different factors and to determine the optimum levels of the parameters to attain the best desirable board properties. The physical and mechanical properties, including the internal bond strength, modulus of rupture, thickness swelling, and formaldehyde emissions, were determined. With the incorporation of nanoparticles into the adhesives, the physical and mechanical properties of the coated particleboard improved, and the formaldehyde emissions reduced accordingly.

Liu et al. [112] identified and characterized the odorous volatile organic compounds (VOC) emitted from wood-based panel products, and larch elements were in their sights with particleboard production again. VOCs with low concentrations, even below the detection limit, can have the negative effects as well. Therefore, VOCs emitted from particleboard and other laminated boards were precisely collected and analyzed, and the main odorous compounds were identified using the odor activity value. Compared with laminated boards, particleboard demonstrated a higher concentration of total VOCs. Halogenated compounds, esters, and aromatic hydrocarbons were the essential contributors to the total VOCs emissions, which could relate to the additive's addition when used for panel manufacturing.

Even Bednarczyk and Boruszewski [113] researched larch particleboard, particularly lightweight particleboard, with the manufacturing modification using a blowing agent from the group of bicarbonates. The low-density particleboard has grown in popularity in recent years due to its facilitated transportation and the lower own mass in the assembled finished products. However, there are still certain restrictions in the general utilization caused by the insufficient, sometimes sorely lacking, mechanical properties. These deficiencies may be controlled by the mechanism of foaming the polymers that bind wood particles in the structure of the boards. The study aimed to determine the possibility of using sodium bicarbonate as a blowing agent of phenolic resin used for bonding wood particles in the technology of lightweight particleboard. It was found that the addition of sodium bicarbonate in the amount of 5% of the dry weight of the phenolic resin substantially increased the internal bond strength of ready-made the particleboard.

The idea of adhesives-free bark panels is also interesting and is addressed in some scientific works, e.g., Wenig et al. [114]. They engaged with adhesives-free bark panels as an alternative application for an accumulated waste material. The motivation for this research was to conserve the bark in a natural common state and to explore alternative processes and applications for its utilization. The common method of tree trunk debarking by peeling was used to obtain, if possible, large bark pieces. The procedure was that two pieces of peeled bark were placed crosswise, approximately with the rhytidome sides (outer bark) facing each other. After diverse conditioning conditions, the bark pieces were pressed with the help of a high temperature to panel products without using any adhesives. The experiment on bark samples of various Central European tree species (including larch) revealed that panel product production with species-dependent properties is basically possible and feasible with certain changes in technology. These are the steps toward the production of the sustainable panel products by the incorporation of the natural waste material, and at the same time, to keep its beneficial structure and its naturally real chemical composition.

Kajita and Imamura [115] carried out a study with chemically modified particleboard and fiberboard made from larch (*Larix leptolepis* G.), mostly to improve the mechanical properties, but also the dimensional stability and biological properties, too. The upgraded technology developed by the authors from the solid wood chemical modification was applied to these targeted reconstituted wood products. Their research report described the means of production and the identified properties of the chemically modified panel boards.

It seems that nowadays larch bark has been more discussed in the research than larch wood. An example is the scientific work by Ninikas et al. [116]. They examined the technical sensuality of low-density insulation particleboard manufacturing. Particleboards were produced from two renewable resources, first tree bark and second hemp fibers (*Cannabis sativa*). Non-toxic methyl cellulose glue was used by them as a binder. Four types of panel products were made, which corresponded to varying mixtures of tree bark (including larch particles) and hemp fibers (tree bark to hemp fibers per-centages of 90:10, 80:20, 70:30, and 60:40). An additional set of panels was produced, consisting only of tree bark as a comparative material. The results achieved pointed out that the hemp fiber addition partially improved the mechanical properties of the boards and an acceptable level of the boards was achieved. However, a negative side effect was that the thermal conductivity increased as the hemp content increased. However, it was gratifying that all the values were still within the acceptable range. Board type 30:70 (with hemp content of 30%) had the highest mechanical properties as well as the optimal thermal conductivity value based on the cluster analysis. So, based on this research, it was concluded that insulation panels with a low density can be successfully produced using these renewable waste raw materials.

The work by Oktoberyani et al. [117] is also worth noting. The authors fabricated the bark particleboard without binders from gelam (*Melaleuca viridiflora* Sol. ex Gaertn.) and other woody species, including larch. They looked at the issue in terms of bark as a waste product. They investigated the influence of the pressing temperature on the particleboard's physical and mechanical properties. The thermal insulation particleboard properties were determined by them as well. Four different temperatures (140 °C, 160 °C, 180 °C, and 200 °C) were used, with the aim to laboratory-produce single-layer binderless bark particleboard with a target density of less than 590 kg/m^3. Their results showed that the pressing temperature, of course, affected the mechanical properties (tensile strength perpendicular to panel surface, modulus of rupture, and modulus of elasticity) and these mechanical properties increased as the pressing temperature increased. The particleboard's physical properties (water absorption and thickness swelling) decreased as the pressing temperature increased. However, the increase in the pressing temperature from 180 °C by 20 °C more did not affect the particleboard's physical or mechanical properties significantly, except for the tensile strength perpendicular to the panel surface. Another positive was that the binderless bark particleboard that was hot-pressed at 200 °C had high water resistance, regardless of its low strength. Its thermal conductivity value was 0.14 W/m·K.

The study by Czajka and Fabisiak [118] determined the radial variation in the content of cellulose and lignin in the cross-section of the chosen softwood species, including larch (*Larix decidua* Mill.). The average values of the content of cellulose and lignin calculated for the entire cross-sections of the larch wood were very similar. In the case of cellulose, for larch wood its average content was 48.29%, and the average content of lignin was 27.93%. Lower values of the cellulose share and higher of the lignin share were obtained for the juvenile larch wood zone compared to the mature larch wood zone.

4.2.4. Overall Evaluation of the Context of Larch Wood's Possibilities in Particleboard Production

It is clear from the mentioned and other unmentioned literature sources that larch wood particles and larch bark particles are suitable prerequisites for their processing into particleboard. Assortments of raw larch wood (*Larix decidua* MILL.) are intended for mechanical processing by sawing technologies as well. Larch is easy to use in this way. Like other types of softwoods, larch cutouts are transformed into lumber by cutting, primarily with frame saws. The mechanical as well as the physical properties of larch wood, but above all its unique color and texture, predetermine it for the production of all products, sometimes more than the industrially more important spruce wood. Its more limited application is due to its lower occurrence in forests compared to spruce. Larch lumber is highly valued in construction and carpentry products, garden architecture products, and exterior and interior floors and coverings. Of course, it achieves the highest market value in the production of all kinds of rustic furniture—sofas, tables, cabinets, and beds. The highest quality assortments of larch wood are used in shipbuilding for structural but mainly decorative purposes. Larch wood dries more slowly than other conifers, and it tends to slightly warp and crack. During natural drying, it is stored with smaller gaps between the sides of the lumber, and thick lintels are not used. Smaller distances are left between the cages, which shade them from the sun. The cages are loaded and protected against rain. For artificial drying, hot-air dryers are mainly used, but also vacuum and condensation dryers. During drying, there is no significant change in the color of the wood. An accompanying phenomenon is the oozing of resin and the release of non-healed knots. For decorative purposes, larch wood is processed in veneer production with cutting machines. These are the highest-quality cutouts. Quality assortments of larch raw wood of a slightly lower quality class are the raw materials for the production of structural veneers or plywood panels. Larch wood of a low quality class is used for the pulp and paper industry, as well as for the production of agglomerated boards, for the production of packaging, and pallets, and for the production of toys, jewelry, and beekeeping supplies. Sometimes larch is mistakenly used in products as an alternative to spruce, fir, and pine wood, but as this literature review clearly shows, this is not correct. Larch has valuable properties and should be considered a first-class and full-fledged woody species [28].

From this collected theoretical information, it appears that larch wood will be usable for particleboard production and for bonding the particles in a mixture with spruce particles into the core layers of particleboard. Although there is less information than for birch wood, in the context of using a mixture of two coniferous species with rather similar chemical compositions (larch and spruce), this board-forming process is probably feasible. In terms of the proportion of larch wood shares in the particleboard core layer (similar to the alder wood and birch wood described in Sections 2 and 3), this will be the subject of future research activities, with which all this collected information will assist.

The latest publication by Akbulut and Ayrılmış [119] confirmed the current optimal technologies for particle production from massive wood and confirmed the optimal procedures for gluing particles into particleboard as well. The information contained in it is also an assumption that modern technologies used for particleboard production are suitable for processing lesser-known and lesser-used wood species, and that no additional investments are required. Wood, as a raw material for particleboard production in 100% representation, will always be the preferred option compared to alternative shares of non-wood raw materials.

5. Conclusions

Lesser-known European species of alder, larch, and birch may be used as additives to major input raw material in the particleboard core layer. This literature search confirmed the diversity of use of these three trees in different fields and unequivocally showed their adequate physical and mechanical properties and chemical composition as usable tree species to cover part of the raw material needs for particleboard.

Of course, the greatest experience in use is with Nordic varieties of birch, which is a very valuable raw material in Northern Europe. However, birch is beginning to gain ground in other parts of the world, and the analysis of its wood and chips in the cited works confirmed this. Birch is also likely to prove useful in the experiments we intend to perform as part of the upcoming research.

Larch is currently a highly researched tree species and not only larch wood but also larch bark has great potential. In particular, the use of larch bark in composites is researched at several research institutes around the world, and the properties of larch composites are tested and refined by various procedures, including experimental ones, to achieve excellent results.

Alder wood has such a beautiful and original color that it is predestined for the manufacture of sliced veneers. Of course, in this case, they must be trees with certain meaningful diameters so that sliced veneers with the necessary dimensions are produced. Smaller diameters of alder wood are very suitable for making particleboard in terms of their chemical and technical properties and their incorporation into particleboard is confirmed by this literature review.

These three tree species provide an eco-friendly and sustainable alternative to the conventional tree species used in the wood-based panel industry. They cannot be a full-fledged replacement, given that their occurrence in forests and wood reserves is not very high, but they can reduce the spruce raw material in particleboard, which is less accessible for the composite industry and there exists a great demand for it in other industrial areas. However, in this respect, fundamental research is still needed to determine the proportional shares of use of these tree species in particleboard for formulating the correct technology shares and rules for their application in the wood-based panel industry. This literature survey confirmed that the potential of alder, birch, and larch for use in the center layer of particleboard is considerable.

Author Contributions: Conceptualization, R.R.; methodology, R.R. and T.P.; validation, L.K. and P.K.; formal analysis, L.K. and T.P.; investigation, R.R. and M.J.; resources, R.R. and L.K.; data curation, L.K.; writing—original draft preparation, R.R. and L.K.; writing—review and editing, R.R. and L.K.; visualization, R.R. and L.K.; supervision, R.R.; project administration, R.R. and L.K.; funding acquisition, R.R. and L.K. All authors have read and agreed to the published version of the manuscript.

Funding: This research was supported by the Slovak Research and Development Agency under contracts No. SK-CZ-RD-21-0100, No. APVV-20-0004, by the Ministry of Education, Science, Research and Sport of the Slovak Republic under the project VEGA 1/0077/24 and by the Ministry of Education, Youth and Sport of the Czech Republic through project No. INTER-EXCELLENCE—LUASK22094 "Analysis of the properties of less-known European wood species in composite materials".

Conflicts of Interest: The authors declare no conflicts of interest.

References

1. Available online: https://www.mordorintelligence.com/industry-reports/particle-board-market (accessed on 1 June 2023).
2. Available online: https://www.researchandmarkets.com/report/particle-board (accessed on 1 June 2023).
3. Available online: https://www.indexbox.io/blog/global-particle-board-market-2019-key-insights/ (accessed on 1 June 2023).
4. Available online: https://www.reportsanddata.com/report-detail/particle-board-market (accessed on 1 June 2023).
5. Barbu, M.C.; Irle, M.; Reh, R. Wood Based Composites, Chapter 1. In *Research Developments in Wood Engineering and Technology*; Engineering Science Reference; Aguilera, A., Davim, P., Eds.; IGI Global: Hershey, PA, USA, 2014; pp. 1–45, ISBN 978-1-4666-4554-7.
6. Available online: https://www.sciencedirect.com/topics/agricultural-and-biological-sciences/particleboards (accessed on 1 June 2023).
7. Niemz, P.; Sandberg, D. Critical wood-particle properties in the production of particleboard. *Wood Mater. Sci. Eng.* **2022**, *17*, 386–387. [CrossRef]
8. Kamke, F.A. Solid wood products/Wood-based composites and panel products. In *Encyclopedia of Forest Sciences*; Burley, J., Ed.; Elsevier: Amsterdam, The Netherlands, 2004; pp. 1338–1345, ISBN 9780121451608. [CrossRef]
9. Available online: https://ec.europa.eu/eurostat/statistics-explained/index.php?title=Wood_products_-_production_and_trade (accessed on 1 June 2023).
10. Youngquist, J.A. Wood-based composites and panel products. In *Wood Handbook: Wood as an Engineering Material*; General Technical Report; FPL–GTR:113; Forest Product Laboratory: Madison, WI, USA, 1999; Chapter 10.
11. Irle, M.A.; Barbu, M.C.; Reh, R.; Bergland, L.; Rowell, R.M. Wood composites. In *Handbook of Wood Chemistry and Wood Composites*; CRC Press: Boca Raton, FL, USA, 2012; pp. 321–411, ISBN 978-1-4398-5380-1.
12. Bentsen, N.S.; Felby, C.; Thorsen, B.J. Agricultural residue production and potentials for energy and materials services. *Prog. Energy Combust. Sci.* **2014**, *40*, 59–73. [CrossRef]
13. Pędzik, M.; Janiszewska, D.; Rogoziński, T. Alternative lignocellulosic raw materials in particleboard production: A review. *Ind. Crops Prod.* **2021**, *174*, 114162. [CrossRef]
14. Mahieu, A.; Vivet, A.; Poilane, C.; Leblanc, N. Performance of particleboards based on annual plant byproducts bound with bio-adhesives. *Int. J. Adhes. Adhes.* **2021**, *107*, 102847. [CrossRef]

15. Barbu, M.C.; Reh, R.; Çavdar, A.D. Non-wood lignocellulosic composites, Chapter 8. In *Research Developments in Wood Engineering and Technology*; Aguilera, A., Davim, P., Eds.; Engineering Science Reference; IGI Global: Hershey, PA, USA, 2014; pp. 281–319, ISBN 978-1-4666-4554-7.
16. Available online: https://8billiontrees.com/trees/how-many-trees-are-in-the-world/ (accessed on 1 June 2023).
17. Available online: https://www.forestportal.sk/odborna-sekcia-i/informacie-o-lesoch/zakladne-informacie-o-lesoch/rozdelenie-lesov/ (accessed on 10 March 2024).
18. Available online: http://www.czechforest.cz/informace-o-lesich#:~:text=Dle%20v%C3%BDsledk%C5%AF%20N%C3%A1rodn%C3%AD%20inventarizace%20les%C5%AF%20p%C5%99eva%C5%BEuj%C3%AD%20v%20%C4%8CR%3A,nejv%C3%ADce%20zastoupen%C3%BDm%20jehli%C4%8Dnanem%20je%20smrk%20ztepil%C3%BD%20%2843%2C1%20%25%29 (accessed on 10 March 2024).
19. Available online: https://www.globalforestwatch.org/dashboards/country/AUT/?category=forest-change&location=WyJjb3VudHJ5IiwiQVVUIl0%3D (accessed on 10 March 2024).
20. Available online: https://ec.europa.eu/eurostat/statistics-explained/index.php?title=Wood_products_-_production_and_trade#Wood-based_industries (accessed on 10 March 2024).
21. Available online: https://jwoodscience.springeropen.com/articles/10.1007/s10086-015-1534-3 (accessed on 10 March 2024).
22. Available online: https://academic.oup.com/forestry/article/83/1/103/546795 (accessed on 10 March 2024).
23. Available online: https://treepursuits.com/where-do-birch-trees-grow-map/?utm_content=cmp-true (accessed on 10 March 2024).
24. Available online: https://openknowledge.fao.org/server/api/core/bitstreams/d27d7ec2-fa9e-46b3-a0e7-529bbc424f6a/content (accessed on 10 March 2024).
25. Available online: https://www.gotreequotes.com/how-many-trees-in-world/ (accessed on 1 June 2023).
26. Available online: https://worldpopulationreview.com/geography/how-many-trees-are-in-the-world (accessed on 1 June 2023).
27. Auer, V.; Rauch, P. Wood supply chain risks and risk mitigation strategies: A systematic review focusing on the Northern hemisphere. *Biomass Bioenergy* **2021**, *148*, 106001. [CrossRef]
28. Klement, I.; Reh, R.; Detvaj, J. *Základné Charakteristiky Lesných Drevín—Spracovanie Drevnej Suroviny V Odvetví Spracovania Dreva*; NLC: Zvolen, Slovakia, 2010; 80p, ISBN 978-80-8093-112-4.
29. Available online: https://www.wood-database.com/european-alder/ (accessed on 5 June 2023).
30. Balabán, K. *Nauka o Dřevě. 1.Část, Anatomie Dřeva*; D-01278; Lesnická knihovna; Státní zemědělské nakladatelství: Praha, Czech Republic, 1955; Volume 16, 216p.
31. Perelygin, L.M. *Náuka o Dreve*; DT 674.02; Slovenské vydavate'stvo technickej literatúry: Bratislava, Slovakia, 1965; 444p.
32. Available online: https://www.woodlandtrust.org.uk/trees-woods-and-wildlife/british-trees/a-z-of-british-trees/alder/ (accessed on 5 June 2023).
33. Coombes, A.J. *Trees*; Dorling Kindersley: New York, NY, USA, 1992; 320p, ISBN 10: 0863188125.
34. Kremer, B.P. *Bäume: Heimische und eingeführte Arten Europas*; Genehmigte Sonderausgabe; Mosaik Verlag: München, Germany, 2000; ISBN 3-572-01086-1.
35. Available online: https://www.hardwoodinfo.com/specifying-professionals/species-guide/species-guide-a-g/alder/ (accessed on 5 June 2023).
36. Požgaj, A.; Chovanec, D.; Kurjatko, S.; Babiak, M. *Štruktúra a Vlastnosti Dreva*; Príroda: Bratislava, Slovakia, 1993; 485p, ISBN 80-07-00600-1.
37. Nečesaný, V. Stavba dřeva. In *Dřevařská Příručka I*; SNTL—Nakladatelství technické literatury: Praha, Czech Republic, 1989; pp. 11–28, ISBN 80-03-00009-2.
38. Forest Products Laboratory. Wood handbook. In *Wood as an Engineering Material*; General Technical Report FPL-GTR-282; Department of Agriculture, Forest Service, Forest Products Laboratory: Madison, WI, USA, 2021; 543p.
39. Nemlį, G. Effects of Some Manufacturing Factors on the Properties of Particleboard Manufactured from Alder (*Alnus glutinosa* subsp. Barbata). *Turk. J. Agric. For.* **2003**, *27*, 6. Available online: https://journals.tubitak.gov.tr/agriculture/vol27/iss2/6 (accessed on 1 June 2023).
40. Tupciauskas, R.; Veveris, A.; Andzs, M.; Gravits, J.; Liitia, T.; Tamminen, T. Investigation of a Grey Alder Particleboard Bonded by Different Industrial Side-Stream Lignins. *Mech. Compos. Mater* **2021**, *57*, 57–68. [CrossRef]
41. Kumaş, İ.; Aras, U.; Kalaycioğlu, H.; Serdar, B.; Yel, H. Effect of growing altitude, particle drying temperature and press temperature on some technological properties of particleboard produced from black alder wood. Drewno. *Pr. Naukowe. Doniesienia. Komun.* **2021**, *64*, 59–71. [CrossRef]
42. Geffert, A.; Geffertová, J.; Dudiak, M.; Výbohová, E. Influence of steaming temperature on chemical characteristics and colour of alder wood. In *Chip and Chipless Woodworking Processes*; Technical University in Zvolen: Zvolen, Slovakia, 2020; Volume 12, pp. 49–56, ISSN 2453-904X.
43. Janceva, S.; Andersone, A.; Spulle, U.; Tupciauskas, R.; Papadopoulou, E.; Bikovens, O.; Andzs, M.; Zaharova, N.; Rieksts, G.; Telysheva, G.; et al. Eco-friendly adhesives based on the oligomeric condensed tannins-rich extract from alder bark for particleboard and plywood production. *Materials* **2022**, *15*, 3894. [CrossRef] [PubMed]

44. Öztürk, H.; Demir, A.; Demirkir, C. Prediction of Optimum Expanded Polystyrene Densities for Best Thermal Insulation Performances of Polystyrene Composite Particleboards by Using Artificial Neural Network. *Drv. Ind.* **2022**, *73*, 385–395. [CrossRef]
45. Öztürk, H.; Çolakoğlu, G.; Demirkir, C. Effect of drying types and polystyrene density on thermal conductivity of polystyrene composite particleboard. *Wood Ind. Eng.* **2019**, *1*, 57–62.
46. Szadkowska, D.; Auriga, R.; Lesiak, A.; Szadkowski, J.; Marchwicka, M. Influence of Pine and Alder Woodchips Storage Method on the Chemical Composition and Sugar Yield in Liquid Biofuel Production. *Polymers* **2022**, *14*, 3495. [CrossRef]
47. Lee, S.H.; Zaidon, A.; Rasdianah, D.; Lum, W.C.; Aisyah, H.A. Alteration in colour and fungal resistance of thermally treated oil palm trunk and rubberwood particleboard using palm oil. *J. Oil Palm. Res.* **2020**, *32*, 83–89.
48. Christy, E.O.; Sumarlan, S.H.; Soehardjono, A. Binderless Bark Particleboard Made from Gelam (*Melaleuca viridiflora* Sol. ex Gaertn.) Bark Waste: The Effect of the Pressing Temperature on Its Mechanical and Physical Properties. *BioResources* **2021**, *16*, 4171. [CrossRef]
49. Dukarska, D.; Pędzik, M.; Rogozińska, W.; Rogoziński, T.; Czarnecki, R. Characteristics of straw particles of selected grain species purposed for the production of lignocellulose particleboards. *Part. Sci. Technol.* **2021**, *39*, 213–222. [CrossRef]
50. Salca, E.A. Black alder (*Alnus glutinosa* L.)—A resource for value-added products in furniture industry under European screening. *Curr. For. Rep.* **2019**, *5*, 41–54. [CrossRef]
51. Torun, S.B.; Cavdar, A.D.; Ozdemir, T. The synergistic effect of intumescent coating containing titanium dioxide and antimony trioxide onto spruce and alder wood species. *J. Build. Eng.* **2020**, *31*, 101407. [CrossRef]
52. Lovaglio, T.; Gindl-Altmutter, W.; Meints, T.; Moretti, N.; Todaro, L. Wetting behavior of alder (Alnus cordata (Loisel) Duby) wood Surface: Effect of thermo-treatment and alkyl ketene dimer (AKD). *Forests* **2019**, *10*, 770. [CrossRef]
53. Bekhta, P.; Pipíška, T.; Gryc, V.; Sedliačik, J.; Král, P.; Ráheľ, J.; Vaněrek, J. Properties of Plywood Panels Composed of Thermally Densified and Non-Densified Alder and Birch Veneers. *Forests* **2023**, *14*, 96. [CrossRef]
54. Bekhta, P.; Krystofiak, T. Performance and Modification of Wood and Wood-Based Materials. *Forests* **2023**, *14*, 963. [CrossRef]
55. Öncel, M.; Vurdu, H.; Aydoğan, H.; Özkan, O.E.; Kaymakci, A. The tensile shear strength of outdoor type plywood produced from fir, alnus, pine and poplar wood. *Wood Res.* **2019**, *64*, 913–920.
56. İlkay, A.T.A.R.; Mengeloğ, F. The effect of ply combination on mechanical and physical properties in laminated veneer lumber (LVL) manufactured from alder and poplar. *Turk. J. For. Sci.* **2022**, *6*, 412–426.
57. Romanenko, M. Influence of Pressing Temperature on Wood Properties. *Mater. Res. Proc.* **2022**, *21*, 32. [CrossRef]
58. Pędzik, M.; Stuper-Szablewska, K.; Sydor, M.; Rogoziński, T. Influence of grit size and wood species on the granularity of dust particles during sanding. *Appl. Sci.* **2020**, *10*, 8165. [CrossRef]
59. Sydor, M.; Mirski, R.; Stuper-Szablewska, K.; Rogoziński, T. Efficiency of Machine Sanding of Wood. *Appl. Sci.* **2021**, *11*, 2860. [CrossRef]
60. Qiu, T.; Zhang, X.; Lu, S.; Zhang, G.; Hou, L. Characterization of combustion behavior of commercial flame-retardant plywoods in fire propagation apparatus. In *IOP Conference Series: Earth and Environmental Science*; IOP Publishing: Bristol, UK, 2022; Volume 983, p. 012046.
61. Sedliačiková, M.; Moresová, M.; Aláč, P.; Malá, D. What is the supply and demand for coloured wood products? An empirical study in Slovakian practice. *Forests* **2021**, *12*, 530. [CrossRef]
62. Available online: https://www.wood-database.com/yellow-birch/ (accessed on 23 April 2024).
63. Available online: https://www.woodlandtrust.org.uk/trees-woods-and-wildlife/british-trees/a-z-of-british-trees/downy-birch/ (accessed on 23 April 2024).
64. Available online: https://www.hardwoodinfo.com/specifying-professionals/species-guide/species-guide-a-g/birch/ (accessed on 23 April 2024).
65. Kollmann, F.F.P.; Kuenzi, E.W.; Stamm, A.J. *Principles of Wood Science and Technology. Vol. 2. Wood Based Materials*; Springer: Berlin/Heidelberg, Germany; New York, NY, USA, 1975; 703p, ISBN 3-540-06467-2.
66. Maloney, T.M. *Modern Particleboard and Dry-Process Fiberboard Manufacturing*; Miller Freeman publications: San Francisco, CA, USA, 1977; 669p, ISBN 0-87930-063-9.
67. Varis, R. *Wood-Based Panels Industry*; Kirjakkari, O., Ed.; Technical University: Zvolen, Slovakia, 2018; 276p, ISBN 978-952-68627-1.
68. Reh, R. *Wood-Based Particle and Fiber Panel Products*; Technical University in Zvolen: Zvolen, Slovakia, 2013; 135p, ISBN 978-80-228-2498-9.
69. Liiri, O. Tutkimuksia lastulevyteollisuuden puuraaka-aineista I. Suomalaiset puulajit lastulevyn raaka-aineina. Investigations on the wood used as a raw material in the particle-board industry. I. Finnish tree species as a raw material for particle board. *Pap. Ja Puu* **1960**, *42*, 467–484.
70. Baharoğlu, M.; Nemli, G.; Sarı, B.; Birtürk, T.; Bardak, S. Effects of anatomical and chemical properties of wood on the quality of particleboard. *Compos. Part B Eng.* **2013**, *52*, 282–285. [CrossRef]
71. Krug, D.; Direske, M.; Tobisch, S.; Weber, A.; Wenderdel, C. Particle-Based Materials. In *Springer Handbook of Wood Science and Technology*; Niemz, P., Teischinger, A., Sandberg, D., Eds.; Springer Handbooks; Springer: Cham, Switzerland, 2023. [CrossRef]
72. Çamlıbel, O.; Aydın, M. The effects of continuous press speed and conditioning time on the particleboard properties at industrial scale. *BioResources* **2022**, *17*, 159–176. [CrossRef]
73. Skurydin, Y.G.; Skurydina, E.M. Digital differential spectrometry in the assessment of the structural characteristics of wood and wooden composite materials. *IOP Conf. Ser. Earth Environ. Sci.* **2021**, *806*, 012030. [CrossRef]

74. Singh, N.; Rana, A.; Badhotiya, G.K. Raw material particle terminologies for development of engineered wood. *Mater. Today Proc.* **2021**, *46*, 11243–11246. [CrossRef]
75. Razinkov, E.M.; Ishchenko, T.L. Analysis of low requirement justification of the current standard for the strength of wood particle boards. In *IOP Conference Series: Earth and Environmental Science*; IOP Publishing: Bristol, UK, 2021; Volume 875, p. 012068. [CrossRef]
76. Tupciauskas, R.; Rizhikovs, J.; Grinins, J.; Paze, A.; Andzs, M.; Brazdausks, P.; Puke, M.; Plavniece, A. Investigation of suberinic acids-bonded particleboard. *Eur. Polym. J.* **2019**, *113*, 176–182. [CrossRef]
77. Jeżo, A.; Wronka, A. Post-extraction birch bark residues as a potential binder in particleboards. Annals of Warsaw University of Life Sciences—SGGW. *For. Wood Technol.* **2022**, *118*, 35–47. [CrossRef]
78. EN 312: 2010. Available online: https://standards.iteh.ai/catalog/standards/cen/310c1aaa-c5c9-4e5d-ad3d-a255c265da52/en-312-2010 (accessed on 10 March 2024).
79. Makars, R.; Godiņa, D.; Rizhikovs, J.; Paze, A.; Tupciauskas, R.; Berzins, R. Investigation of Furfural Formation and Mechanical Properties of Suberinic Acids-Bonded Particleboards Depending on their Preparation Parameters. *KEM* **2021**, *903*, 235–240. [CrossRef]
80. Makars, R.; Rizikovs, J.; Godina, D.; Paze, A.; Merijs-Meri, R. Utilization of Suberinic Acids Containing Residue as an Adhesive for Particle Boards. *Polymers* **2022**, *14*, 2304. [CrossRef]
81. Tupciauskas, R.; Meile, K.; Godina, D.; Rizhikovs, J.; Syrpas, M.; Venskutonis, P.R. Qualitative Differences and Emission Persistence of Volatile Organic Compounds from Bio-Based Particleboards. *Materials* **2022**, *15*, 5278. [CrossRef]
82. Dudík, R.; Borůvka, V.; Zeidler, A.; Holeček, T.; Riedl, M. Influence of Site Conditions and Quality of Birch Wood on Its Properties and Utilization after Heat Treatment. Part II—Surface Properties and Marketing Evaluation of the Effect of the Treatment on Final Usage of Such Wood. *Forests* **2020**, *11*, 556. [CrossRef]
83. Kowaluk, G.; Jeżo, A. Compression strength-focused properties of wood composites induced by structure. Annals of Warsaw University of Life Sciences—SGGW. *For. Wood Technol.* **2021**, 131–140.
84. Dumitrascu, A.E.; Lunguleasa, A.; Salca, E.A.; Ciobanu, V.D. Evaluation of selected properties of oriented strand boards made from fast growing wood species. *BioResources* **2020**, *15*, 199–210. [CrossRef]
85. Xu, W. *Horizontal Density Distribution of Particleboard: Origin and Implications*; University of British Columbia: Vancouver, BC, Canada, 1994. Available online: https://open.library.ubc.ca/collections/ubctheses/831/items/1.0088293 (accessed on 10 March 2024).
86. Carre, J. Use of forest biomass for the manufacture of particleboard. *Bull. Rech. Agron. Bemloux.* **1978**, *13*, 287–309.
87. Pedieu, R.; Riedl, B.; Pichette, A. Properties of white birch (*Betula papyrifera*) outer bark particleboards with reinforcement of coarse wood particles in the core layer. *Ann. For. Sci.* **2008**, *65*, 701. [CrossRef]
88. Available online: https://www.wood-database.com/european-larch/ (accessed on 23 April 2024).
89. Available online: https://www.woodlandtrust.org.uk/trees-woods-and-wildlife/british-trees/a-z-of-british-trees/european-larch/ (accessed on 23 April 2024).
90. Available online: https://leafyplace.com/larch-trees/ (accessed on 23 April 2024).
91. Ragan, B.; Kačíková, D.; Paul'uro, M. Influence of physical and chemical characteristics of selected tree species on mass loss and rate of burning after exposure to radiant heating. *Acta Fac. Xylologiae Zvolen* **2016**, *58*, 121–131. [CrossRef]
92. Chauzov, K.; Varankina, G.; Tambi, A. Investigation on gluing larch wood by modified glue. In *9th International Scientific Conference on Production Engineering Development and Modernization of Production*; Tehnički fakultet Bihač: Bosna i Herzegovina, 2013. Available online: https://api.semanticscholar.org/CorpusID:208221259 (accessed on 10 March 2024).
93. Pazio, B.; Boruszewski, P. Analysis of the influence of larch fibers and particles on selected properties of fiber- and particleboards. Annals of WULS—SGGW. *For. Wood Technol.* **2020**, *111*, 43–52. [CrossRef]
94. Muhcu, S.; Nemli, G.; Ayrilmis, N.; Bardak, S.; Baharoğlu, M.; Sarı, B.; Gerçek, Z. Effect of log position in European Larch (*Larix decidua* Mill.) tree on the technological properties of particleboard. *Scand. J. For. Res.* **2015**, *30*, 357–362. [CrossRef]
95. Bardak, S.; Nemli, G.; Bardak, T. The quality comparison of particleboards produced from heartwood and sapwood of European larch. *Maderas. Cienc. Y Tecnol.* **2019**, *21*, 4. [CrossRef]
96. Lee, P.W.; Chung, G. Effect of press time and temperature on the physical properties of larch particle board. *J. Korean For. Soc.* **1984**, 12–20.
97. Matsumoto, A.; Nishikawa, K.; Anazawa, T.; Nunomura, A. Effects of the flake size on the properties of particle boards [from larch]. *J. Hokkaido For. Prod. Res. Inst.* **1982**.
98. Dix, B.; Roffael, E. Mechanical and technological properties of particleboards made of larch sapwood and heartwood. *Holz Als Roh-Und. Werkst.* **1994**.
99. Shupe, T.F.; Hse, C.Y.; Choong, E.T.; Groom, L.H. Effect of Silvicultural Practice and Wood Type on Loblolly Pine Particleboard and Medium Density Fiberboard Properties. *Holzforsch. Gruyter* **1999**, *53*, 215–222. [CrossRef]
100. Simatupang, M.H.; Lange, H.; Neubauer, A. Einfluß der Lagerung von Pappel, Birke, Eiche und Lärche sowie des Zusatzes von SiO2-Feinstaub auf die Biegefestigkeit zementgebundener Spanplatten. *Holz Als Roh-Und. Werkst.* **1987**, *45*, 131–136. [CrossRef]
101. Moslemi, A.A.; Garcia, J.F.; Hofstrand, A.D. Effect of Various Treatments and Additives on Wood-Portland Cement-Water Systems. *Wood Fiber Sci.* **1983**, *15*, 164–176.

102. Simatupang, M.H.; Geimer, R.L. Inorganic binder for wood composites: Feasibility and limitations. In Proceedings of the Wood Adhesives Symposium, Madison, WI, USA, 16–18 May 1990; Forest Products Laboratory, USDA Forest Service, Forest Products Research Society: Madison, WI, USA, 1990.
103. Keegan, C.E.; Charles, E.; Blatner, K.A.; Wichman, D.P. Use and value of western larch as a commercial timber species. *Ecol. Manag. Larix For. A Look Ahead; Gen. Tech. Rep. GTR-INT-319* **1995**, 155–157.
104. Gou, J.; Chang, J.; Huang, Y.; Han, Y.; Ren, X. Characterization of Nitrogen Forms in Larch Wood Particle Board. In Proceedings of the 5th International Conference on Bioinformatics and Biomedical Engineering, Wuhan, China, 10–12 May 2011; pp. 1–4. [CrossRef]
105. Liu, Y.; Shen, J.; Zhu, X.D. Influence of processing parameters on VOC emission from particleboards. *Environ. Monit. Assess.* **2010**, *171*, 249–254. [CrossRef]
106. Lo, K.S.; Liew, K.C. Effect of extraction solvent on mechanical properties of particleboard produced from cultivated Acacia hybrid. *J. Indian Acad. Wood Sci.* **2011**, *8*, 46–49. [CrossRef]
107. Tudor, E.M.; Barbu, M.C.; Petutschnigg, A.; Reh, R.; Kristak, L. Analysis of Larch-Bark Capacity for Formaldehyde Removal in Wood Adhesives. *Int. J. Environ. Res. Public Health* **2020**, *17*, 764. [CrossRef] [PubMed]
108. Kain, G.; Tudor, E.M.; Barbu, M.C. Bark Thermal Insulation Panels: An Explorative Study on the Effects of Bark Species. *Polymers* **2020**, *12*, 2140. [CrossRef] [PubMed]
109. Tudor, E.M.; Dettendorfer, A.; Kain, G.; Barbu, M.C.; Reh, R.; Kristak, L. Sound-Absorption Coefficient of Bark-Based Insulation Panels. *Polymers* **2020**, *12*, 1012. [CrossRef]
110. Kain, G.; Stratev, D.; Tudor, E.; Lienbacher, B.; Weigl, M.; Barbu, M.C.; Petutschnigg, A. Qualitative investigation on VOC-emissions from spruce (*Picea abies*) and larch (*Larix decidua*) loose bark and bark panels. *Eur. J. Wood Prod.* **2020**, *78*, 403–412. [CrossRef]
111. Liu, Y.; Shen, J.; Zhu, X.D. Evaluation of mechanical properties and formaldehyde emissions of particleboards with nanomaterial-added melamine-impregnated papers. *Eur. J. Wood Prod.* **2015**, *73*, 449–455. [CrossRef]
112. Liu, Y.; Zhu, X.; Qin, X.; Wang, W.; Hu, Y.; Yuan, D. Identification and characterization of odorous volatile organic compounds emitted from wood-based panels. *Env. Monit Assess* **2020**, *192*, 348. [CrossRef]
113. Bednarczyk, D.; Boruszewski, P. Lightweight particleboards—Manufacturing modification using a blowing agent from the group of bicarbonates. Annals of Warsaw University of Life Sciences—SGGW. *For. Wood Technol.* **2022**, *117*, 55–62. [CrossRef]
114. Wenig, C.; Reppe, F.; Horbelt, N.; Spener, J.; Berendt, F.; Cremer, T.; Frey, M.; Burgert, I.; Eder, M. Adhesives free bark panels: An alternative application for waste material. *PLoS ONE* **2023**, *18*, e0280721. [CrossRef] [PubMed]
115. Kajita, H.; Imamura, Y. Chemically Modified Particleboards. In *Recent Research on Wood and Wood-Based Materials*; Shiraishi, N., Kajita, H., Norimoto, M., Eds.; Elsevier: Amsterdam, The Netherlands, 1993; pp. 67–74, ISBN 9781483178219. [CrossRef]
116. Ninikas, K.; Mitani, A.; Koutsianitis, D.; Ntalos, G.; Taghiyari, H.R.; Papadopoulos, A.N. Thermal and Mechanical Properties of Green Insulation Composites Made from Cannabis and Bark Residues. *J. Compos. Sci.* **2021**, *5*, 132. [CrossRef]
117. Tudor, E.M.; Scheriau, C.; Barbu, M.C.; Réh, R.; Krišťák, L'.; Schnabel, T. Enhanced resistance to fire of the bark-based panels bonded with clay. *Appl. Sci.* **2020**, *10*, 5594. [CrossRef]
118. Czajka, M.; Fabisiak, E. Radial variation of the content of cellulose and lignin in selected softwood species. *Ann. Wars. Univ. Life Sci.—SGGW For. Technol.* **2017**, *100*, 233–238. Available online: https://journals.indexcopernicus.com/api/file/viewByFileId/349531 (accessed on 10 March 2024).
119. Akbulut, T.; Ayrılmış, N. *Yongalevha Endüstrisi (Particleboard Industry). İstanbul Universitesi-Cerrahpaşa;* IUC University Press: Yokohama, Kanagawa, 2024; 126p, ISBN 978-605-7880-49-9.

polymers

MDPI

Article

Influence of Pressing Temperatures on Physical–Mechanical Properties of Wood Particleboards Made with Urea-Formaldehyde Adhesive Containing Al_2O_3 and CuO Nanoparticles

Luana Cristal Lirya Silva [1,*], Felipe Oliveira Lima [2], Victor Almeida De Araujo [2,3], Herisson Ferreira Dos Santos [4], Francisco Antonio Rocco Lahr [5], André Luis Christoforo [3], Higor Rogério Favarim [2] and Cristiane Inácio de Campos [1,2,*]

1 Science and Engineering Faculty, São Paulo State University, Guaratinguetá 12516-410, Brazil
2 Science and Engineering Institute, São Paulo State University, Itapeva 18409-010, Brazil
3 Exact Sciences & Technology Center, Federal University of São Carlos, São Carlos 13565-905, Brazil
4 Campus of Ariquemes, Federal Institute of Education, Science and Technology of Rondônia, Ariquemes 76870-000, Brazil
5 São Carlos School of Engineering, University of São Paulo, São Carlos 13566-590, Brazil
* Correspondence: luana.cristal@unesp.br (L.C.L.S.); cristiane.campos@unesp.br (C.I.d.C.)

Abstract: Particleboards have gained attention in the global market. Understanding their physical–mechanical behavior in the current technological context is essential due to adhesive polymerization, which depends on variables such as pressing time and temperature. Today, the use of nanoparticles has become a plausible option for improving the properties of polymers used in wood-based composites. This study evaluates the influences of the addition of non-commercial 0.5% aluminum oxide (Al_2O_3) and aluminum oxide copper (CuO) nanoparticles using a greener route with a lower environmental impact obtaining a urea-formaldehyde (UF)-based polymeric adhesive to manufacture particle composites of *Eucalyptus urophylla* var. *grandis* wood. Regarding characterizations, the resin properties analyzed were viscosity, gel time, and pH, as well as panel properties, including density, moisture content, thickness swelling, modulus of elasticity, modulus of rupture, and thermal conductivity. The results were compared with scientific publications and standards. The addition of nanoparticles interfered with viscosity, and all treatments indicated a basic pH. It was not possible to determine the gel time after 10 min. Nanoparticles added to the polymers in the internal layer did not cause an improvement in the swelling properties in terms of thickness, with no significant statistical difference for density and moisture content. The increase from 150 °C to 180 °C may have caused an improvement in all physical–mechanical properties, indicating that the higher temperature positively influenced the polymerization of the formaldehyde-based adhesive. Therefore, the additions of both nanoparticles (0.5% in each condition) led to a limitation in the material influence with respect to physical–mechanical performance.

Keywords: wood-based panel; density; thermal property; physical property; mechanical property; pressing temperature; nanomaterials; polymeric adhesive

check for updates

Citation: Silva, L.C.L.; Lima, F.O.; De Araujo, V.A.; Dos Santos, H.F.; Lahr, F.A.R.; Christoforo, A.L.; Favarim, H.R.; de Campos, C.I. Influence of Pressing Temperatures on Physical–Mechanical Properties of Wood Particleboards Made with Urea-Formaldehyde Adhesive Containing Al_2O_3 and CuO Nanoparticles. *Polymers* **2024**, *16*, 1652. https://doi.org/10.3390/polym16121652

Academic Editors: Ľuboš Krišťák, Roman Reh, Eugenia Mariana Tudor and Marius Cătălin Barbu

Received: 15 May 2024
Revised: 4 June 2024
Accepted: 5 June 2024
Published: 11 June 2024

1. Introduction

Particleboards have been increasingly used, with a favorable trend for the coming decades [1,2]. Due to the increase in the market competitiveness for bio-based particulate composites, understanding their physical and mechanical behavior is paramount in the current technological context [3].

Particleboard (PB) is a composite manufactured with wood particles aggregated by synthetic adhesives through reconstitution and consolidation of the material using heating and pressure in a forming press. Urea-formaldehyde is one of the most usable resins in particleboard production [4].

Therefore, it is important to understand that the performance of a wood composite is directly related to the polymerization of the adhesive, which can depend on variables such as pressing time and temperature [3,5–7].

Thus, the composite properties can still be made worse by inefficient heat transfer, as the physical–mechanical performance is directly related to an efficient constituent interaction consolidated by heat transfer from the heated press along the panel thickness [6,8]. It is worth highlighting that pressing is one of the different manufacturing stages in panel production, because the reduction of costs involved in the production must be prioritized as a leading objective of every manufacturer.

The low thermal conductivity of wood is responsible for problems in heat transfer from pressed panels under the influence of temperature. In this sense, new technological trends verify the possibility of better performance with the use of nanometric-scale materials. Nanomaterials can present new behaviors and properties when compared to materials on a macroscopic scale [9]. Thus, the heat-conducting nature of nanoparticles produced with metals can be used to improve heat transfer to the innermost layer of the panel mat, which contributes to the polymerization of the resin and, consequently, impacts the properties of the composite [10–12]. Aluminum oxide nanoparticles have been used as reinforcement in polymer composites for multiple purposes [13]. Reinforcement with aluminum oxide nanoparticles in various thermosetting resins has been researched for the development of value-added products [14]. Researchers from different countries have been studying the application of nanoparticles in polymeric adhesives for the manufacture of wood composites. Some studies present the addition of nanomaterials to urea-formaldehyde resins, reporting improvements in mechanical properties, directly affecting the swelling of panels when in contact with water, and reducing the formaldehyde emission [15–17].

According to Cadermatori et al. [15], thermodynamic analysis revealed that Al_2O_3 is an effective additive for urea-formaldehyde (UF), maintaining crucial curing parameters such as the vitrification point and gel temperature. Additionally, Al_2O_3 nanoparticles basically reduced formaldehyde emissions during the UF resin curing process at elevated temperatures and achieved up to a 14% reduction in thickness at room temperature.

Zhang et al. [18] concluded that aluminum oxide nanoparticles, in percentages of 0% to 4%, when used together with phenol-formaldehyde adhesive in plywood pressed at 140 °C, showed an ability to accelerate and optimize the curing process, which increases manufacturing efficiency and reduces energy consumption in the production process.

Gupta et al. [19] studied a mixture of aluminum oxide with urea-formaldehyde resin, in percentages of 0.5%, 1%, and 1.5%, and they verified a reduction in curing time due to the greater thermal conductivity of the medium-density fiberboard matrix, as well as the significant improvement of the physical and mechanical properties with the addition of Al_2O_3 nanoparticles. Taghiyari and Bibalan [20] studied the use of Cu nanoparticles in particleboards with urea-formaldehyde adhesive, in proportions of 100 and 150 mL/kg based on the weight of dry particles, and better polymerization of the resin in the inner layer of the panel was verified due to improved heat transfer from copper nanoparticles and a very significant reduction in hot-pressing time.

Unlike other reported works that utilize commercial nanoparticles (Al_2O_3, CuO, and ZnO) and formaldehyde-based adhesives commonly utilized in panel production, our main goal in this study involves the use of nanoparticles from an alternative route for nanoparticle synthesis with lower costs and environmental impacts.

For this, this work analyzed two pressing temperature levels (150 °C to 180 °C) for wood particleboards and the effects on the curing of polymeric resin (UF) reinforced with (0.5%) and without aluminum oxide (Al_2O_3) and copper oxide (CuO) nanoparticles, verifying their physical–mechanical properties for a more environmentally friendly option.

2. Materials and Methods

2.1. Panel Configurations and Characterization of Nanoparticles

2.1.1. Materials

The following materials related to panels and nanoparticles were included:

- Particles of Eucalyptus urophylla var grandis;
- Urea-formaldehyde adhesive, with 68% solid content and pH 8 (Hexion, Curitiba, Brazil);
- Ammonium sulfate;
- Al_2O_3 nanoparticles (own laboratory production, Itapeva, Brazil);
- CuO nanoparticles (own laboratory production, Itapeva, Brazil);
- X-ray diffraction equipment (Bruker AXS D8, Billerica, MA, USA);
- Pneumatic press (Hidralmaq HMP 80 ton model, Araraquara, Brazil);
- K thermocouple (Hikari HKP01 model, São Paulo, Brazil);
- Gluing rotational machine (Marconi MA686, Piracicaba, Brazil);
- Device for data acquisition and computer.

2.1.2. Methods

For panels, particles of *Eucalyptus urophylla* var. *grandis* were used as the main raw material, urea-formaldehyde adhesive with 65% solid content as the gluing agent, and ammonium sulfate as the catalyst with 20% solid content, as well as 0.5% aluminum oxide nanoparticles (Al_2O_3) and 0.5% copper oxide (CuO) nanoparticles. Water was utilized in the resin preparation, as well as in the water absorption test.

In the characterization of nanoparticles, the calcined material was analyzed by X-ray diffraction. Subsequently, microstructural analysis was carried out with a scanning electron microscope (SEM) to determine the shape of the synthesized material, as well as the size of the particles before the panel production. Scherrer's analysis was carried out.

The panels were manufactured rigorously according to the methodology previously utilized by Silva et al. [8] and Lima [21]. All stages of the panel manufacturing process in the laboratory and the characterization procedures followed the recommendations given by the ABNT NBR 14810-3:2018 [22].

Regarding materials, wood, adhesive, and additives were obtained by confidential donations through Brazilian manufacturers. To produce each panel with a calculated nominal density of 720 kg/m^3 considering nominal dimensions (40 cm $\times$ 40 cm $\times$ 1.3 cm), 1500 g of *Eucalyptus urophylla* var. *grandis* wood particles was used, which were properly classified according to 5, 9, 16, 35, and 60 mesh. Of these proportions, 5, 9, and 16 mesh were used for the single inner layer, while 35 and 60 mesh were used for the two outer layers.

Urea-formaldehyde (UF) adhesive with a solids content of 65% was prepared under a dosage based on the weight of the particles, using 10% for the external layers and 8% for the internal layer, where 0.5% nanoparticles were utilized in the inner layer of the panel. The nanoparticles were obtained in partnership with other researchers and were produced by the sol–gel protein method. A protein precursor was dissolved at 40 °C together with the metal salts in a stoichiometric proportion. The solution was dried in an oven until it formed a spongy structure, which was then burned and calcined to form the nanomaterial, as performed by Silva et al. [8].

Wood particles were dried in a chamber (103 °C $\pm$ 2) to reach 3% moisture content and properly weighed in a proportion of 25–50–25% between layers. The resin was sprayed onto the outer- and inner-layer particles on a rotary gluing machine (Figure 1a).

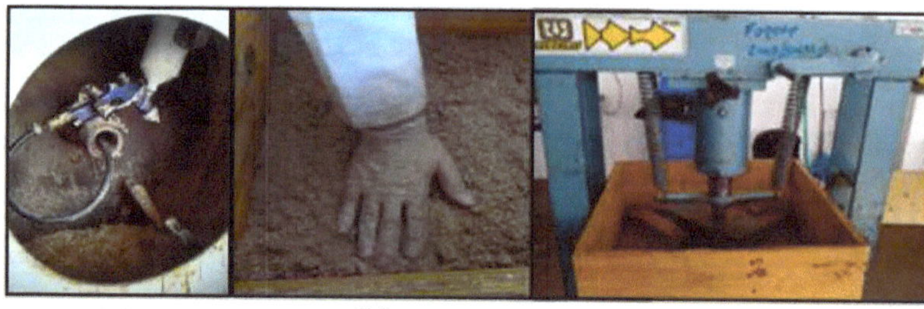

(**a**) (**b**) (**c**)

Figure 1. Panel production stages: (**a**) resin spraying, (**b**) formation of mat with wood particles, and (**c**) pre-pressing.

The particle mat was assembled in a forming box (40 cm × 40 cm) (Figure 1b), and the formation was finalized through cold pre-pressing in a pneumatic press (Figure 1c) for 300 s at 0.3 MPa. After the pre-pressing stage, the analysis of temperature variation was performed by a type K thermocouple inserted into the innermost layer of the panel.

This thermocouple was coupled to a data acquisition device (DAQ) to obtain the temperature x time graph.

From pre-pressed mats (Figure 2a), the hot-pressing process was started with a total cycle of 600 s, with two pressure releases with a duration of 30 s each (Figure 2b). The pressure was set at 4 MPa with variations in temperature, at 150 °C and 180 °C. Thus, particleboards were successfully obtained, with an average thickness of 13 mm, under different manufacturing conditions (Figure 2c), including three material compositions (with Al_2O_3, with CuO, and without nanoparticles) and two temperature pressing levels.

(**a**) (**b**) (**c**)

Figure 2. Panel production stages: (**a**) pre-pressed mat, (**b**) hot pressing, and (**c**) particleboard.

2.2. *Characterization of Resin*

2.2.1. Materials

The following materials related to resin characterization were used:

- Ford cup (Marte N4, São Paulo, Brazil);
- Urea-formaldehyde adhesive;
- Al_2O_3 nanoparticles (own laboratory production, Itapeva, Brazil);
- CuO nanoparticles (own laboratory production, Itapeva, Brazil);
- Meter device (Digimed DM22, São Paulo, Brazil).

2.2.2. Methods

Viscosity testing was conducted using a Ford cup with a 5.20 mm orifice under ASTM D1200-2005 [23] (Standard Test Method for Viscosity by Ford Viscosity Cup).

The pH values of all resin samples were determined through direct measurements using a pH meter device (Digimed DM22).

To determine the gel time of each sample, 5 g of adhesive was utilized. Each adhesive was placed in a test tube and immersed in glycerin at 130 °C. A glass rod was continuously moved through the adhesive to induce gelation. The time taken for gelation in each sample was recorded as the gel time value.

2.3. Characterization of Physical and Mechanical Properties and Statistical Analysis

2.3.1. Materials

The following materials related to physical–mechanical tests were used:

- Pachymeter (Mitutoyo Absolute 150 mm, Jundiaí, Brazil);
- Micrometer (Mitutoyo Digital 25 mm, Jundiaí, Brazil);
- Analytical digital scale (Marte 0.2 g precision, São Paulo, Brazil);
- Universal testing machine (EMIC 300T, Curitiba, Brazil);
- Muffle furnace (Marconi MA035, Piracicaba, Brazil).

2.3.2. Methods

Regarding the physical and mechanical tests carried out to characterize these panels, the evaluation was based on density, moisture content, thickness swelling, static bending, and perpendicular tensile strength according to specifications given by the ABNT NBR 14810-3:2018 [22], including the dimensions and quantities of specimens to be characterized. Thus, 10 specimens were used for each test. In turn, a thermal conductivity test was carried out following the recommendations of the ASTM E1530:2011 [24]. For this, panel specimens 50 mm in diameter were used, with a total of three samples for each treatment.

After the characterization of particleboards, the results were statistically analyzed with an analysis of variance (ANOVA) at a 5% significance level to test for the existence of a significant difference between the means of the respective results.

Tukey's test was then performed, also with 5% significance, to analyze the interactions. This analysis was carried out through R software version 3.2.0. In the result tables with statistics, the letter "A" denotes the group with the highest average value, "B" is the group with the second-highest average value, and so on.

3. Results and Discussion

3.1. Characterization of Nanoparticles

The analysis of sizes and structures of the nanoparticles involved the characterization through X-ray diffraction. Figure 3 presents the diffractogram obtained from the analysis of Al_2O_3 and CuO nanoparticles. The diffractogram indicates the characteristic reflections of the γ-alumina phase, verifying the typical peaks of this cubic structure presented by the ICDD (International Center of Diffraction Data) with the reference number JCPDS 010-0425, which confirms the formation of crystalline γ-alumina.

The peaks have higher intensities. However, they are slightly broadened due to the nature of the crystallite size of these materials. Scherrer equation analysis of the average crystallite size for γ-alumina—from the full width at half maximum (FWHM) for the (311), (400), and (440) peaks found 7, 15, and 26 nm sizes, respectively—proves the synthesis of a nanocrystalline material. Likewise, by analyzing the ICSD database (Inorganic Crystal Structure Database), the presence of a single phase for copper II oxide with a monoclinic structure (ICSD reference number 16025) was verified. Thereby, Scherrer's equation indicates a size of 37 nm, which also proves the synthesis of a nanocrystalline material.

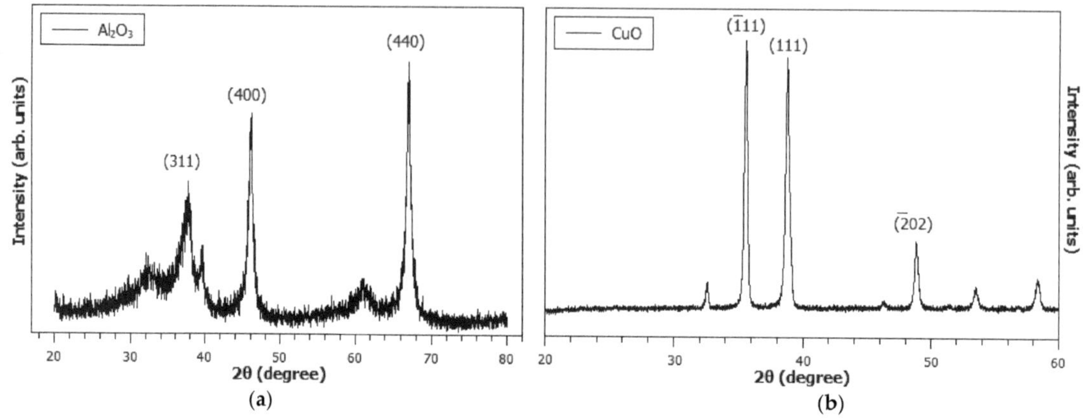

Figure 3. Diffractograms according to nanoparticles: (**a**) Al_2O_3 and (**b**) CuO.

3.2. Characterization of Resin

The pH, viscosity, and gel time properties of the resin are presented in Table 1.

Table 1. pH, viscosity, and gel time properties.

Resin	Obtained Results	
	pH	Viscosity (Cp)
Urea-formaldehyde (UF)	8.51 ^{A}	435.60 ^{A}
UF + nano Al_2O_3	8.31 ^{A}	572.19 ^{B}
UF + nano CuO	8.64 ^{A}	638.88 ^{B}

^{A}: group with the highest average value; ^{B}: group with the second-highest average value.

From the results presented in Table 1, it is possible to state that the pH does not show a significant statistical difference with the addition of the nanomaterial. As verified in some previous studies, the resin presented a basic pH [6,18]. Gelatinization after 10 min was not found for any of the variations studied. This fact has already been observed by Silva et al. [25]. Regarding viscosity, the addition of nanoparticles increased the resin viscosity, which influenced the interaction between the matrix and reinforcement.

3.3. Characterization of Physical Properties

After the nanometric characterization of raw materials and physical tests of the panels, the results were compared with the standard values of ABNT NBR 14810:2018 (class P2) [26], EN 312:2003 [27], and ANSI 208-1:2016 (class M1) [28] for density, moisture, and thickness swelling. Non-structural panels measuring 13 to 20 mm thick were taken as a reference, for internal uses. All panels below this value were considered references and were classified as medium-density particles (MDPs). As shown in Table 2, the results obtained for panel density with respective coefficients of variation are presented with the ranges of values given by these standards valid in Brazil, Europe, and North America. Our results remained within the ranges from Brazilian and European standard documents, although they slightly exceeded the maximum average value stipulated for the North American market.

Table 2. Density: results and comparison with values given by standards.

Temperature Pressing	150 °C	180 °C
Without nanoparticles	661.36 [1,A] (5.98 [2])	667.43 [1,A] (6.45 [2])
Al_2O_3 nanoparticle	674.72 [1,A] (6.03 [2])	707.57 [1,A] (5.28 [2])
CuO nanoparticle	682.64 [1,A] (6.59 [2])	665.68 [1,A] (7.48 [2])
Brazil ABNT [26]	551 to 750 [1]	
Europe EN [27]	500 to 800 [1]	
North America ANSI [28]	640 to 800 [1]	

[1] panel density (in kg/m^3); [2] coefficient of variation; [A]: group with the highest average value.

From Table 2, the density results also demonstrate some similarity to other studies, although they were lower than the average values reported by the literature. For example, Lima et al. [12], when manufacturing medium-density particleboards made with zinc oxide nanoparticles (ZnO), glued with urea-formaldehyde resin, and pressed at 180 °C in different proportions (0.5% and 1%), achieved average densities of 720.31 kg/m^3 and 749.98 kg/m^3, with no statistical difference between their results. It is worth highlighting that the experimental density of panels was lower than the nominal density (as indicated in the methodology), as manufacturing losses were verified in the particle gluing and mat formation stages.

Following the same presentation of our results and the ranges from standards as exemplified in Table 2, Table 3 analyzes the moisture content of particleboards pressed in two temperature levels and configured in three material compositions (with Al_2O_3, with CuO, and without nanoparticles).

Table 3. Moisture content: results and comparison with values prescribed by standards.

Temperature Pressing	150 °C	180 °C
Without nanoparticles	6.49 [1,A] (4.93 [2])	6.41 [1,A](4.70 [2])
Al_2O_3 nanoparticle	6.65 [1,A] (2.11 [2])	6.51 [1,A] (5.53 [2])
CuO nanoparticle	6.56 [1,A] (2.90 [2])	6.15 [1,A] (5.85 [2])
Brazil ABNT [26]	5 to 13 [1]	
Europe EN [27]	-	
North America ANSI [28]	-	

[1] moisture content (in %); [2] coefficient of variation; [A]: group with the highest average value.

Regarding the moisture analysis of the particleboards developed in this study, it was possible to verify that all treatments met the values referenced by the ABNT [26], with no statistical difference among the results from Table 3. All results were close to the minimum moisture content suggested by this same standard document.

These results were also similar to those reported by other scientific studies. Lima et al. [12] presented moisture contents of 6.18 and 6.24% for wood particle panels with 0.5% and 1% ZnO (urea-formaldehyde resin and 180 °C pressing temperature), while Silva et al. [8] obtained an average of 6.61% moisture under the same conditions. These authors did not find any statistical difference.

Similar to the organization of the previous tables, Table 4 discloses the obtained results for thickness swelling after 24 h. For thickness swelling after 24 h, none of the panels met the specifications from Brazilian and European standards. Particleboards with aluminum oxide nanoparticles at 150 °C reached the worst performance in this property, while panels without nanoparticles and at 180 °C showed the best results, statistically differing from the others (Table 4).

Table 4. Thickness swelling: results and comparison with values prescribed by standards.

Temperature Pressing	150 °C	180 °C
Without nanoparticles	36.31 [1,B] (16.33 [2])	25.57 [1,C] (10.60 [2])
Al_2O_3 nanoparticle	52.18 [1,A] (9.37 [2])	44.95 [1,A,B] (1.91 [2])
CuO nanoparticle	43.21 [1,A,B] (13.75 [2])	34.33 [1,B] (11.97 [2])
Brazil ABNT [26]	<22 [1]	
Europe EN [27]	<16 [1]	
North America ANSI [28]	-	

[1] thickness swelling after 24 h (in %); [2] coefficient of variation; [A]: group with the highest average value; [B]: group with the second-highest average value; [C]: group with the third-highest average value.

For medium-density particleboards manufactured with 4% SiO_2 nanoparticles (urea-formaldehyde resin and 160 °C pressing temperature), Valle [10] reached an average value of 36.46% for this same property after 24 h. In nanocellulose-treated particleboards (urea-formaldehyde resin and 160 °C pressing temperature), Cardoso et al. [29] verified thickness swelling values of 30.35% and 53.68% for 2% and 3% nanocellulose, respectively. They still confirmed a negative influence with the addition of nanoparticles.

Taghiyari and Bibalan [20] also obtained an increase in the 24 h swelling of wood particle panels (urea-formaldehyde resin and 200 °C pressing temperature) when adding 150 mL/kg of copper nanoparticles, which increased this property from 19.52% to 22.68%.

Figure 4 illustrates the interactions between the treatments as well as possible trends presented by the addition of the nanoparticles.

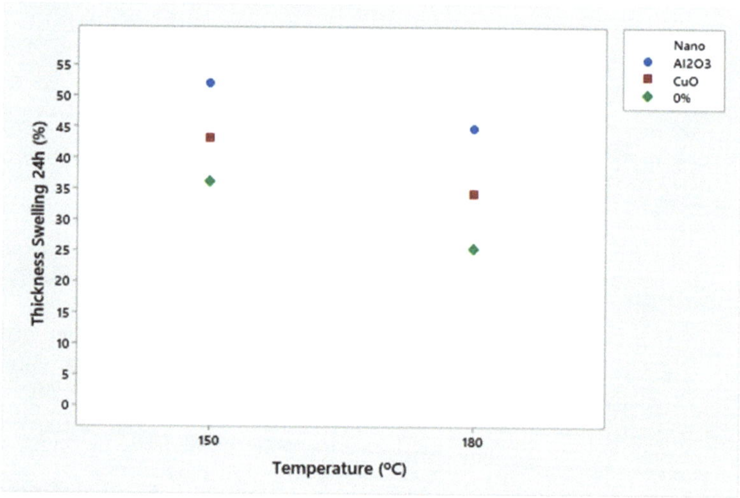

Figure 4. Interaction graph of results for thickness swelling after 24 h according to each panel composition.

Analyzing the interactions (Figure 4), it was possible to observe that—although the values for CuO were lower than for Al_2O_3—the addition of nanomaterials exhibited the same effect on the particleboards for this property, in which the pressing temperature was the greatest influential variable. Thus, a higher pressing temperature favors the curing of the polymeric adhesive, reducing thickness swelling values. It is also noteworthy that the percentage of nanoparticles used was smaller than in other research studies—for example, Zhang et al. [18] used up to 4% aluminum oxide nanoparticles.

Figure 4 illustrates the decrease in thickness swelling after water immersion for 24 h in all panels (without nanoparticles, with copper oxide, and with aluminum oxide). The greater the difference in values between the points for the same treatment, the greater the interaction between the factors. Occurred at a temperature of 180 °C, this reduction can be explained by the greater polymerization of the resin at higher temperatures, improving the performance of the panels in contact with water, that is, allowing better interaction of the resin with the wood particles, as confirmed by Lima et al. [12].

3.4. Characterization of Mechanical Properties

The same category of non-structural panel for internal uses, with a thickness between 13 and 20 mm, was considered for this analysis. The results were compared with the standard values of ABNT NBR 14810:2018 (class P2) [26], EN 312:2003 [27], and ANSI 208-1:2016 (class M1) [28] for commercial use in dry environments concerning the MOE and MOR.

Table 5 presents the results obtained for the modulus of elasticity (MOE), including respective coefficients of variation, and the ranges of values given by standard documents from Brazil, Europe, and North America. Only the particleboard pressed at 180 °C and produced with CuO nanoparticles did not reach the minimum expected for the European standards. However, it met the expected modulus of elasticity for the Brazilian and North American markets (Table 5). All configurations produced at a 150 °C pressing temperature exceeded the minimum standardized conditions for the modulus of elasticity.

Table 5. Modulus of elasticity: results and comparison with values given by standards.

Temperature Pressing	150 °C	180 °C
Without nanoparticles	1831 ^{1,B} (18.12 ^{2})	1840 ^{1,B} (13.22 ^{2})
Al_2O_3 nanoparticle	1880 ^{1,B} (12.01 ^{2})	2316 ^{1,A} (11.47 ^{2})
CuO nanoparticle	1845 ^{1,B} (8.35 ^{2})	1703 ^{1,B} (19.56 ^{2})
Brazil ABNT [26]	>1600 ^{1}	
Europe EN [27]	>1800 ^{1}	
North America ANSI [28]	>1550 ^{1}	

[1] modulus of elasticity (in MPa); [2] coefficient of variation; [A]: group with the highest average value; [B]: group with the second-highest average value.

Although almost all treatments were superior to the minimum modulus of elasticity specified by these standards (Table 5); only particleboards made with Al_2O_3 nanoparticles pressed at 180 °C differed statistically.

Obtained in static bending test, the modulus of rupture (MOR) is analyzed in Table 6. All panels pressed at 180 °C exceeded the minimum modulus of rupture given by the standards under consideration. In contrast, all panels pressed at 150 °C only met the minimum requirements for the North American region. All particleboards are close to the minimum performance expected by Brazilians, Europeans, and North Americans.

Table 6. Modulus of rupture: results and comparison with values given by standards.

Temperature Pressing	150 °C	180 °C
Without nanoparticles	10.24 ^{1,B} (4.18 ^{2})	11.97 ^{1,A,B} (17.46 ^{2})
Al_2O_3 nanoparticle	10.42 ^{1,B} (15.54 ^{2})	13.64 ^{1,A} (14.74 ^{2})
CuO nanoparticle	10.70 ^{1,A,B} (16.92 ^{2})	12.24 ^{1,A,B} (14.71 ^{2})
Brazil ABNT [26]	>11.00 ^{1}	
Europe EN [27]	>11.50 ^{1}	
North America ANSI [28]	>10.00 ^{1}	

[1] modulus of rupture (in MPa); [2] coefficient of variation; [A]: group with the highest average value; [B]: group with the second-highest average value.

It is known that there is a relationship between mechanical resistance and density, and this is not influenced by the addition of nanoparticles. Thus, it was already expected that there would not be visible changes in the modulus of rupture values. The results showed a small difference between the treatments, while the reduced percentage of nanoparticles added to the adhesive can be considered as previously mentioned.

It is possible to highlight that the panels produced with copper oxide at 150 °C were statistically equivalent to the panels produced without nanoparticles at 180 °C, suggesting that the use of this nanomaterial allows a reduction in the pressing temperature. This gain is both economically and environmentally favorable in terms of reducing energy consumption in pressing, which is the most expensive stage in panel manufacturing. In future studies with higher percentages of nanoparticles, better results are expected.

Through the interactions, it is possible to observe that the modulus of elasticity for CuO nanoparticles showed a decreasing trend with increasing temperature (Figure 5a).

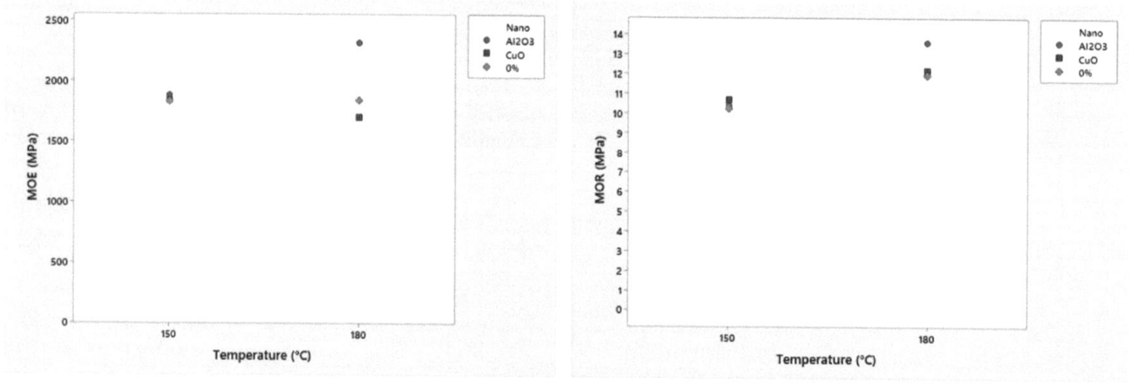

(a) (b)

Figure 5. Interaction graph of results according to each panel composition for (**a**) modulus of elasticity and (**b**) modulus of rupture.

Also, Taghiyari and Bibalan [20] obtained a reduction in the MOE in panels produced with urea-formaldehyde at 200 °C, as the addition of 150 mL/kg of copper nanoparticles decreased this property from 1830 MPa to 1775 MPa. In Figure 5, the interaction graph shows differences in behavioral trends with increasing temperature for the three treatments analyzed. For copper nanoparticles, there was a decrease, although it was not significant (Table 5). The better performance verified in the treatment with aluminum nanoparticles can be justified due to the increase in thermal conductivity with the addition of the nanomaterial, which was also observed by Gupta et al. [19]. The effect of temperature increasing on the MOE for the treatment without nanoparticles does not present a significant change—this is justified by the fact that UF resin can cure efficiently from 150 °C [5].

Regarding the modulus of rupture (Figure 5b), all treatments exhibited an upward trend with increasing temperature, with a more notable effect observed for Al_2O_3 nanoparticles. As observed for the MOE, the increase in thermal conductivity resulting from the use of Al_2O_3 nanoparticles promoted a significant increase in the MOR property. Other treatments demonstrated non-significant increases.

Comparing our results to other studies, Valle [10] obtained MOR values close to 12.25 MPa, finding no statistical difference, whereas Lima et al. [5] reached an average value of 13.3 MPa with ZnO added to the urea-formaldehyde adhesive. Neither study showed statistical differences. Also, Taghiyari and Bibalan [20] did not obtain an increase in the modulus of rupture for particulate panels produced with urea-formaldehyde and pressed at 200 °C when adding 100 mL/kg of copper nanoparticles, although the 150 mL/kg addition showed an increase from 11.56 MPa to 12.43 MPa in this same mechanical property.

3.5. Analysis of the Variation in Pressing Temperature

Sequentially to the nanometric evaluation of the material and physical–mechanical characterizations of the panels reported in the previous subitems, Figure 6a,b shows the graphs obtained from the average panel data after hot pressing at 150 °C and 180 °C.

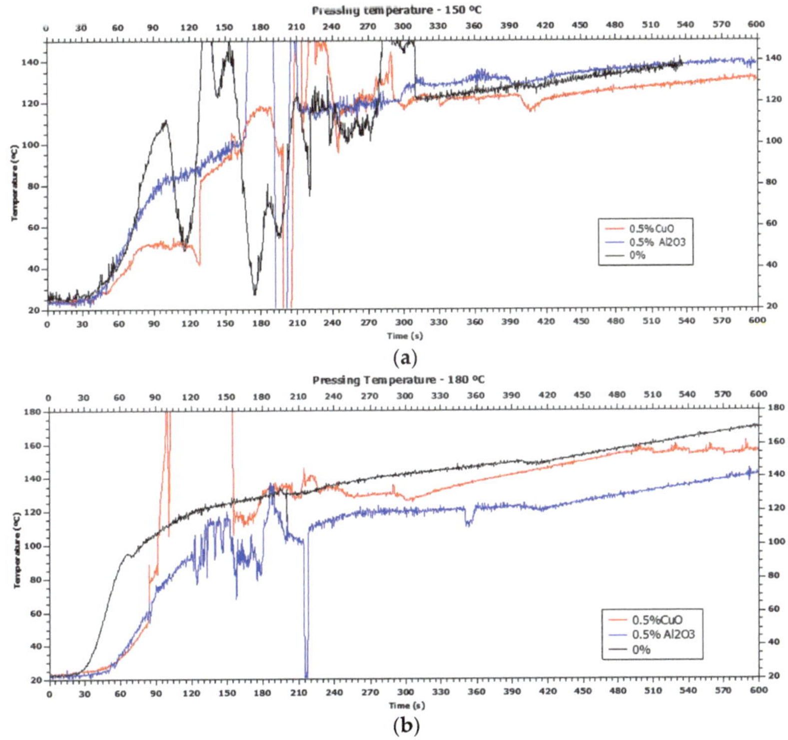

Figure 6. Graphs of variation in pressing temperature at (**a**) 150 °C and (**b**) 180 °C.

It was possible to observe that both the particleboards produced without nanoparticles and the panels manufactured with the addition of nanoparticles (aluminum oxide and copper oxide) did not reach the expected pressing temperatures (150 °C in Figure 5a and 180 °C in Figure 5b), which could influence in the polymerization of the resin, as well as affect the adequate curing of the eucalypt wood panels developed in this study.

When adding copper oxide nanoparticles, the pressing temperature in the innermost layer of the panel was reduced in this exact location of thermocouple measuring, which did not occur when adding aluminum oxide nanoparticles (Figure 5a). In turn, both additions of nanoparticles influenced the protocols for manufacturing different panel compositions, as the presence of these nanoparticles reduced the pressing temperature in Figure 4.

Aluminum oxide has greater thermal insulation capacity as this treatment has lower thermal conductivity than copper oxide. According to Barea [30], when the temperature of aluminum oxide increases, its thermal conductivity decreases and, therefore, it may lead to a 5 °C difference between the final pressing temperatures for two analyses (Figures 4 and 5).

At 150 °C, it is noted that particleboards produced with aluminum oxide reached the highest temperature. At 180 °C, the panel without nanoparticles reached the highest temperature, which approached the ideal temperature. In this case, this nanoparticle-free treatment also obtained the best thickness swelling properties. At 180 °C, it is also possible to observe that the panels produced with nanoparticles presented slower heating, which is justified by the refractory effect present in the oxides. The added nanomaterials may also have interfered with the physical and mechanical properties by retaining greater heat and, therefore, may have affected the polymerization of the urea-formaldehyde adhesive.

Regarding copper oxide nanoparticles, when comparing the two graphs in Figure 5, practically the same variation in final pressing temperature is observed in particleboards without nanoparticles (25 and 30 °C, respectively), presenting intermediate performances in their physical and mechanical properties compared to other treatments.

From effects of copper oxide nanoparticle addition, Taghiyari and Bibalan [20] concluded that, to obtain better heat transfer performance of copper nanoparticles in wood particleboards, it is desirable to increase their quantity. They noticed a reduction in the modulus of elasticity, justifying that it may have been caused by a negative influence on the bonding between the wood particles and the polymeric resin.

For higher levels of nanoparticles, better physical–mechanical properties and lower energy consumption were confirmed by [18]. Therefore, the observed effects can be attributed to the temperature gradient between both surfaces and the core in the particleboard during hot pressing, as explained by Zhang et al. [6], where the type and relative balance of covalent and ionic bonds in the polymeric resin structure may differ in these different regions of composite panels, that is, externally and internally.

3.6. Analysis of Thermal Conductivity

As shown in Table 7, thermal conductivity was analyzed. The added nanoparticles led to greater values for the lower temperature level at 150 °C, a condition opposite to the panel without nanoparticles. Particleboard containing copper oxide showed higher thermal conductivity than panels with aluminum oxide.

Table 7. Thermal conductivity: results.

Temperature Pressing	150 °C	180 °C
Without nanoparticle	0.137 ^{1,B} (11.67 ^{2})	0.153 ^{1,A,B} (3.27 ^{2})
Al_2O_3 nanoparticle	0.162 ^{1,A} (3.08 ^{2})	0.153 ^{1,A,B} (9.15 ^{2})
CuO nanoparticle	0.164 ^{1,A} (5.49 ^{2})	0.158 ^{1,A,B} (1.93 ^{2})

[1] thermal conductivity (in W/mK); [2] coefficient of variation; [A]: group with the highest average value; [B]: group with the second-highest average value.

Thermal conductivity ranged from 0.137 W/mK to 0.164 W/mK for our particleboards under evaluation (Table 7).

Our results also suggest that tested panels may serve as efficient thermal insulating materials, as they are in accordance with the observations of Bonduelle [31]. Panels with insulating performance are recommended for construction uses, where indoor spaces can be more pleasant than the external environment. Values between 0.09 W/mK and 0.197 W/mK were found by Çavuş et al. [32] in their study with different wood species, while Binici et al. [33] identified values between 0.075 W/mK and 0.1588 W/mK for corn-straw panels. This better thermal performance was possible due to the addition of nanoparticles. The added nanomaterials increased the panel conductivity, as it was 20% higher in panels produced at 150 °C (Table 7). Therefore, particleboards produced without nanoparticles require a higher pressing temperature to achieve the same effect provided by the nanomaterials, especially copper oxide. In this way, the nanoparticles allow the use of a lower pressing temperature for better thermal conduction.

4. Conclusions

As shown by our results obtained in physical tests to characterize the particleboards, the nanoparticles did not provide a significant improvement in thickness swelling properties, and they did not interfere with other properties. This can be justified by the 0.5% content of nanoparticles, which led to a limited influence on the material. For mechanical tests, the best performances are related to the increase in temperature, where a higher pressing temperature contributes to the heat transfer during pressing and assists the polymerization of the adhesive and curing of panels, which results in more efficient chemical bonds and the improvement of mechanical properties.

Specifically, we concluded the following:

- For both temperatures, our particleboards were suitable for internal uses, as they were in accordance with Brazilian and European standards for particleboards.
- An increase in temperature from 150 °C to 180 °C allowed better physical and mechanical results for all treatments, reaffirming that the innermost region of the panel reached a higher temperature for polymerization and curing.
- The thermal conductivity of panels with nanoparticles was higher than panels without them, although the increase in temperature reduced this effect. Oxide nanoparticles may have provided a refractory effect common to ceramic materials, which makes it preferable to opt for lower pressing temperatures, especially aluminum oxide, which is positive in terms of reducing energy consumption in a costly stage of panel production.
- Thermal conductivity and pressing temperature evaluations suggest that, in certain situations, the nanoparticles studied can contribute to better heat transfer to the innermost region of the panel. However, an impaired interaction between the wood particles and the polymeric formaldehyde-based adhesive was observed, provided by the addition of more environmentally friendly nanomaterials, which did not allow significant improvements in physical and mechanical properties. This effect can be justified by the increase in the viscosity of the resin when the nanomaterial is added, indicating a poor interaction of the aluminum oxide and copper oxide nanoparticles with the polymers.
- The percentage of 0.5% nanoparticles produced by the sol–gel protein methodology in use was lower than studies found in the literature, as these publications obtained better results in the physical–mechanical performance of panels with commercial nanoparticle contents ranging from 2 and 4%. The lower performance obtained can be justified by the higher viscosity of the adhesive when nanoparticles were added, which impaired the interaction between the particles and the polymeric adhesive.

In light of the above conclusions, different levels of nanoparticles and percentages of nanoparticles between external and internal layers are suggested for future studies, as these topics can be regarded to identify better heat transfers to allow a more significant influence of nanoparticles. In addition, new alternatives for the application of nanoparticles ought to be studied to obtain better results, including the insertion of other nanomaterials, different pressing temperature levels, structural adhesives, and other composite products such as plywood, fiberboards, structural particleboards, and glued engineered wood products.

Author Contributions: Conceptualization, L.C.L.S. and C.I.d.C.; methodology, L.C.L.S., C.I.d.C., F.O.L. and H.R.F.; software, L.C.L.S. and H.R.F.; validation, V.A.D.A., A.L.C. and F.A.R.L.; formal analysis, L.C.L.S., C.I.d.C., F.O.L. and H.R.F.; resources, F.O.L. and H.F.D.S.; writing—original draft preparation, all authors; writing—review and editing, L.C.L.S., V.A.D.A., F.O.L., C.I.d.C. and A.L.C.; visualization, C.I.d.C., A.L.C. and V.A.D.A.; supervision, C.I.d.C.; funding acquisition, C.I.d.C. and H.F.D.S. All authors have read and agreed to the published version of the manuscript.

Funding: This research was funded by FAPESP (Fundação de Amparo à Pesquisa do Estado de São Paulo -São Paulo Research Foundation) with grant numbers 2019/02926-4 and 2015/04660-0, and by CAPES (Coordenação de Aperfeiçoamento de Pessoal de Nível Superior—Coordination of Superior Level Staff Improvement) with grant number 001. The APC was fully funded by PROPESP-IFRO (Pró-reitoria de Pesquisa do Instituto Federal de Educação, Ciência e Tecnologia de Rondônia—Dean of Research of the Federal Institute of Education, Science and Technology of Rondônia).

Institutional Review Board Statement: Not applicable, as this study does not involve humans or animals.

Data Availability Statement: Data are contained within the article.

Conflicts of Interest: The authors declare no conflicts of interest.

References

1. UNECE/FAO. COFFI Market Forecasts. 2023. Available online: https://unece.org/forests/coffi-market-forecasts (accessed on 25 March 2024).
2. De Araujo, V.; Vasconcelos, J.; Lahr, F.; Christoforo, A. Timber forest products: A way to intensify global bioeconomy from bio-materials. *Acta Fac. Xylologiae Zvolen* **2022**, *64*, 99–111. [CrossRef]
3. Sakai, S.; Chen, S.; Matsuo-Ueda, M.; Umemura, K. Curing behavior of sucrose with p-toluenesulfonic acid. *Polymers* **2023**, *15*, 4592. [CrossRef] [PubMed]
4. Iwakiri, S. *Painéis de Madeira Reconstituída [Panels of Reconstituted Wood]*; FUPEF: Curitiba, Brazil, 2005; pp. 1–247. (In Portuguese)
5. Lima, F.O.; Silva, L.C.L.; Campos, C.I.; Christoforo, A.L.; Lahr, F.A.R. Pressing time influences on physical and mechanical properties of MDP panels. *Sci. For.* **2018**, *46*, 387–393. [CrossRef]
6. Zhang, B.; Chen, X.; Pizzi, A.; Petrissans, M.; Dumarcay, S.; Petrissans, A.; Zhou, X.; Du, G.; Colin, B.; Xi, X. Highly branched tannin-tris(2-aminoethyl)amine-urea wood adhesives. *Polymers* **2023**, *15*, 890. [CrossRef]
7. Galdino, D.S.; Kondo, M.Y.; De Araujo, V.A.; Ferrufino, G.L.A.A.; Faustino, E.; Santos, H.F.; Christoforo, A.L.; Luna, C.M.R.; Campos, C.I. Thermal and gluing properties of phenol-based resin with lignin for potential application in structural composites. *Polymers* **2023**, *15*, 357. [CrossRef] [PubMed]
8. Silva, L.C.L.; Lima, F.O.; Favarim, H.R.; Campos, C.I. Heat transfer and physical-mechanical properties analysis of particleboard produced with ZnO nanoparticles addition. *BioResources* **2019**, *4*, 9904–9915. [CrossRef]
9. Taghiyari, H.R.; Rangavar, R.; Bibalan, O.F. Effect of nano-silver on reduction of hot-pressing time and improvement in physical and mechanical properties of particleboard. *BioResources* **2011**, *6*, 4067–4075. [CrossRef]
10. Valle, A.C.M. *Análise das Propriedades Físicas e Mecânicas de Painéis MDP de Madeira de Eucalipto Com Adição de Nanopartículas de Sílica ao Adesivo Uréia-Formaldeído [Analysis of the Physical and Mechanical Properties of MDP of Eucalypt Wood with Addition of Silica Nanoparticles to Urea-Formaldehyde Adhesive]*; São Paulo State University: Itapeva, Brazil, 2015; pp. 1–181. (In Portuguese)
11. da Silva, A.P.S.; Ferreira, B.S.; Favarim, H.R.; Silva, M.F.F.; Silva, J.V.F.; Azambuja, M.A.; Campos, C.I. Physical properties of medium density fiberboard produced with the addition of ZnO nanoparticles. *BioResources* **2019**, *14*, 1618–1625. [CrossRef]
12. Lima, F.O.; Silva, L.C.L.; Ferreira, B.S.; Morais, C.A.G.; Bertolini, M.S.; Barreiros, R.M.; Azambuja, M.A.; Caraschi, J.C.; Favarim, H.R.; Campos, C.I. Influence of the addition of Al_2O_3 nanoparticles and the duration of pressing on the physical properties of OSB panels. *BioResources* **2022**, *17*, 3014–3024. [CrossRef]
13. Mallakpour, S.; Khadem, E. Recent development in the synthesis of polymer nanocomposites based on nano-alumina. *Prog. Polym. Sci.* **2015**, *51*, 74–93. [CrossRef]
14. Yu, J.; Huang, X.; Wang, L.; Peng, P.; Wu, C.; Wu, X.; Jiang, P. Preparation of hyperbranched aromatic polyamide grafted nanoparticles for thermal properties reinforcement of epoxy composites. *RSC Polym. Chem.* **2011**, *2*, 1380–1388. [CrossRef]
15. Cademartori, P.H.G.; Artner, M.A.; Freitas, R.A.; Magalhães, W.L.E.M. Alumina nanoparticles as formaldehyde scavenger for urea-formaldehyde resin: Rheological and in-situ cure performance. *Compos. Part B Eng.* **2019**, *176*, 107281. [CrossRef]
16. Roumeli, E.; Papadopoulou, E.; Pavlidou, E.; Vourlias, G.; Bikiaris, D.; Paraskevopoulos, K.M.; Chrissafis, K. Synthesis, characterization and thermal analysis of urea–formaldehyde/nanoSiO_2 resins. *Thermochim. Acta* **2012**, *527*, 33–39. [CrossRef]
17. Tian, H.; Zhang, Q.; Pizzi, A.; Lei, H.; Wang, J.; Xi, X. Adhesion properties and formaldehyde emissions of MnO_2/UF nanocomposite adhesives. *Int. J. Adhes. Adhes.* **2023**, *128*, 103527. [CrossRef]
18. Zhang, R.; Jin, X.; Wen, X.; Chen, Q.; Qin, D. Alumina nanoparticle modified phenol-formaldehyde resin as a wood adhesive. *Int. J. Adhes. Adhes.* **2018**, *81*, 79–82. [CrossRef]
19. Gupta, A.; Kumar, A.; Sharma, K.V.; Gupta, R. Application of high conductive nanoparticles to enhance the thermal and mechanical properties of wood composite. *Mater. Today Proc.* **2018**, *5*, 3143–3149. [CrossRef]
20. Taghiyari, H.R.; Bibalan, O.F. Effect of copper nanoparticles on permeability, physical and mechanical properties of particleboard. *Eur. J. Wood Wood Prod.* **2013**, *71*, 69–77. [CrossRef]
21. Lima, Z.M.F. Nova Rota de Síntese de Nanopartículas de $NiMn_2O_4$ Usando o Método Sol-Gel Protéico [New Route of Synthesis Nanoparticles $NiMn_2O_4$ Method Using Sol-Gel Protein]. Master's Thesis, Federal University of Ceará, Fortaleza, Brazil, 2011. (In Portuguese).
22. *NBR 14810-3:2013*; Chapas de Madeira Aglomerada—Parte 3: Métodos de Ensaio [Wood Particleboard—Part 3: Test Methods]. Brazilian Association of Technical Standards: Rio de Janeiro, Brazil, 2013. (In Portuguese)
23. *ASTM D1200:2005*; Standard Test Method for Viscosity by Ford Viscosity Cup. ASTM Standards: West Conshohocken, PA, USA, 2023.
24. *ASTM E1530:2011*; Standard Test Method for Evaluating the Resistance to Thermal Transmission of Materials by the Guarded Heat Flow Meter Technique. ASTM Standards: West Conshohocken, PA, USA, 2011.
25. Silva, G.C.; Lelis, R.C.C.; de Oliveira, G.L.; da Silva, B.C.; da Lossano, W.C.S.; dos Abreu, H.S. Propriedades de adesivo aplicado em painéis a partir da substituição por lignossulfonato do processo sulfito. *Ciênc. Florest.* **2019**, *29*, 322–335. [CrossRef]
26. *NBR 14810-2:2018*; Chapas de Madeira Aglomerada—Parte 2: Requisitos [Wood Particleboard—Part 2: Requirements]. Brazilian Association of Technical Standards: Rio de Janeiro, Brazil, 2018. (In Portuguese)
27. *BS EN312:2003*; Particleboards. Specifications. British Standards Institution: London, UK, 2003.
28. *ANSI A208.1:2016*; Particleboard. American National Standards Institute: Gaithersburg, MD, USA, 2016.

29. Cardoso, G.V.; Teixeira, F.P.; Ferreira, E.S. Nanocelulose como catalisador de ureia-formaldeído para produção de painéis aglomerados [Nanocellulose as a urea-formaldehyde catalyst for particleboard production]. In Proceedings of the 15th Brazilian Conference of Woods and Timber Structures (XV Ebramem), Curitiba, Brazil, 9–11 March 2016. (In Portuguese).
30. Barea, R.; Belmonte, M.; Osendi, M.I.; Miranzo, P. Thermal conductivity of Al_2O_3/SiC platelet composites. *J. Eur. Ceram. Soc.* **2003**, *23*, 1773–1778. [CrossRef]
31. Bonduelle, G.M. *Propriedades Térmicas da Madeira [Thermal Properties of Wood]*; UFPR: Curitiba, Brazil, 2017.
32. Çavuş, V.; Şahin, S.; Esteves, B.; Ayata, Ü. Determination of thermal conductivity properties in some wood species obtained from Turkey. *BioResources* **2019**, *14*, 6709–6715. [CrossRef]
33. Binici, H.; Aksogan, O.; Demirhan, C. Mechanical, thermal and acoustical characterizations of an insulation composite made of bio-based materials. *Sustain. Cities Soc.* **2016**, *20*, 17–26. [CrossRef]

polymers

MDPI

Article

Molded Plywood with Proportions of Beech Bark in Adhesive Mixtures: Production on an Industrial Scale

Roman Reh [1,*], Lubos Kristak [1], Jan Sedliacik [1], Pavlo Bekhta [1,2], Anita Wronka [3] and Grzegorz Kowaluk [3]

[1] Faculty of Wood Science and Technology, Technical University in Zvolen, 96001 Zvolen, Slovakia; kristak@tuzvo.sk (L.K.); sedliacik@tuzvo.sk (J.S.); bekhta@tuzvo.sk (P.B.)
[2] Department of Wood-Based Composites, Cellulose and Paper, Ukrainian National Forestry University, 79057 Lviv, Ukraine
[3] Institute of Wood Sciences and Furniture, Warsaw University of Life Sciences, 02-776 Warsaw, Poland; anita_wronka@sggw.edu.pl (A.W.); grzegorz_kowaluk@sggw.edu.pl (G.K.)
* Correspondence: reh@tuzvo.sk

Abstract: Molded plywood is used for furniture components such as seats, backrests, or integral seat shells, and it must be durable and harmless to health. Molded plywood is made with urea-formaldehyde (UF) adhesives; therefore, the issue of the fillers used in them is important. The potential of using ground beech (*Fagus sylvatica* L.) bark as an eco-friendly additive in UF adhesives for molded plywood manufacturing was investigated in this work. Wheat flour was used as a reference filler. The beech bark (BB) level as a filler was 10%, a value verified under laboratory conditions. Nine-layer flat and molded plywood were produced under industrial conditions from beech veneers bonded with a UF adhesive mixture. The mechanical (bending strength and bonding quality) and physical (swelling and absorbency values after 2 and 24 h) properties of the industrially fabricated molded plywood were evaluated and compared with the European standard requirements (EN 310 and EN 314-2). The mechanical properties of the molded plywood with the addition of BB in the adhesive mixture were acceptable and met these standards' requirements. The positive effect of BB in the UF adhesive mixture on a reduction in formaldehyde emissions from the molded plywood was also confirmed. BB, considered to be wood-processing industry waste or a by-product, has significant potential to be used as a filler in UF resins for molded plywood production, providing an environmentally friendly, inexpensive solution for the industrial valorization of bark as a bio-based formaldehyde scavenger.

Keywords: molded beech plywood; UF; ground beech bark; eco-friendly fillers; free formaldehyde emission

Citation: Reh, R.; Kristak, L.; Sedliacik, J.; Bekhta, P.; Wronka, A.; Kowaluk, G. Molded Plywood with Proportions of Beech Bark in Adhesive Mixtures: Production on an Industrial Scale. *Polymers* **2024**, *16*, 966. https://doi.org/10.3390/polym16070966

Academic Editor: George Z. Papageorgiou

Received: 23 February 2024
Revised: 19 March 2024
Accepted: 1 April 2024
Published: 2 April 2024

1. Introduction

Molded plywood is usually a three-dimensional product that is created from multiple beech (birch) veneer layers that are glued together using adhesive, heat, and pressure in a press. The veneer layers are arranged crosswise at an angle of 90°. Molded plywood is used for furniture components such as seats, backrests, or integral seat shells [1,2].

The main task of molded plywood is to maintain the shape in which it was made in the long term to properly serve its purpose. This is made possible by the fact that a grain-interlocking pattern of the adjacent veneers is applied, which greatly reduces swelling and shrinking and generates greater surface resistance [3,4].

The technological processes in the production of molded plywood are the same as in the manufacture of flat plywood; the difference is in the shape of the press plates used. In the manufacture of molded plywood, pressing molds, a matrix, and a patrix of the necessary shape are used and molded plywood is pressed between the surfaces. The condition is that both surfaces of the matrix and the patrix maintain the same distance when pressing when measured to the common perpendiculars, i.e., one surface must be equidistant from the other. Attention must be paid to the projection of the pressing pressure. The pressure in the surfaces perpendicular to its direction is manifested in full, but the pressure decreases with a deviation from the perpendicular direction about the angle forming the direction of the pressed surface. This is from a perpendicular direction to the direction of pressure [5,6]. These are facts of fundamental importance. This paper focuses on the composition of adhesive mixtures used to produce molded plywood, which must be suitable for the machinery and technological operations indicated. They must be such that they make a significant contribution to the long-term stabilization of the manufactured shape of the plywood and that they do not degrade over time [5,7].

In terms of applications of shaped pressing technologies of veneer materials, molded plywood is a product with a long history [5,8–12]. Therefore, the correct composition of the adhesive mixture used for molded plywood pressing is essential. Urea-formaldehyde (UF) adhesives are used for common shapes in molded plywood manufacturing because chairs made using them are used indoors [10,11,13–17].

UF adhesives create a sufficient bonding strength of veneers for molded plywood [9,15,18]. However, they are not used as one-component adhesives; there is always an adhesive mixture present in molded plywood production and, therefore, the resulting adhesive effect is also dependent on the proportion of other components of the UF adhesive mixture used [2,15,17].

Today's modern UF adhesives are of significantly better quality than those produced and offered to molded plywood manufacturers in the 20th century. Formaldehyde emissions are notably already at relatively small levels, corresponding with strict technical standards.

The common method used to decrease free formaldehyde in the adhesive mixtures of molded plywood is the modification of the adhesive mixture using formaldehyde scavengers, whereby the scavengers are directly applied to the adhesive mixtures by mixing and by the related reaction [19–22].

The issue of minimum formaldehyde emission values from the adhesive mixtures used in molded plywood must be completely resolved before production in light of today's strict technical standards because of the length of time pupils and students sit on chairs made from molded plywood during their daily school attendance. This is so constant that the health of pupils and students could be severely tested and endangered over the years [23–26]. Therefore, the use of scavengers in UF adhesive mixtures is of fundamental importance.

This paper describes research into the properties of molded (and flat) plywood produced in a factory. Both plywood types were pressed using disintegrated beech bark (BB) as a filler for UF mixtures. The industrial production of plywood was preceded by the thorough laboratory research testing of forest tree barks as fillers as well as their fractions and shares. This took place from 2013 to 2022 at the Technical University in Zvolen, Slovakia, mostly within its development workshops and laboratories. Only after rigorous laboratory tests and achieving optimal laboratory results was it decided to carry out the industrial tests in a full production operation. The industrial production of plywood using disintegrated beech bark as a filler for UF adhesive mixtures took place at the industrial plant BEKER-MP, s.r.o., Prešov, Slovakia (www.beker.sk, accessed on 23 February 2024).

Various flours are used in UF glue mixtures as fillers and they work well. The viscosity of such glue mixtures is reasonable, the industrial application of the adhesive mixtures is reasonably good, and the cost is not high [2,15,27]. Some scientists and researchers have attempted to replace flour with another type of filler that has all the effective properties of flour, but with the additional effect of effectively capturing the leakage of harmful formaldehyde from the adhesives. Some disintegrated bark of forest trees seems to fulfill this role [21,22,28,29].

Beech is a high-quality raw material used for plywood production and its bark is widely available in plywood mills because of debarking. Thus, this bark is present at the place of wood processing, but it is considered to be ineligible for further processing. The proposal to use it as a filler has given it a new purpose. Thus, BB was selected to produce molded plywood under industrial conditions (fully fledged production in a full industrial operation). BB has been proved to be an effective filler of adhesive mixtures to produce flat plywood under laboratory conditions and its chemical and technological parameters as well as collection, drying, and grinding methods have already been published [22,27,30–33].

2. Materials and Methods

2.1. Materials

Rotary cut beech (*Fagus sylvatica* L.) veneers with an average thickness of 1.30 ± 0.05 mm, 6% ± 1% moisture content (MC), and dimensions of 500 × 500 mm^2 were used to produce the plywood. The raw wood material, which was used for peeling, came from the region of Jasov (Slovenské Rudohorie Mountains) in Slovakia. The veneers were made using a four-foot rotary lathe from KSB (Královopolská strojírna Brno, Czech Republic) at the industrial plant of DYHA TIROLA, s.r.o., Moldava and Bodvou, Slovakia (www.dyhatirola.sk, accessed on 23 February 2024).

As a binder, an industrial UF adhesive (Kronores CB 1100 F; DIAKOL Strážske s.r.o., Strážske, Slovakia) with about 67% dry content was used with an ammonium nitrate (NH_4NO_3) water solution (47%) as a hardener to reach the curing time to glue the mass at 100 °C in about 88 s [28]. The hardener was added at a ratio of 10 parts by weight (pbw) per 100 pbw of adhesive, according to common industrial formulations. The average adhesive viscosity to bond the veneers was 1220 mPa.s and its pH value was 8.6. Wheat flour (WF) was used as the reference filler (REF).

UF adhesive with 10 pbw per 100 pbw of WF based on a liquid UF resin was used as a reference sample. This concentration of fillers for molded plywood production was selected based on our long-term research; it achieved optimal values in the results of previous tests [21,22] and was consistent with the observations of other authors examining the issue of fillers in UF adhesive mixtures [27,29,34–36].

2.2. Methods

The beech bark was properly and finely ground (to a fraction of bark with grains smaller than 0.125 mm) (Figure 1). This was utilized as an alternative to WF; the same fraction of flour with grains smaller than 0.125 mm is used as a filler with a share of 10% in UF adhesives for plywood manufacturing. A higher bark content (15–20%) in adhesive mixtures has not been tested under industrial conditions because under laboratory conditions, it was found that the viscosity of such a mixture became problematic from the point of view of the perfect application of the adhesive mixture to the veneer surface.

Figure 1. Beech bark in a dry state.

Nine-layer flat (18 pieces) and molded (18 pieces) plywood samples were produced under industrial conditions with the use of the bonding mixtures listed in Table 1. The following parameters were applied during the plywood preparation: a bonding mixture spread of 150 g/m^2, the adhesive mixture was then applied using a roller spreader, and a maximum press pressure of 1.4 MPa was used for the hydraulic press (Hydroma RM-3, Uherský Brod, Czech Republic). The pressing times are given in Table 1. The resulting thickness of all plywood after pressing was 10.0 ± 0.2 mm. All the produced flat and molded plywood samples were subjected to conditioning at 20 °C/65% ± 1% relative humidity (RH) to obtain a constant weight before further testing.

Table 1. Compositions of adhesive mixtures, plywood molds, and selected pressing parameters.

Variant Label	Filler	Filler Content (pbw [1] per 100 pbw of Solid Resin)	Pressing Temperature (°C)	Pressing Time (min)
REF 10 flat plywood	Wheat flour	10	110	5
BB 10 flat plywood	Beech bark	10	110	5
REF 10 molded plywood	Wheat flour	10	110	5
BB 10 molded plywood	Beech bark	10	110	5

[1] Pbw: parts by weight.

Flat plywood was produced as a REF product on identical presses to the molded plywood. Their press plates were not molded; in this case, they were flat. The flatness of the plywood—and, therefore, their entire area—allowed us to produce a larger range of test samples than in the case of molded plywood production, where suitable places to produce test sample bodies were limited because of plywood roundness. Molded plywood was produced using molded press plates (Figure 2).

Figure 2. Molded plywood in the molded press during the pressing process.

The following tests were completed for the produced flat plywood (REF): density (8 repetitions), bending strength (MOR) in a parallel direction to the grains of the face veneer layer (16 repetitions), MOR in a cross-direction to the grains of the face veneer layer (8 repetitions), bonding quality (8 repetitions), TS after 2 h (8 repetitions), TS after 24 h (8 repetitions), WA after 2 h (8 repetitions), and WA after 24 h (8 repetitions) [37,38]. We respected and observed the international conventions and procedures for the selection of samples from the panels. Every part of the panel had an equal chance of becoming a sample, but we omitted the edge of the board with a width of 50 mm and we performed this as objectively as possible [37,38].

The following tests were completed for the produced flat plywood (REF): density (8 repetitions), modulus of rupture (MOR) in a parallel direction to the grains of the face veneer layer (16 repetitions), MOR in a cross-direction to the grains of the face veneer layer (8 repetitions), bonding quality (8 repetitions), thickness swelling after 2 h (8 repetitions), thickness swelling after 24 h (8 repetitions), water absorption after 2 h (8 repetitions), water absorption after 24 h (8 repetitions), and free formaldehyde emissions [37,38].

The following tests were completed for the produced molded plywood: density (6 repetitions), MOR in a parallel direction to the grains of the face veneer layer (6 repetitions), MOR in a cross-direction to the grains of the face veneer layer (4 repetitions), bonding quality (6 repetitions), thickness swelling after 2 h (6 repetitions), thickness swelling after 24 h (6 repetitions), water absorption after 2 h (6 repetitions), water absorption after 24 h (6 repetitions), and free formaldehyde emissions [37,38]. The molded plywood properties were crucial in this research because we were studying the development of a new product using an alternative adhesive mixture; thus, the properties of the flat plywood produced using the same process and with the same adhesive mixture as the molded plywood served as the REF.

All mechanical tests were performed using a computer-controlled universal testing machine (TIRA 2200 Heckert Testing Machine, Schalkau, Germany) at the Technical University in Zvolen, Slovakia (within its own development workshops and laboratories). Physical tests were performed at the Technical University in Zvolen, Slovakia, as well. The free formaldehyde emissions were measured using the chamber method according to the standard EN 717-1 [39]. A UviLine SI 5000 spectrophotometer (SI Analytics, Plains, NY, USA) at 412 nm was used to determine the total formaldehyde content.

2.3. Statistical Analysis

An analysis of variance (ANOVA) and *t*-test calculations were used to test ($\alpha = 0.05$) for significant differences between factors and levels, where appropriate. A comparison of the means was performed by employing the Duncan test when the ANOVA indicated a significant difference. The statistically significant differences for the achieved results are given in the Results and Discussion paragraphs, where the data are evaluated.

3. Results and Discussion

3.1. Density of Plywood Panels Produced under Industrial Conditions

As industrially produced plywood is usually produced in larger volumes and emphasis is placed on the production economics, some production operations can take place at a higher pace and less precisely than laboratory plywood production. Thus, in the first step, we produced the plywood under laboratory conditions with exact technological conditions and parameters and we then used these proven parameters in the production of the molded plywood in the factory. The achieved density results of the flat and molded plywood were reasonable and they are presented in Table 2.

Table 2. The density of flat and molded plywood panels.

Variant Label	Filler	Filler Content (pbw [1] per 100 pbw of Solid Resin)	Densities (kg/m^3)
REF 10 flat plywood [1]	Wheat flour	10	758 (11) *
REF 10 flat plywood [2]	Wheat flour	10	764 (6)
BB 10 flat plywood [1]	Beech bark	10	751 (10)
BB 10 flat plywood [2]	Beech bark	10	755 (6)
REF 10 molded plywood [1]	Wheat flour	10	761 (21)
REF 10 molded plywood [2]	Wheat flour	10	730 (22)
BB 10 molded plywood [1]	Beech bark	10	754 (11)
BB 10 molded plywood [1]	Beech bark	10	735 (5)

[1] Parallel direction to the grains of the face veneer layer. [2] Cross-direction to the grains of the face veneer layer. * Standard deviations are in parentheses.

3.2. Bending Strength of Plywood Panels

The results of the MOR tests are presented in Table 3.

Table 3. Bending strength of flat and molded plywood panels.

Variant Label	Filler	Filler Content (pbw[1] per 100 pbw of Solid Resin)	MOR (MPa)
REF 10 flat plywood [1]	Wheat flour	10	95 (9) *
REF 10 flat plywood [2]	Wheat flour	10	62 (3)
BB 10 flat plywood [1]	Beech bark	10	93 (5)
BB 10 flat plywood [2]	Beech bark	10	69 (3)
REF 10 molded plywood [1]	Wheat flour	10	153 (22)
REF 10 molded plywood [2]	Wheat flour	10	58 (2)
BB 10 molded plywood [1]	Beech bark	10	93 (8)
BB 10 molded plywood [2]	Beech bark	10	60 (4)

[1] Parallel direction to the grains of the face veneer layer. [2] Cross-direction to the grains of the face veneer layer.
* Standard deviations are in parentheses.

The plywood MOR results were more than adequate in all cases. They were higher than the stated MOR values for standard plywood available on the market [40–43]. If we considered the type, corresponding plywood thicknesses, and number of veneer layers in the plywood, the achieved MOR values should have been at the level of 70–75 MPa in the case of the parallel direction to the grains of the face veneer layer and in the case of the cross-direction to the grains of the face veneer layer, at the level of 52–54 MPa. The achieved MOR values of the plywood industrially produced in the case of the parallel direction to the grains of the face veneer layer were higher by ±30% and in the case of the cross-direction to the grains of the face veneer layer, higher by ±15%. This applied to both the technical flour filler and ground BB filler. These were quite understandable values because plywood manufacturers tend to achieve higher values of mechanical strength of plywood than the minimum required. The values achieved in this research demonstrated that the tested adhesive mixtures had been properly developed and were suitable for the production of molded plywood. Statistically significant deviations did not occur for about 10% of filler content. Both tested fillers behaved in flat and molded plywood in much the same way in terms of MOR and were, therefore, suitable for the application. An increase in the filler content of about 5–10% significantly increased the viscosity of the binder; the mixture became problematic and possible problems with the even-spreading of the binder over the veneer surface could negatively influence the MOR of the plywood panels. Due to the results of previous laboratory research, we did not experiment with increasing the proportion of fillers in the adhesive mixtures in the plant. This avoided possible difficulties with applying adhesive mixtures or equipment damage.

3.3. Bonding Quality of Plywood Panels

The results of the bonding quality tests are presented in Table 4.

As in the case of MOR and also in the case of bonding quality, we compared the achieved values with those that were published as standards. Our results were more than adequate in all cases. They were higher than the stated bonding quality values for standard plywood available on the market [40–43] or in research papers [21,22,28,35,44,45]. If we considered the type, corresponding plywood thicknesses, and number of veneer layers in the plywood, the achieved bonding quality values should have been at the level of 0.8–1.4 MPa. The achieved bonding quality values of the plywood produced at the plant as part of this research were higher by ±65% in the case of a parallel direction to the grains of the face veneer layer. Both tested fillers behaved in much the same way in the flat and molded plywood in terms of the bonding quality and were, therefore, suitable for application. The fundamental differences in the break zones of the flat and molded plywood samples bonded with both types of filler pressed using the hot press process after the shear strength test did not occur. This could be explained by the fact that the proportion of filler of 10% in the adhesive was not large and, therefore, there was almost no difference between WF and disintegrated BB in the structure of the adhesive mixture used. This was also consistent with the observations of other authors who investigated the mechanical properties of plywood materials after modifications to the adhesive mixtures used to glue veneers or adhesive penetration in wood (veneers) [40,41,46–55].

Table 4. Bonding quality of flat and molded plywood panels.

Variant Label	Filler	Filler Content (pbw [1] per 100 pbw of Solid Resin)	Bonding Quality (MPa)
REF 10 flat plywood [1]	Wheat flour	10	2.5 (0.3) *
REF 10 flat plywood [2]	Wheat flour	10	2.4 (0.1)
BB 10 flat plywood [1]	Beech bark	10	2.8 (0.3)
BB 10 flat plywood [2]	Beech bark	10	2.8 (0.3)
REF 10 molded plywood [1]	Wheat flour	10	2.9 (0.0)
REF 10 molded plywood [2]	Wheat flour	10	2.8 (1.5)
BB 10 molded plywood [1]	Beech bark	10	3.1 (0.4)
BB 10 molded plywood [1]	Beech bark	10	3.0 (0.2)

[1] Parallel direction to the grains of the face veneer layer. [2] Cross-direction to the grains of the face veneer layer.
* Standard deviations are in parentheses.

3.4. *Thickness Swelling and Water Absorption of Plywood Panels*

The tested plywood was designed for indoor environments, yet the thickness swelling and water absorption values after 2 and 24 h were authoritative and informative.

The results of TS and WA of the flat and molded plywood panels after 2 and 24 h are presented in Tables 5 and 6.

Table 5. Thickness swelling of flat and molded plywood panels after 2 and 24 h.

Variant Label	Filler	Filler Content (pbw [1] per 100 pbw of Solid Resin)	TS (%) (after 2/24 h)
REF 10 flat plywood [1]	Wheat flour	10	1.4 (0.1)/3.3 (0.2) *
REF 10 flat plywood [2]	Wheat flour	10	1.6 (0.1)/3.6 (0.1)
BB 10 flat plywood [1]	Beech bark	10	1.6 (0.3)/3.6 (0.7)
BB 10 flat plywood [2]	Beech bark	10	1.6 (0.1)/4.5 (0.3)
REF 10 molded plywood [1]	Wheat flour	10	2.6 (0.9)/5.6 (1.0)
REF 10 molded plywood [2]	Wheat flour	10	2.0 (0.1)/5.0 (1.0)
BB 10 molded plywood [1]	Beech bark	10	2.3 (0.2)/4.7 (0.5)
BB 10 molded plywood [1]	Beech bark	10	1.9 (0.1)/4.2 (0.3)

[1] Parallel direction to the grains of the face veneer layer. [2] Cross-direction to the grains of the face veneer layer.
* Standard deviations are in parentheses.

From Tables 5 and 6 it is evident that TS and WA of both the flat and molded plywood after 2 and 24 h reached the standard values achieved by conventional plywood of this category and these thicknesses. The achieved values of WA and TS for the examined flat and molded plywood produced under industrial conditions did not significantly differ from each other, which corresponded with the idea that the industrially tested concentrations of adhesive mixtures and their composition were the results of long-term laboratory measurements and that there were no unexpected fluctuations in the physical properties tested [3,48,49]. Under the influence of water, the structure and properties of the adhesive mixtures in the cured state within the produced flat and molded plywood did not change. The extraction from the adhesive mixtures of water-soluble ingredients (mainly fillers, but also plasticizers, stabilizers, etc., in the adhesive) was at a reasonably suitable level [1,2,17,18,56–58].

Table 6. Water absorption of flat and molded plywood panels after 2 and 24 h.

Variant Label	Filler	Filler Content (pbw [1] per 100 pbw of Solid Resin)	WA (%) (after 2/24 h)
REF 10 flat plywood [1]	Wheat flour	10	14.7 (1.2)/33.0 (2.3) *
REF 10 flat plywood [2]	Wheat flour	10	14.1 (1.0)/31.9 (1.5)
BB 10 flat plywood [1]	Beech bark	10	12.8 (1.3)/28.8 (1.0)
BB 10 flat plywood [2]	Beech bark	10	22.4 (1.3)/41.1 (0.4)
REF 10 molded plywood [1]	Wheat flour	10	26.2 (2.6)/40.7 (2.0)
REF 10 molded plywood [2]	Wheat flour	10	20.6 (1.0)/39.5 (2.1)
BB 10 molded plywood [1]	Beech bark	10	15.6 (2.3)/37.8 (2.2)
BB 10 molded plywood [1]	Beech bark	10	17.1 (2.8)/35.7 (3.6)

[1] Parallel direction to the grains of the face veneer layer. [2] Cross-direction to the grains of the face veneer layer.
* Standard deviations are in parentheses.

3.5. Formaldehyde Emissions from Plywood Panels: Laboratory Tests at the Technical University in Zvolen

The results of the formaldehyde emission tests of plywood samples manufactured under industrial conditions are shown in Table 7. The obtained results clearly demonstrated that replacing wheat flour with beech bark as a UF adhesive filler led to a significant reduction in formaldehyde emissions. The reduction in formaldehyde emissions was approximately 24% compared with the control samples and approximately 66% compared with the requirements of the standard EN 636 [59]. It was encouraging that the use of both wheat flour and beech bark provided formaldehyde emissions for the E1 class of plywood as well as E0.5 class samples according to that standard, in alignment with the requirements of the German Chemikalien-Verbotsverordnung [60] for wood-based materials. The reduction in free formaldehyde emissions could be attributed to the presence of lignin and polyphenolic extractives in the chemical composition of tree bark [61]. Lignin can react with formaldehyde in an acidic medium [62], while polyphenolic extractives can react with formaldehyde even at an ambient temperature [63]. The potential of using tree bark to reduce the formaldehyde emissions from plywood panels has also been confirmed by other authors [21,22,28,31,44,46,63].

Table 7. Formaldehyde emissions of molded plywood panels.

Class of Formaldehyde Emission	Formaldehyde Emissions Using a UF Filler				Requirement According to EN 636 (mg/m^3)	Requirement According to Chemikalien-Verbotsverordnung	
	Wheat Flour (mg/m^3)	Beech Bark (mg/m^3)	Wheat Flour (ppm)	Beech Bark (ppm)		(mg/m^3)	(ppm)
E1	0.055	0.042	0.044	0.034	≤0.124	-	-
E0.5	0.055	0.042	0.044	0.034	-	0.062	0.050

The indisputable advantage of using bark as a filler is that the bark is a residue from debarking logs. Moreover, the replacement of technical wheat flour in UF adhesive mixtures also allows for a wider use of food flour in the food industry, thus eliminating the a threat to the food security of the population.

3.6. Formaldehyde Emissions from Plywood Panels: Official Authorized Test in an Accredited Foreign Independent Laboratory

The official authorized test of formaldehyde emissions from molded plywood produced under industrial conditions using BB as a filler in the adhesive mixture took place in an accredited foreign independent laboratory. The results of the test were that molded plywood produced under industrial conditions using BB as a filler in the adhesive mixture achieved the following values:

1. According to EN 636: 2012 + A1: 2015, Class E1, the resulting value was 0.058 mg/m^3 The limit is up to 0.124 mg/m^3 [39,59].

2. According to the German Chemikalien-Verbotsverordnung standards for wood-based materials (enforcement dated 1 January 2020), the resulting value was 0.092 ppm. The limit is up to 0.1 ppm (the concentration of the EN 717-1 test was multiplied by a factor of 2) [60].

The fulfillment of the requirements according to both methods was satisfactory, with a large and clear margin from the limit value (decision rule: simple acceptance according to ILAC-G8: 09/2019, 4.2.1 [64]). The results of the test in an independent laboratory confirmed the laboratory tests of free formaldehyde emission measurements and the reliably low repeated measured values from the laboratories of the Technical University in Zvolen.

4. Conclusions

Beech bark (*Fagus sylvatica* L.) that is properly and finely ground (the fraction of bark with grains smaller than 0.125 mm) can be effectively utilized as an eco-friendly alternative to WF as a filler with a share of 10% in UF adhesives for plywood manufacturing. The determined values of the industrially produced flat and molded plywood panels with 10% bark content for MOR, bonding quality, and thickness swelling and water absorption after 2 and 24 h were equal to or higher than the mechanical and physical properties of commercially produced plywood and met the European standard requirements. A higher bark content (15–20%) in the adhesive mixtures was not tested under industrial conditions because under laboratory conditions, it was found that the viscosity of such a mixture became problematic from the point of view of the perfect application of the adhesive mixture to the veneer surface (due to high viscosity).

The positive effect of adding BB to the UF adhesive mixture on reducing harmful formaldehyde emissions in industrially produced plywood panels for two methods was also confirmed. After 669 h (four weeks), the detected value of free formaldehyde emissions from the tested molded plywood in an accredited foreign independent laboratory reached just under 47% of the permitted limit, according to EN 636: 2012 + A1: 2015. It also met the requirements of the German Chemikalien-Verbotsverordnung for wood-based materials by a sufficient margin. The results of the laboratory tests measuring emissions from molded plywood at the Technical University in Zvolen were similarly positive and the plywood clearly and reliably met the formaldehyde emissions requirements of E1 and, at the same time, met the formaldehyde emissions requirements of E0.5.

Based on the results presented, it can be concluded that BB, considered to be wood-processing industry waste or a by-product, has significant potential to be used as a filler in UF resins for molded plywood production, providing an environmentally friendly, inexpensive solution for the industrial valorization of bark as a bio-based formaldehyde scavenger. For academic researchers, the validation of the technology developed in industrial practice is encouraging, even more so when it comes to the successful development of a new product that is also ecological and renewable. It is sometimes difficult to find a willing partner in the industry to support operational testing. In this case, it succeeded, and many thanks go to the plant that is considering the permanent introduction of this technology and is currently developing a system of harvesting, drying, and grinding beech bark so that it is a process that is unpretentious and economically inexpensive. The assumptions and indications for this exist.

Author Contributions: Conceptualization, R.R. and L.K.; methodology, R.R.; validation, J.S., P.B. and G.K.; formal analysis, R.R.; investigation, A.W.; resources, R.R.; data curation, A.W.; writing—original draft preparation, R.R.; writing—review and editing, R.R.; supervision, P.B. All authors have read and agreed to the published version of the manuscript.

Funding: This research was supported by the Slovak Research and Development Agency under contract No. APVV-19-0269 and APVV-22-0238, and by the Ministry of Education, Science, Research and Sport of the Slovak Republic under the project VEGA 1/0077/24.

Data Availability Statement: Data are contained within the article.

Acknowledgments: Experience regarding the application of bark products in plywood production was collected during the realization of the project "BarkBuild", funded by Horizon 2020 Research and Innovation of the European Union under grant agreement No. 773324, has been used in the Methodology and Results sections. This work was also supported by the NextGenerationEU through the Recovery and Resilience Plan for Slovakia under project No. 09I03-03-V01-00124. Our thanks go to the production factory BEKER-MP, s.r.o., Prešov. This research was supported by the Student Furniture Scientific Group (Koło Naukowe Meblarstwa), Faculty of Wood Technology, Warsaw University of Life Sciences, SGGW, Warsaw, Poland.

Conflicts of Interest: The authors declare no conflicts of interest.

References

1. Youngquist, J.A. Wood based composites and panel products. In *Wood Handbook—Wood as an Engineering Material*; General Technical Report, FPL–GTR:113; Forest Product Laboratory: Madison, WI, USA, 1999; Chapter 10.
2. Sellers, T. *Plywood and Adhesive Technology*; CRC Press: Boca Raton, FL, USA, 1985; ISBN 978-0-8247-7407-3.
3. Heebink, B.G. *Fluid-Pressure Molding of Plywood*; Report No. R1624; US FS Forest Products Laboratory: Madison, WI, USA, 1953; 25p.
4. Nickum, C.C. The Implication of Molded Plywood. *Plan B Pap.* **1962**, *195*, 39. Available online: https://thekeep.eiu.edu/plan_b/195 (accessed on 22 February 2024).
5. Obst, J. *Hospodárná Výroba Překližek*; Průmyslové Vydavatelství: Praha, Czech Republic, 1952; 72p.
6. Irle, M.A.; Barbu, M.C.; Réh, R.; Bergland, L.; Rowell, R.M. Wood composites. In *Handbook of Wood Chemistry and Wood Composites*; CRC Press: Boca Raton, FL, USA, 2012; pp. 321–411. ISBN 978-1-4398-5380-1.
7. Available online: https://www.futuremarketinsights.com (accessed on 2 May 2023).
8. Available online: https://www.expertmarketresearch.com/reports/plywood-market (accessed on 2 May 2023).
9. Available online: https://www.imarcgroup.com/plywood-market (accessed on 3 May 2023).
10. Available online: https://www.marketsandmarkets.com/Market-Reports/plywood-market-233250253.html (accessed on 3 May 2023).
11. Available online: https://www.globenewswire.com/news-release/2022/01/20/2370402/0/en/Plywood-Market-Outlook-European-Import-Prices-Spike-26-IndexBox.html (accessed on 3 May 2023).
12. Drápela, J. *Výroba Nábytku—Technologie*; SNTL—Státní Nakladatelství Technické Literatury: Praha, Czech Republic, 1980; 486p.
13. Available online: https://www.plycollection.com/modern-school-chairs-plycollection/ (accessed on 4 May 2023).
14. Available online: https://mdd.eu/en/seating/new-school-chair/ (accessed on 4 May 2023).
15. Pizzi, A. *Wood Adhesives: Chemistry and Technology*; CRC Press: Boca Raton, FL, USA, 2018. [CrossRef]
16. Barbu, M.C.; Irle, M.; Réh, R. Wood Based Composites, Chapter 1. In *Research Developments in Wood Engineering and Technology*; Aguilera, A., Davim, P., Eds.; Engineering Science Reference; IGI Global: Hershey, PA, USA, 2014; pp. 1–45.
17. Dunky, M. Adhesives in the wood industry. In *Handbook of Adhesive Technology*, 2nd ed; Pizzi, A., Mittal, K.L., Eds.; Marcel Dekker Inc.: New York, NY, USA, 2003; p. 71. [CrossRef]
18. Frihart, C.R.; Hunt, C.G. Adhesives with wood materials, bond formation and performance. In *Wood Handbook—Wood as an Engineering Material*; General Technical Report FPL-GTR-190; U.S. Department of Agriculture, Forest Service, Forest Products Laboratory: Madison, WI, USA, 2010; Chapter 10.
19. Kristak, L.; Antov, P.; Bekhta, P.; Lubis, M.A.R.; Iswanto, A.H.; Reh, R.; Sedliacik, J.; Savov, V.; Taghiyari, H.R.; Papadopoulos, A.N.; et al. Recent progress in ultra-low formaldehyde emitting adhesive systems and formaldehyde scavengers in wood-based panels: A review. *Wood Mater. Sci. Eng.* **2023**, *18*, 763–782. [CrossRef]
20. Barbu, M.C.; Lohninger, Y.; Hofmann, S.; Kain, G.; Petutschnigg, A.; Tudor, E.M. Larch bark as a formaldehyde scavenger in thermal insulation panels. *Polymers* **2020**, *12*, 2632. [CrossRef] [PubMed]
21. Réh, R.; Igaz, R.; Krišťák, Ľ.; Ružiak, I.; Gajtanska, M.; Božíková, M.; Kučerka, M. Functionality of beech bark in adhesive mixtures used in plywood and its effect on the stability associated with material systems. *Materials* **2019**, *12*, 1298. [CrossRef] [PubMed]
22. Ružiak, I.; Igaz, R.; Krišťák, Ľ.; Réh, R.; Mitterpach, J.; Očkajová, A.; Kučerka, M. Influence of ureaformaldehyde adhesive modification with beech bark on chosen properties of plywood. *BioResources* **2017**, *12*, 3250–3264. [CrossRef]
23. Available online: https://www.hermanmiller.com/en_lac/products/seating/side-chairs/eames-molded-plywood-chairs/design-story/ (accessed on 10 May 2023).
24. Available online: https://www.plycollection.com/about/ (accessed on 12 May 2023).
25. Available online: https://decor-plywood.com/en/plywood.html (accessed on 12 May 2023).
26. Available online: https://licit.eu (accessed on 8 May 2023).
27. Račko, V.; Čunderlík, I. Thickness and bark proportion of selected hardwood logs. *Acta Fac. Xylologiae Zvolen* **2007**, *49*, 19–25.
28. Reh, R.; Kristak, L.; Sedliacik, J.; Bekhta, P.; Bozikova, M.; Kunecova, D.; Vozarova, V.; Tudor, E.M.; Antov, P.; Savov, V. Utilization of Birch Bark as an Eco-Friendly Filler in Urea-Formaldehyde Adhesives for Plywood Manufacturing. *Polymers* **2021**, *13*, 511. [CrossRef]

29. Oh, Y.S. Evaluation of Chestnut Shell and Coffee Waste with Phenol-Formaldehyde Resin for Plywood Filler X1—Avaliação de Cascas de Castanha e Resíduos de Café Como Material de Enchimento Do Adesivo Fenol-Formaldeído Para a Produção de Compensados. *Ciência Florest* **2021**, *31*, 1991–2001. [CrossRef]
30. Cetin, N.S.; Ozmen, N.; Narlioglu, N.; Cavus, V. Effect of bark flour on the mechanical properties of HDPE composites. *J. Mater. Sci.* **2014**, *1*, 23–32.
31. Aydin, I.; Demirkir, C.; Colak, S.; Colakoglu, G. Utilization of bark flours as additive in plywood manufacturing. *Eur. J. Wood Wood Prod.* **2017**, *75*, 63–69. [CrossRef]
32. Tudor, E.M.; Barbu, M.C.; Petutschnigg, A.; Reh, R. Added-value for wood bark as a coating layer for flooring tiles. *J. Clean. Prod.* **2018**, *170*, 1354–1360. [CrossRef]
33. Liu, J.; Li, Y.; Mo, H.; Xie, E.; Fang, J.; Gan, W. Current Utilization of Waste Biomass as Filler for Wood Adhesives: A Review. *J. Ind. Eng. Chem.* **2022**, *115*, 48–61. [CrossRef]
34. Sanghvi, M.R.; Tambare, O.H.; More, A.P. Performance of Various Fillers in Adhesives Applications: A Review. *Polym. Bull.* **2022**, *79*, 10491–10553. [CrossRef]
35. Kawalerczyk, J.; Dziurka, D.; Mirski, R.; Trociński, A. Flour Fillers with Urea-Formaldehyde Resin in Plywood. *BioResources* **2019**, *14*, 6727–6735. [CrossRef]
36. Mohamed Abdoul-Latif, F.; El Montassir, Z.; Ainane, A.; Gharby, S.; Sakar, E.H.; Merito, A.; Mohamed, J.; Ainane, T. Use of Thymus Plants as an Ecological Filler in Urea-Formaldehyde Adhesives Intended for Bonding Plywood. *Processes* **2022**, *10*, 2209. [CrossRef]
37. *EN 314-1*; Plywood—Bonding Quality—Part 1: Test Methods. European Committee for Standardization: Brussels, Belgium, 2004.
38. *EN 310*; Wood-Based Panels. Determination of Modulus of Elasticity in Bending and of Bending Strength. European Committee for Standardization: Brussels, Belgium, 1993.
39. *EN 717-1*; Wood Based Panels—Determination of Formaldehyde Release—Part 1: Formaldehyde Emission by the Chamber Method. European Committee for Standardization: Brussels, Belgium, 2004.
40. Samek, J. Velkolplošné Dřevní Materiály. In *Dřevařská Příručka I*; Kafka, E., Ed.; SNTL—Nakladatelství Technické Literatury: Praha, Czech Republic, 1989; pp. 163–208. ISBN 80-03-00009-2.
41. Available online: https://apawood-europe.org/official-guidelines-3/apa-plywood-main-eu-standards/performance-characteristics-standards/ (accessed on 15 May 2023).
42. Available online: https://www.engineeringtoolbox.com/timber-mechanical-properties-d_1789.html (accessed on 15 May 2023).
43. Available online: https://www.euroguide.org/structural-timber-design/plywood.html (accessed on 15 May 2023).
44. Ježo, A.; Wronka, A.; Dębiński, A.; Kristak, L.; Reh, R.; Rizhikovs, J.; Kowaluk, G. Influence of Upcycled Post-Treatment Bark Biomass Addition to the Binder on Produced Plywood Properties. *Forests* **2023**, *14*, 110. [CrossRef]
45. Liu, J.; Zhang, H.; Wang, S.; Wang, M.; Lu, X. Study of compressed plywood structure and density variances. *Wood Res.* **2013**, *58*, 663–670.
46. Jorda, J.; Kain, G.; Barbu, M.-C.; Haupt, M.; Krišťák, Ľ. Investigation of 3D-Moldability of Flax Fiber Reinforced Beech Plywood. *Polymers* **2020**, *12*, 2852. [CrossRef]
47. Oikawa, E.; Umeda, H. Design Development of Molded Plywood Stool. In Proceedings of the Annual (61st) Conference of Japanese Society for the Science of Design; Japanese Society for the Science of Design: Tokyo, Japan, 2014; p. 216.
48. Zamirian, L. Process Improvement for Plywood Product Manufacturing Using Design of Experiments. Master's Thesis, Concordia University, Montreal, QU, Canada, 2020; 88p.
49. Graham, P.H. Commercial molded plywood methods much improved. *Veneers Plywood* **1950**, *44*, 12–14.
50. Sandeep, C.; Sirish, N. In Situ Investigation of the Kinematics of Ply Interfaces During Composite Manufacturing. *J. Manuf. Sci. Eng.* **2021**, *143*, 021006.
51. Matsuda, S.; Oshima, K.; Hosaka, M.; Satokawa, S. Effect of annealing on the separation of resin from CFRP cross-ply laminate via electrical treatment. *Compos. Struct.* **2020**, *234*, 111665. [CrossRef]
52. Bekhta, P.; Salca, E.A.; Lunguleasa, A. Some properties of plywood panels manufactured from combinations of thermally densified and non-densified veneers of different thicknesses in one structure. *J. Build. Eng.* **2020**, *29*, 101116. [CrossRef]
53. Bekhta, P.; Hiziroglu, S.; Shepelyuk, O. Properties of plywood manufactured from compressed veneer as building material. *Mater. Des.* **2009**, *30*, 947–953. [CrossRef]
54. Talaei, A.; Ashori, A.; Heydari, V. A Comparative Study on the Mechanical and Physical Properties of Plywood Panels Prepared by Chitosan as Bio-Adhesive. *J. Polym. Environ.* **2022**, *30*, 4263–4270. [CrossRef]
55. Auriga, R.; Gumowska, A.; Szymanowski, K.; Wronka, A.; Robles, E.; Ocipka, P.; Kowaluk, G. Performance properties of plywood composites reinforced with carbon fibers. *Compos. Struct.* **2020**, *248*, 112533. [CrossRef]
56. Kamke, F.; Lee, J. Adhesive penetration in wood—A review. *Wood Fiber Sci.* **2007**, *39*, 205–220.
57. Barboutis, I.; Kamperidou, V. Properties of two different thicknesses 3-ply plywood of tree-of-heaven veneers. In Proceedings of the 22nd International Scientific Conference Wood Is Good—EU Preaccession Challenges of the Sector, Proceedings, Zagreb, Croatia, 21 October 2011; pp. 9–16, ISBN 978-953-292-022-2.
58. Benthien, J.T.; Ohlmeyer, M. Thickness swelling and water absorption of WPC after immersion in cold and boiling water. *Eur. J. Wood Prod.* **2013**, *71*, 437–442. [CrossRef]
59. *EN 636:2012+A1:2015*; Plywood—Specifications. European Committee for Standardization: Brussels, Belgium, 2015.

60. Available online: https://www.umweltbundesamt.de/en/formaldehyde (accessed on 19 May 2023).
61. Tudor, E.M.; Barbu, M.C.; Petutschnigg, A.; Réh, R.; Krišťák, Ľ. Analysis of Larch-Bark Capacity for Formaldehyde Removal in Wood Adhesives. *Int. J. Environ. Res. Public Health* **2020**, *17*, 764. [CrossRef]
62. Van Der Klashorst, G.H.; Strauss, H.F. Polymerization of lignin model compounds with formaldehyde in acidic aqueous medium. *J. Polym. Sci. Part A Polym. Chem.* **1986**, *24*, 2143–2169. [CrossRef]
63. Bekhta, P.; Sedliačik, J.; Noshchenko, G.; Kačík, F.; Bekhta, N. Characteristics of beech bark and its effect on properties of UF adhesive and on bonding strength and formaldehyde emission of plywood panels. *Eur. J. Wood Wood Pract.* **2021**, *79*, 423–433. [CrossRef]
64. Available online: https://ilac.org/publications-and-resources/ilac-guidance-series/ (accessed on 22 February 2024).

polymers

MDPI

Article

Selected Properties of Bio-Based Layered Hybrid Composites with Biopolymer Blends for Structural Applications

Aneta Gumowska [1], Eduardo Robles [2], Arsene Bikoro [2], Anita Wronka [1,*] and Grzegorz Kowaluk [1,*]

[1] Department of Technology and Entrepreneurship in Wood Industry, Warsaw University of Life Sciences, Nowoursynowska Str. 159, 02-776 Warsaw, Poland

[2] University of Pau and the Adour Region, E2S UPPA, CNRS, Institute of Analytical and Physicochemical Sciences for the Environment and Materials (IPREM-UMR 5254), 403 Rue de Saint Pierre, 40004 Mont de Marsan, France

* Correspondence: anita_wronka@sggw.edu.pl (A.W.); grzegorz_kowaluk@sggw.edu.pl (G.K.)

Abstract: In this study, layered composites were produced with different biopolymer adhesive layers, including biopolymer polylactic acid (PLA), polycaprolactone (PCL), and biopolymer blends of PLA + polyhydroxybutyrate (PHB) (75:25 *w*/*w* ratio) with the addition of 25, 50% microcrystalline cellulose (MCC) and 3% triethyl Citrate (TEC) for these blends, which acted as binders and co-created the five layers in the elaborated composites. Modulus of rupture (MOR), modulus of elasticity (MOE), internal bonding strength (IB), density profile, differential scanning calorimetry (DSC), thermogravimetric analysis (TGA), and scanning electron microscopy (SEM) analysis were obtained. The results showed that among the composites in which two pure biopolymers were used, PLA obtained the best results, while among the produced blends, PLA + PHB, PLA + PHB + 25MCC, and PLA + PHB + 25MCC + 3TEC performed best. The mechanical properties of the composites decreased with increases in the MCC content in blends. Therefore, adding 3% TEC improved the properties of composites made of PLA + PHB + MCC blends.

Keywords: composite; mechanical properties; blends; polylactide; polycaprolactone; polyhydroxybutyrate; microcrystalline cellulose; triethyl citrate

Citation: Gumowska, A.; Robles, E.; Bikoro, A.; Wronka, A.; Kowaluk, G. Selected Properties of Bio-Based Layered Hybrid Composites with Biopolymer Blends for Structural Applications. *Polymers* **2022**, *14*, 4393. https://doi.org/10.3390/polym14204393

Academic Editors: Ľuboš Krišťák, Réh Roman, Marius Cătălin Barbu and Eugenia Mariana Tudor

Received: 2 October 2022
Accepted: 15 October 2022
Published: 18 October 2022

1. Introduction

The continuous development of science and technology increases the demand for environmentally friendly products of natural origin and the increased reuse of forestry and agricultural byproducts that are treated mostly as waste [1,2]. Due to concerns related to the depletion of petroleum and greenhouse gas emissions resulting from the production of petroleum products, the use of renewable, recyclable, and compostable raw materials is becoming increasingly desirable [3]. In recent years, environmental regulations have forced producers of wood-based composites to think about using sustainable materials in producing new products. This is one of the reasons why the use of alternative raw materials in wood-based composites is of growing interest [4,5]. New political strategies aim to improve efficiency and reduce impacts on health and the environment, and are also an essential element in promoting competitiveness and the idea of sustainable development [6]. Trends in wood-based products for commodities show an increase in demand, which corresponds to a combination of different factors such as aesthetic aspects, awareness of the origin of products, service life, and performance. However, some of these factors can be contradictory, as products from natural origin have a shorter service life than petroleum products, and products with enhanced performance usually imply a more complex disposal of the residues. On the other hand, there is pressure for new products to perform perfectly during use without harming health and the environment, and, at the end of their life cycle, to be easily reduced, reused, recycled, composted, or recovered as energy. The elaborate production of biocomposites, defined as materials composed of at least two distinct constituent materials (with at least one being naturally derived), using biopolymers and natural fibers is the answer to the current market needs. Unlike petroleum products, natural products can be entirely recovered for further applications [7,8].

Currently, adhesives based on petrochemical products contain volatile organic compounds (VOCs), and toxic chemicals are commonly used in wood-based materials, particleboards, fiberboards, and layered panels [9]. Synthetic adhesives dominate the production process for wood-based materials due to their low manufacturing cost, superior bonding properties and board reliability [10]. However, adhesives based on formaldehyde come from non-renewable petrochemical sources. Despite many efforts to develop low-formaldehyde emission binders, as well as the use of formaldehyde scavengers [11], these have the disadvantage of free formaldehyde emissions during manufacturing and during the use of the final products, which have a detrimental effect on health and the environment [12,13]. In addition, urea-formaldehyde (UF) and other amine resins are a severe concern for human health due to their pathogenic character, which has been extensively studied and demonstrated. Factors driving the development of renewably sourced adhesives as a promising alternative to synthetic adhesives are the regulation of formaldehyde emissions from wood-based industries and market interest in bio-adhesives [14]. Soubam et al. [10] produced plywood with biopolymers as an adhesive; in this case, elaboration and tests of layered composites were performed based on two biopolymers: natural rubber latex (NRL) and dimethylol dihydroxy ethylene urea (DMDHEU) cross-linked with rice starch. Researchers proved that synthetic-based adhesives could be sustainably replaced with selected blends of biopolymers. Based on the results, the mechanical properties of plywood bonded with 10 g NRL + 10 g cross-linked rice starch DMDHEU showed MOR, MOE, and IB values that met the standards.

The use of polymer blends to bond fibers for wood-based products seems an attractive substituent because of the low cost and suitability for a wide range of applications. In this sense, polymer blends with either lignocellulosic fibers or wood flour have been studied for several years. However, the current polymer consumption per human has provoked increasing concern regarding waste management and environmental impact. Post-service-life processes associated with plastics, such as recycling, reducing, and reusing, have become necessary and commendatory, and several laws have been made for this purpose worldwide. Therefore, the use of polymers from renewable materials that represent less harm to the environment has increased in research and industry. There is a steadily growing interest in developing non-toxic and environmentally friendly adhesives that have adhesion properties similar to synthetic commercial resins. The most common biodegradable synthetic polymers are aliphatic polyesters such as polylactic acid (PLA), polyglycolic acids (PGA), polycaprolactone (PCL), and polyhydroxyalkanoates (PHA), among others. Above all, PLA is widely used in commodities for many applications, including grocery and composting bags, automobile panels, textiles, and bio-absorbable medical materials [15–17]. The thermal properties of this material make it an attractive option for thermoplastic processes such as extrusion, injecting molding, blow molding, thermoforming, sheet forming, and film forming [18,19]. PLA is a semi-crystalline aliphatic polyester with a low glass-transition temperature and high transparency. On the other hand, it can be highly brittle [20] and has low heat resistance [21] and poor barrier properties [22], although these properties can be improved with the addition of polymers with higher crystallinity. PLA may not biodegrade at room temperature, but it does biodegrade in controlled composting systems [23]. Polycaprolactone (PCL) is widely used in applications including biomedical materials [24], hot melt adhesives [25], and degradable plastics [26]. The advantages include its excellent biodegradability, high flexibility, biocompatibility, and processibility [27].

PHB is one of the most used polyhydroxyalkanoates; it can be synthesized by controlled bacterial fermentation [28], which has attracted increasing attention. PHB presents high crystallinity [29]; nevertheless, its main drawback is its melting temperature (Tm) which is approximately 170–180 °C, close to its degradation temperature of approximately

270 °C, yielding a small processing range for melt extrusion [30]. Blending two or more polymers with different properties to produce composite materials is a well-known strategy to obtain specific physical properties without the need for complex polymeric systems [31]. PLA and PHB blends have been intensively studied because of the good synergy they can form, with PHB increasing the crystallinity of the blend [32–34] and PLA increasing the stiffness [35]. The mechanical blending of these two polymers can be achieved in a melt state because of their similar melting temperatures, which allows for better blending. Many studies concluded that the best blend is achieved by blending 75 wt% PLA with 25 wt% PHB [32,35–37].

Polymers reinforced with natural fibers are replacing synthetic fiber-reinforced plastics in different industrial sectors, including the automotive industry, packaging, and furniture production, providing lighter materials with better thermal properties [38]. This investigation aimed to assess the impact of natural biopolymer binders on selected mechanical and physical properties of a five-layer lignocellulosic composite produced with different biopolymer layers (including, for example, PLA and PCL), using biopolymer blends as an adhesive and to co-create the five layers in the composites. In the light of the state of art mentioned above, this study fills the gap between the development of formaldehyde-free binders for wood, made of renewable resources, and that of wood-based layered composites modified by additional layers made of biopolymers and their blends.

2. Materials and Methods

2.1. Materials

This study produced layered composites from beech (*Fagus sylvatica* L.) veneers. The nominal dimensions of the commercial veneers were 2500 mm × 200 mm × 0.60 mm, length × width × thickness, respectively. The veneers were cut into 200 mm × 200 mm sheets. The moisture content of ca. 5% of every veneer was measured using an ultrasonic moisture control device. Pure, laboratory-purpose polylactide (PLA, Sigma-Aldrich, product no. 38534, Burlington, MA, USA), polycaprolactone (PCL, Sigma-Aldrich, product no. 704105) in drops with a diameter of 3 mm, and five variants of biopolymer blends, obtained under laboratory conditions, were used as binders. The following components were used to achieve biopolymer blends: PLA was provided by Futerro (Belgium) at extrusion grade; Polyhydroxybutyrate (PHB, P309E) was provided by Biomer (Germany); Sigma-Aldrich provided microcrystalline cellulose (MCC); triethyl citrate (TEC) was provided by Acros Organics.

2.2. Biopolymer Blend Elaboration

The PLA–PHB masterbatch (MB) was blended in a 75:25 w/w ratio according to recommendations from the literature [29,33], and composites were elaborated by mixing MB with different contents of MCC and TEC, as shown in Table 1. Blends were manufactured using a twin-screw extruder (M250, LabTech Engineering, Thailand) with a screw speed of 30–200–100–100 rpm (feed–mix–extrusion) and a temperature profile process of 180–185–190–195 °C. Composites were extruded twice to guarantee dispersion and then granulated into pellets.

Table 1. Composition and shortcode for every elaborated biopolymer blend.

Matrix	PLA [%]	PHB [%]	MCC [%]	TEC [%]
PLA + PHB (75:25)	75.0000	25.0000	0.0000	0.0000
PLA + PHB (75:25)	54.5625	18.1875	24.2500	0.0000
PLA + PHB (75:25)	55.5000	18.5000	25.0000	3.0000
PLA + PHB (75:25)	37.5000	12.5000	50.0000	0.0000
PLA + PHB (75:25)	36.3750	12.1250	48.5000	3.0000

2.3. Manufacturing of Biopolymer Adhesive Layers

The biopolymer adhesive layers were manufactured with an intended thickness of 1 mm. The granules were manually spread onto a frame mold with an average total of 62 g (1550 g m^{-2}) over PTFE mats and placed on pressing steel plates. The granules were evenly distributed over the entire surface, which was limited by the frame (Figure 1a). The first stage involved heating slightly above the melting point of the binder; therefore, only the bottom steel plate with biopolymers/blends spread on the PTFE mat was placed in the press. Adequately thick spacer bars allowed the press shelves to be closed without the upper shelf contacting the granules, while simultaneously trying to provide heat from both sides. This treatment with the bottom steel plate and spacer bars made it possible to control the melting of these materials. The heating time was 10 min, enough to reach the melting point of the blends (Figure 1b). When the biopolymers and blends achieved a molten consistency, they were covered from above with a second PTFE mat and a steel plate and re-inserted between the press shelves to be pressed to a thickness of 1 mm. The temperature of the press was 185 °C. A water bath was used to cool the layers after removal from the press, and then the obtained adhesive layers were cut along the sides of the frame to obtain an adhesive sheet (Figure 1c).

(**a**)

(**b**)

(**c**)

Figure 1. (**a**) Overheating phase of the granules of biopolymers and biopolymer blends; (**b**) molten pellets (PLA in this case); (**c**) adhesive biopolymer layer.

2.4. Layered Composite Manufacturing

Five-layer composites (alternating layers of veneer with biopolymer layers) were manufactured with dimensions 200 mm × 200 mm with an average thickness of 3 mm. The middle layer grain directions were oriented 90° relative to the surface veneers. As a result, seven types of composites were produced with different adhesive layers (subsequently denoted by the shortcodes listed in Table 2), and no less than four layered panels of each binder type. The composites were pressed for 5 min without pressure, after which they were overheated and the adhesive layers were allowed to melt between the veneers; then, 0.6 MPa pressure was applied for 1 min, followed by 1.5 MPa for 1 min. The total pressing time was 7 min. The press temperature was 185 °C. Following the research plan, the produced composites were conditioned in ambient conditions (20 °C; 65% R.H.) to a constant weight for seven days before being cut.

Table 2. Shortcodes for every elaborated composite.

Biopolymer Layers	Sample
Polylactide	PLA
Polycaprolactone	PCL
PLA + PHB	PLA + PHB
PLA + PHB + 25% microcrystalline cellulose (MCC)	PLA + PHB + 25MCC
PLA + PHB + 25% MCC + 3% triethyl citrate (TEC)	PLA + PHB + 25MCC + 3TEC
PLA + PHB + 50% MCC	PLA + PHB + 50MCC
PLA + PHB + 50% MCC + 3% TEC	PLA + PHB + 50MCC + 3TEC

2.5. Mechanical and Physical Properties

The physical and mechanical properties were tested according to European standards. The moduli of rupture (MOR) and moduli of elasticity (MOE) were determined according to EN 310 [39] and were reported as the average of ten measurements. Internal bond (IB) was measured according to EN 319 [40]. Five samples of the layered lignocellulosic composite for each binder variant were used for bond quality. All mechanical properties were examined with an INSTRON 3369 (Instron, Norwood, MA, USA) standard laboratory testing machine. The density profile (DP) of samples was analyzed using a DA-X measuring instrument (GreCon, Alfeld, Germany). Measurement based on direct scanning X-ray densitometry was carried out with a speed of 0.05 mm s^{-1} across the panel thickness with a sampling step of 0.02 mm. The nominal dimensions of all samples were 50 mm × 50 mm. Graphs of the density distribution for each composite type were obtained based on three replicates. Differential Scanning Calorimetry (DSC) tests were performed using a DSC Q20 Instrument (TA Instruments, New Castle, DE, USA). The measurements were carried out at a heating rate of 10 °C min^{-1} under an inert gas (nitrogen) atmosphere with a flow rate of 50 mL min^{-1}. Samples of 5 mg were tested with two repetitions. Thermogravimetric analysis (TGA) was performed on a Q500 (TA Instruments, New Castle, DE, USA) apparatus in air (flow rate 40 mL min^{-1}) in a temperature range of 50–600 °C at a heating rate of 10 °C min^{-1}. Samples of 5–10 mg were tested with two repetitions. Scanning electron microscopy images (SEM) were obtained with a Quanta 200 (FEI, Hillsboro, OH, USA) scanning electron microscope. Pictures of the cross-sections of the manufactured five-layer composites were obtained with a NIKON SMZ 1500 (Kabushiki-gaisha Nikon, Minato, Tokyo, Japan) optical microscope.

2.6. Statistical Analysis

Analysis of variance (ANOVA) and *t*-test calculations were used to test (α = 0.05) for significant differences between factors and levels, where appropriate, using IBM SPSS statistical software (IBM, SPSS 20, Armonk, NY, USA). In addition, a comparison of the means was performed by employing the Duncan test when the ANOVA indicated a significant difference.

3. Results and Discussion

TGA and DSC analysis was carried out for more exhaustive characterization of all produced biopolymer adhesive layers (Figures 2 and 3). The obtained TGA results give information regarding the thermal resistance of the tested materials. The results for samples of PLA and PCL were widely analyzed and compared to the findings in Gumowska et al. (2021) [41]. TGA analysis recorded the degradation temperature, which should not be exceeded during further tests with these materials. TGA analysis of the biopolymers and blends showed one main thermal degradation stage at temperatures of 280–430 °C, with a mass loss of approximately 90%. Figure 2 presents two characteristic curve profiles. The first is a smooth transition (PLA and PCL) and the second shows a deflection characteristic of materials consisting of two or more components (biopolymer blends). The data in Table 3 display the thermal stability established for 50% and 80% weight loss of the tested samples. The higher temperature obtained is related to higher thermal resistance. The highest thermal stability was noted for PCL. The rest of the samples showed similar thermal stability between them. The endothermal melting process of PLA and PCL was similar to that found in published data. In Figure 3 multiple melting peaks can be seen, which were previously reported for PHB and copolymers. These multiple peaks could be caused by melting, recrystallization, and remelting during heating; polymorphism; different molecular weight species; different lamellar thickness, perfection, or stability; and physical aging or relaxation of the rigid amorphous fraction, among others [42]. For example, melting–recrystallization–remelting was considered the cause of complex double melting in PHBV [43]. A similar mechanism may be supposed for the PHB reference, but the occurrence of different molecular weight species due to chain scission during melt processing or the presence of β crystals in addition to the common α form of PHB should not be excluded [44].

The results of the density profiles are summarized in Figure 4. The average densities of all composites ranged from 980 to 1080 kg m^{-3}. The density profiles for individual samples were symmetrical to the middle of the thickness of the composites; therefore, the graph presents the density profiles for their axis of symmetry to facilitate analysis. The graph shows the estimated boundaries of the layers in the composites. The tested composites consisted of five layers, alternating between veneer layers and biopolymer layers. Regardless of the biopolymer or blend layer used, the shapes of the profiles did not differ significantly. The most remarkable information corresponds to PCL, which is usually less dense than PLA and its derived blends. Every determined profile shows an increase in density characteristic of the layer materials, precisely in the bonding zones, which is related to compaction of the resin due to resistance of the wood to impregnation [45,46]. This image justifies calling the produced composites multi-layered composites, as they could be perceived as five-layer panels, of which the biopolymer layer acted as a binder, and separate layers were visible even to the naked eye. The bonding lines are accurately shown on the graphs; they were flat over the entire section of each sample's width. The highest average values of bonding line density were recorded for biopolymer blends with 50% MCC due to the density of cellulose (around 1500 kg m^{-3}). Another remarkable fact is the uneven distribution of densities at the 90° veneer (right side of the plots), which corresponds to an inverted section in comparison to the face veneer, thus presenting a different structure along the section with a lower density in the zones with tracheids.

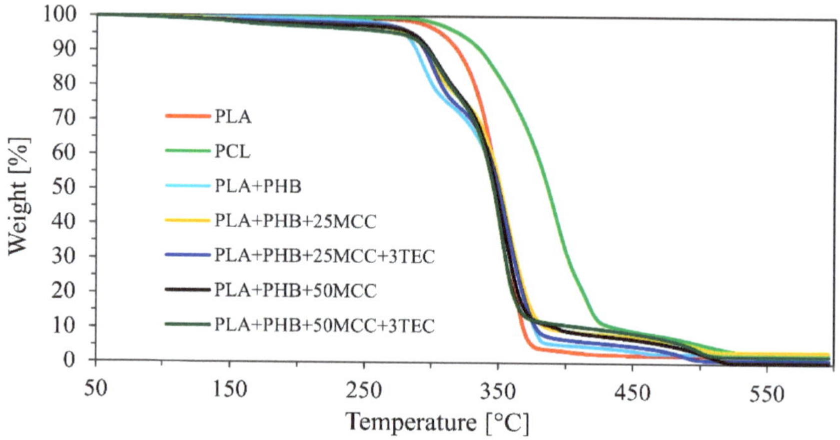

Figure 2. Thermogravimetric analysis (TGA).

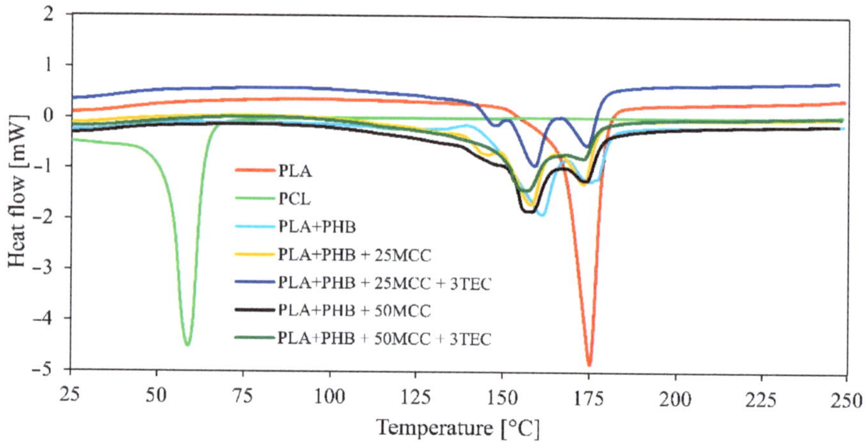

Figure 3. Differential scanning calorimetry (DSC).

Table 3. Thermal stability estimators for the investigated biopolymers and biopolymer blend samples.

Tested Materials	Mass Loss	
	50%	80%
	°C	
PLA	349.3	363.3
PCL	387.4	413.8
MB	350.3	369.7
MB + C1	352.2	370.5
MB + C2	350.6	368.3
MB + C3	348.5	364.9
MB + C4	346.8	361.0

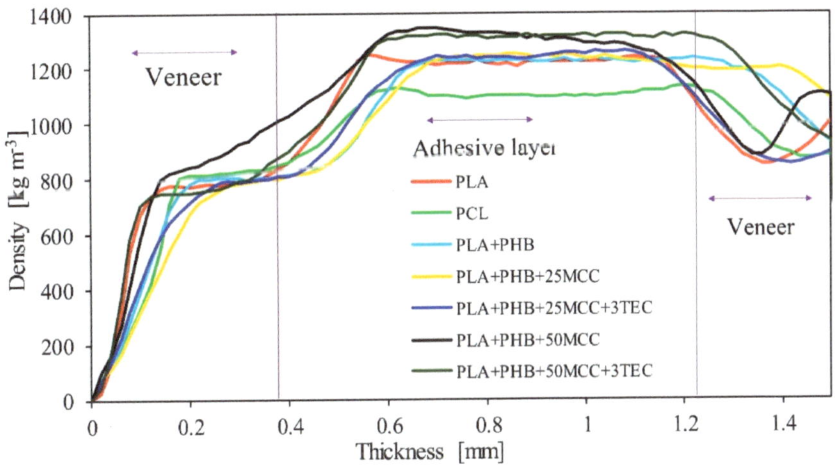

Figure 4. Density profiles of the produced five-layer composites.

The average values of modulus of rupture (MOR) and modulus of elasticity (MOE) under three-point bending stress for the tested composites are presented in Figure 5. The highest average value of MOR (153 N mm^{-2}) was found for PLA + PHB samples, while the lowest (93 N mm^{-2}) was found for PLA + PHB + 50MCC. In the case of MOE, the highest average value was achieved for PLA + PHB (13,718 N mm^{-2}) and MOR. The lowest was for PCL (11,437 N mm^{-2}). Adding MCC to PLA + PHB reduced MOR and MOE, while adding 3% TEC increased MOR and MOE in composites with 25 and 50% MCC in biopolymer blends. The tasks of plasticizers are, among others, to enhance polymer chain mobility [47] and to reduce intermolecular interactions, thus giving the material greater flexibility and a plasticization effect [48]. The addition of 3% TEC to blends with 25% and 50% MCC caused the bonding to anchor deeper in the wood structure than blends without added TEC, which was confirmed by the density profile graph for these samples and potentially also affected the mechanical properties [16]. Based on statistical analysis, there were no statistically significant differences between the average MOR values for PLA, PLA + PHB, PLA + PHB + 25MCC, PLA + PHB + 25MCC + 3TEC, PLA + PHB + 50MCC + 3TEC, or between PCL and PLA + PHB + 50MCC binders.

The results of internal bonding (IB) tests are presented in Figure 6. The outcomes show that the highest average value of IB was that of PLA (8.67 N mm^{-2}), and the lowest value was found for PLA + PHB + 50MCC (3.29 N mm^{-2}). When analyzing the biopolymer blends, it was noticed that composites with 25% MCC fared better than 50% MCC. Increasing to 25% and 50% MCC in blends resulted in decreases in internal bonding of more than 4% and 37%, respectively. This occurred because the mixture of PLA, PHB, and cellulose resulted in packed phases rather than a new polymer or copolymer, potentially generating contact zones prone to failure due to bad mixing. Even if the blends with PLA and PHB showed good properties overall, bad blending at the extruder or a shorter pre-melting time during the first steps of pressing might generate interstices in which fracture could occur. This was further noted in samples containing MCC as the powder also generated surface–surface defects, as shown before [18].

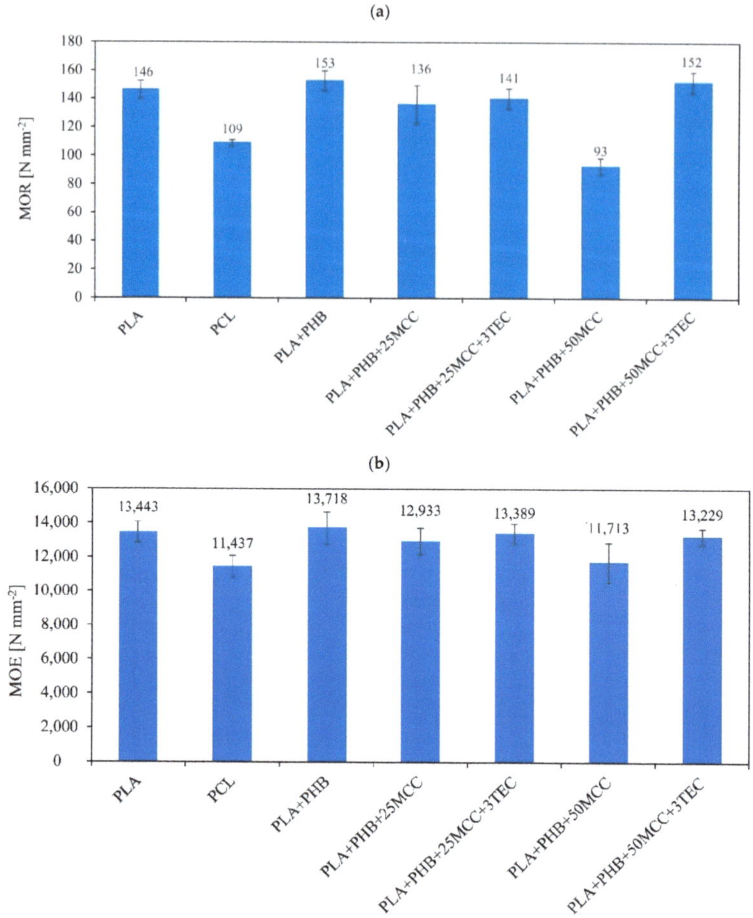

Figure 5. (**a**) Modulus of rupture (MOR) and (**b**) modulus of elasticity (MOE) of the tested five-layer composites.

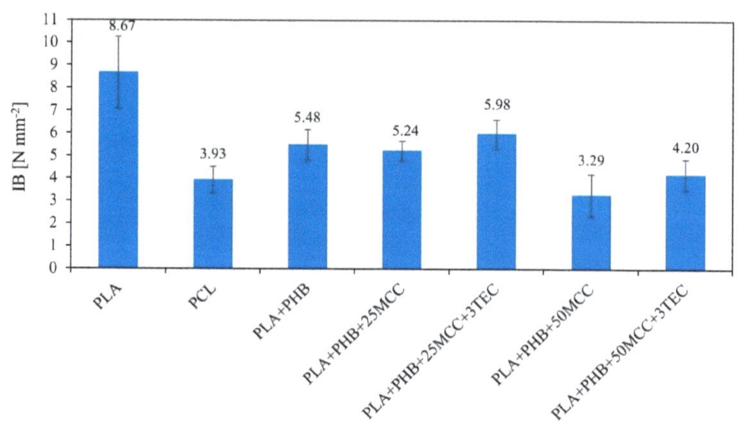

Figure 6. Internal bond (IB) results of tested five-layer composites.

On the other hand, adding 3% TEC to the blends with 25% MCC increased IB by 14%, whereas adding 3% TEC to 50% cellulose blends increased IB by 28%, thus proving the positive contribution of TEC to composite blends. Based on statistical analysis, statistically-significant differences existed between the average values of IB for PLA and the rest of the samples. There were no statistically significant differences between PCL, PLA + PHB + 50MCC, PLA + PHB + 50MCC + 3TEC, and PLA + PHB, PLA + PHB + 25MCC, and PLA + PHB + 25MCC + 3TEC. The two major forms of damage to the samples resulting from the IB test are presented in Figure 7. The first type of damage occurred in the near-surface zone, along with partial destruction in the wood structure obtained as a biopolymer adhesive. The second group includes samples in which the damage took place in the near-surface zone along with destruction on the surface of the adhesive layers.

(a) (b)

Figure 7. The two major forms of damage after the IB test: (**a**) near-surface zone along with partial destruction in the structure of the wood; (**b**) near-surface zone along with destruction in the adhesive layers.

SEM analysis and optical microscopy observation showed the penetration of the biopolymers and biopolymer blends into the wood structure (Figure 8.). There was a recognizable difference in the adhesion zone between the adhesive layers and the wood, where partial penetration of the binder into the pores of the wood could be seen. Penetration into the wood structure is visible in the optical microscope photographs; for an example, see the images of PLA + PHB + 25MCC and the same binder with the addition of 3% TEC (Figure 8e,f). It can be seen that the veneers were not 100% impregnated with the binder, which was also confirmed by the density profiles of the manufactured five-layer composites and which allows for the presence of an exposed face of the veneer with no traces of any polymer resin, which in turn permits either a raw or a finished appearance on the surface face of the elaborated composites. In addition, the addition of TEC affected the penetration depth into the pores of the wood, which could be the origin of the better mechanical properties shown by these composites, in which the wood structure acted as the skeleton of the solidified polymer resin, resulting in a tougher composite.

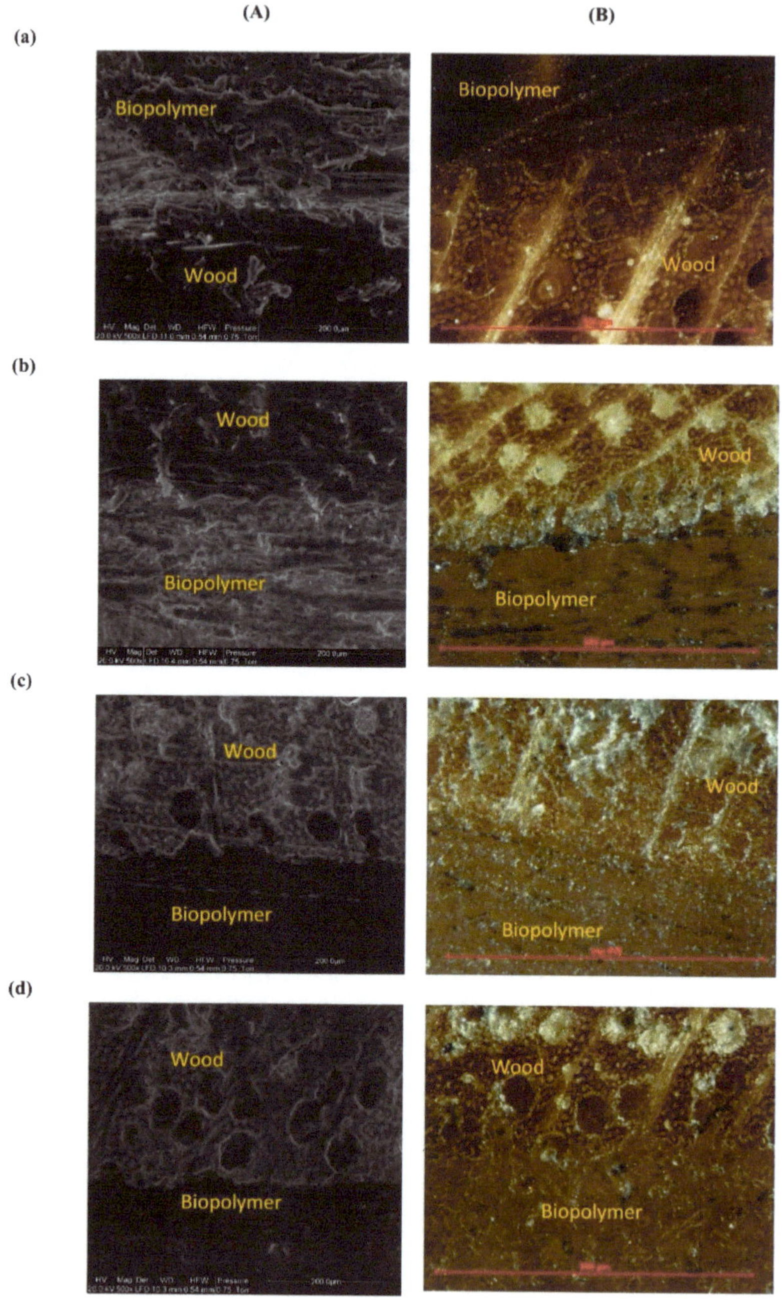

Figure 8. *Cont.*

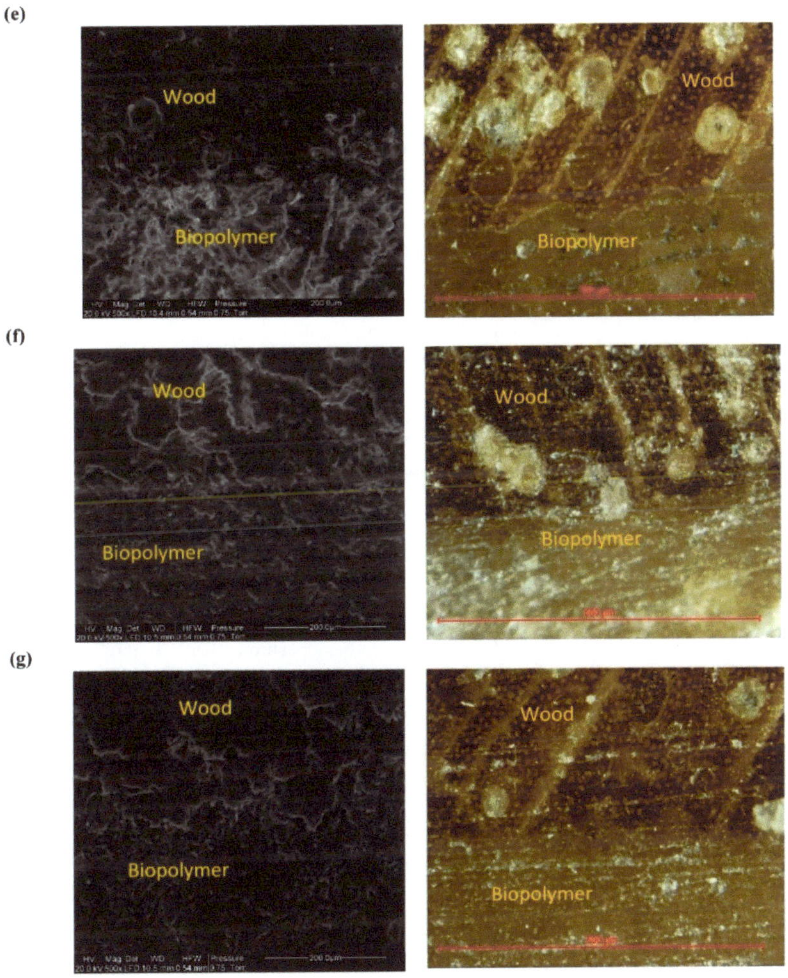

Figure 8. (**A**) SEM pictures (×500) and (**B**) optical microscope photos of the cross-section of composites (10×): (**a**) PLA, (**b**) PCL, (**c**) PLA + PHB, (**d**) PLA + PHB + 25MCC, (**e**) PLA + PHB + 25MCC + 3TEC, (**f**) PLA + PHB + 50MCC, (**g**) PLA + PHB + 50MCC + 3TEC. The wood and binder layers are indicated in the pictures.

4. Conclusions

In the above study, five-layer composites were produced in which biopolymer adhesive layers (PLA, PCL, blends of PLA + PHB (75:25 *w*/*w* ratio) with or without the addition of 25% or 50% MCC, and 3% TEC for these blends) were used as binders. The density profiles of the composites manufactured with biopolymers and biopolymer blends as binders did not deviate from the characteristic profiles of plywood, with increased densities at the bonding line. The produced layers acted as binders and co-created the five layers in the composites. The mechanical properties of the composites decreased with increases in the amount of MCC in blends. MOR, MOE, and IB values for PLA + PHB + 25MCC and PLA + PHB + 50MCC decreased by 11% and 39%, 6% and 15%, and 4% and 40%, respectively. Therefore, adding TEC improved the properties of composites made of PLA + PHB + MCC blends. Among the composites in which two pure biopolymers were used, PLA obtained the best results, while among the produced blends, PLA + PHB, PLA + PHB + 25MCC, and PLA + PHB + 25MCC + 3TEC performed best.

The results achieved herein, regarding an attempt to produce layered wood-based composites with different biopolymers and their blends as special properties layers and binders, allow for the conclusion that it is possible to create a formaldehyde-free wood-based layered composite that enhances the properties of both materials—wood and biopolymer. However, additional work should be performed in the field of biopolymer blend composition to improve adhesion to wood.

Author Contributions: A.G.: Formal analysis, investigation, writing—original draft, funding acquisition; E.R.: Conceptualization, methodology, investigation, resources, writing—original draft, funding acquisition; A.B.: Investigation, A.W.: Investigation, resources, funding acquisition, writing—review and editing; G.K.: Conceptualization, methodology, resources, supervision, project administration, funding acquisition, writing—review and editing. All authors have read and agreed to the published version of the manuscript.

Funding: The PROM International Scholarship Exchange for Ph.D. supported some of the work. Students and Academic Staff were co-financed by the European Social Fund under the Operational Programme Knowledge Education Development, a non-competitive project entitled International Scholarship Exchange for Ph.D. Students and Academic Staff, under the contract POWR.03.03.00-00-PN13/18. Furthermore, the presented study was co-financed by the Polish Ministry of Education and Science from the state budget under the program "Student scientific groups create innovations" (*"Studenckie koła naukowe tworzą innowacje"*)—agreement no. SKN/SP/495845/2021. Some research was completed under the PHC Polonium project, acronym LaLiCoBio, co-financed by the Polish National Agency for Academic Exchange (agreement no. PPN/BFR/2020/1/00042/U/00001). E.R. wishes to acknowledge the tenure track position "BOIS", part of E2S UPPA supported by the "Investissements d'Avenir", a French program managed by ANR (ANR-16-IDEX-0002).

Institutional Review Board Statement: Not applicable.

Data Availability Statement: The data presented in this study are available on request from the corresponding author.

Acknowledgments: The authors would like to acknowledge the members of the Student Furniture Scientific Group (*Koło Naukowe Meblarstwa*), Warsaw University of Life Sciences (SGGW) for support in the realization of the selected research task.

Conflicts of Interest: The authors declare no conflict of interest.

References

1. Brito, F.M.S.; Bortoletto Júnior, G.; Paes, J.B.; Belini, U.L.; Tomazello-Filho, M. Technological characterization of particleboards made with sugarcane bagasse and bamboo culm particles. *Constr. Build. Mater.* **2020**, *262*, 120501. [CrossRef]
2. Aydin, I.; Demirkir, C.; Colak, S.; Colakoglu, G. Utilization of bark flours as additive in plywood manufacturing. *Eur. J. Wood Wood Prod.* **2017**, *75*, 63–69. [CrossRef]
3. Rajeshkumar, G.; Seshadri, S.A.; Devnani, G.L.; Sanjay, M.R.; Siengchin, S.; Maran, J.P.; Al-Dhabi, N.A.; Karuppiah, P.; Mariadhas, V.A.; Sivarajasekar, N.; et al. Environment friendly, renewable and sustainable poly lactic acid (PLA) based natural fiber reinforced composites—A comprehensive review. *J. Clean. Prod.* **2021**, *310*, 127483. [CrossRef]
4. Pędzik, M.; Janiszewska, D.; Rogoziński, T. Alternative lignocellulosic raw materials in particleboard production: A review. *Ind. Crops Prod.* **2021**, *174*, 114162. [CrossRef]
5. Lee, S.H.; Lum, W.C.; Boon, J.G.; Kristak, L.; Antov, P.; Pędzik, M.; Rogoziński, T.; Taghiyari, H.R.; Lubis, M.A.R.; Fatriasari, W.; et al. Particleboard from agricultural biomass and recycled wood waste: A review. *J. Mater. Res. Technol.* **2022**, *20*, 4630–4658. [CrossRef]
6. França, W.T.; Barros, M.V.; Salvador, R.; de Francisco, A.C.; Moreira, M.T.; Piekarski, C.M. Integrating life cycle assessment and life cycle cost: A review of environmental-economic studies. *Int. J. Life Cycle Assess.* **2021**, *26*, 244–274. [CrossRef]
7. Hammiche, D.; Boukerrou, A.; Azzeddine, B.; Guermazi, N.; Budtova, T. Characterization of polylactic acid green composites and its biodegradation in a bacterial environment. *Int. J. Polym. Anal. Charact.* **2019**, *24*, 236–244. [CrossRef]
8. Couret, L.; Irle, M.; Belloncle, C.; Cathala, B. Extraction and characterization of cellulose nanocrystals from post-consumer wood fiberboard waste. *Cellulose* **2017**, *24*, 2125–2137. [CrossRef]
9. Haag, A.P.; Maier, R.M.; Combie, J.; Geesey, G.G. Bacterially derived biopolymers as wood adhesives. *Int. J. Adhes. Adhes.* **2004**, *24*, 495–502. [CrossRef]

10. Soubam, T.; Gupta, A.; Sharma, S.; Shima Jamari, S. Mechanical property study of plywood bonded with dimethylol dihydroxy ethylene urea crosslinked rice starch-natural rubber latex-based adhesive. *Mater. Today Proc.* **2022**, *in press*. [CrossRef]
11. Kristak, L.; Antov, P.; Bekhta, P.; Lubis, M.A.R.; Iswanto, A.H.; Reh, R.; Sedliacik, J.; Savov, V.; Taghiyari, H.R.; Papadopoulos, A.N.; et al. Recent progress in ultra-low formaldehyde emitting adhesive systems and formaldehyde scavengers in wood-based panels: A review. *Wood Mater. Sci. Eng.* **2022**, 1–20. [CrossRef]
12. Yu, C.W.F.; Crump, D.R. Testing for formaldehyde emission from wood-based products—A review. *Indoor Built Environ.* **1999**, *8*, 280–286. [CrossRef]
13. Antov, P.; Savov, V.; Neykov, N. Sustainable bio-based adhesives for eco-friendly wood composites a review. *Wood Res.* **2020**, *65*, 51–62. [CrossRef]
14. Islam, M.N.; Rahman, F.; Das, A.K.; Hiziroglu, S. An overview of different types and potential of bio-based adhesives used for wood products. *Int. J. Adhes. Adhes.* **2022**, *112*, 102992. [CrossRef]
15. Halász, K.; Hosakun, Y.; Csóka, L. Reducing Water Vapor Permeability of Poly(lactic acid) Film and Bottle through Layer-by-Layer Deposition of Green-Processed Cellulose Nanocrystals and Chitosan. *Int. J. Polym. Sci.* **2015**, *2015*, 954290. [CrossRef]
16. Arrieta, M.P.; López, J.; Hernández, A.; Rayón, E. Ternary PLA-PHB-Limonene blends intended for biodegradable food packaging applications. *Eur. Polym. J.* **2014**, *50*, 255–270. [CrossRef]
17. Shi, G.; Cai, Q.; Wang, C.; Lu, N.; Wang, S.; Bei, J. Fabrication and Biocompatibility of Cell Scaffolds of Poly(L-lactic acid) and Poly(L-lactic-co-glycolic acid). *Polym. Adv. Technol.* **2002**, *13*, 227–232. [CrossRef]
18. Robles, E.; Urruzola, I.; Labidi, J.; Serrano, L. Surface-modified nano-cellulose as reinforcement in poly(lactic acid) to conform new composites. *Ind. Crops Prod.* **2015**, *71*, 44–53. [CrossRef]
19. Auras, R.; Harte, B.; Selke, S. An Overview of Polylactides as Packaging Materials. *Macromol. Biosci.* **2004**, *4*, 835–864. [CrossRef]
20. Auras, R.A.; Harte, B.; Selke, S.; Hernandez, R. Mechanical, Physical, and Barrier Properties of Poly(Lactide) Films. *J. Plast. Film Sheeting* **2003**, *19*, 123–135. [CrossRef]
21. Kose, R.; Kondo, T. Size effects of cellulose nanofibers for enhancing the crystallization of poly(lactic acid). *J. Appl. Polym. Sci.* **2013**, *128*, 1200–1205. [CrossRef]
22. Martino, V.P.; Jiménez, A.; Ruseckaite, R.A.; Avérous, L. Structure and properties of clay nano-biocomposites based on poly(lactic acid) plasticized with polyadipates. *Polym. Adv. Technol.* **2011**, *22*, 2206–2213. [CrossRef]
23. Fortunati, F.; Armentano, I.; Iannoni, A.; Barbale, M.; Zaccheo, S.; Scavone, M.; Visai, L.; Kenny, J.M. New Multifunctional Poly(lactide acid) Composites:Mechanical, Antibacterial, and Degradation Properties. *J. Appl. Polym. Sci.* **2012**, *124*, 87–98. [CrossRef]
24. Malikmammadov, E.; Tanir, T.E.; Kiziltay, A.; Hasirci, V.; Hasirci, N. PCL and PCL-based materials in biomedical applications. *J. Biomater. Sci. Polym. Ed.* **2018**, *29*, 863–893. [CrossRef] [PubMed]
25. Cheng, H.N.; Ford, C.V.; He, Z. Evaluation of polyblends of cottonseed protein and polycaprolactone plasticized by cottonseed oil. *Int. J. Polym. Anal. Charact.* **2019**, *24*, 389–398. [CrossRef]
26. Mina Hernandez, J.H. Effect of the incorporation of polycaprolactone (Pcl) on the retrogradation of binary blends with cassava thermoplastic starch (tps). *Polymers* **2021**, *13*, 38. [CrossRef]
27. Ilyas, R.A.; Zuhri, M.Y.M.; Norrrahim, M.N.F.; Misenan, M.S.M.; Jenol, M.A.; Samsudin, S.A.; Nurazzi, N.M.; Asyraf, M.R.M.; Supian, A.B.M.; Bangar, S.P.; et al. Natural Fiber-Reinforced Polycaprolactone Green and Hybrid Biocomposites for Various Advanced Applications. *Polymers* **2022**, *14*, 182. [CrossRef] [PubMed]
28. Chee, J.-Y.; Yoga, S.-S.; Lau, N.; Ling, S.; Abed, R.M.M.; Sudesh, K. Bacterially Produced Polyhydroxyalkanoate (PHA): Converting Renewable Resources into Bioplastics. *Curr. Res. Technol. Educ. Top. Appl. Microbiol. Microb. Biotechnol.* **2010**, *2*, 1395–1404.
29. Calvão, P.S.; Chenal, J.-M.; Gauthier, C.; Demarquette, N.R.; Bogner, A.; Cavaille, J.Y. Understanding the mechanical and biodegradation behaviour of poly(hydroxybutyrate)/rubber blends in relation to their morphology. *Polym. Int.* **2012**, *61*, 434–441. [CrossRef]
30. Malinová, L.; Brožek, J. Mixtures poly((R)-3-hydroxybutyrate) and poly(l-lactic acid) subjected to DSC. *J. Therm. Anal. Calorim.* **2011**, *103*, 653–660. [CrossRef]
31. Mendizábal, E.; Candia, J.M.; Jasso, C.F.; Cruz, L. Comparison of the time-temperature curing dependence of industrial and dry blend modified plastisols. *J. Vinyl Addit. Technol.* **1992**, *14*, 202–206. [CrossRef]
32. Zhang, M.; Thomas, N.L. Blending polylactic acid with polyhydroxybutyrate: The effect on thermal, mechanical, and biodegradation properties. *Adv. Polym. Technol.* **2011**, *30*, 67–79. [CrossRef]
33. Tsuyoshi, F.; Harumi, S.; Rumi, M.; Jianming, Z.; Yong-Xin, D.; Isao, N.; Shukichi, O.; Yukihiro, O. Structure, Dispersibility, and Crystallinity of Poly(hydroxybutyrate)/Poly(l-lactic acid) Blends Studied by FT-IR Microspectroscopy and Differential Scanning Calorimetry. *Macromolecules* **2005**, *38*, 6445–6454. [CrossRef]
34. Arrieta, M.P.; Castro-López, M.D.M.; Rayón, E.; Barral-Losada, L.F.; López-Vilariño, J.M.; López, J.; González-Rodríguez, M.V. Plasticized Poly(lactic acid)–Poly(hydroxybutyrate) (PLA–PHB) Blends Incorporated with Catechin Intended for Active Food-Packaging Applications. *J. Agric. Food Chem.* **2014**, *62*, 10170–10180. [CrossRef]
35. Vogel, C.; Siesler, H.W. Thermal Degradation of Poly(ε-caprolactone), Poly(L-lactic acid) and their Blends with Poly(3-hydroxybutyrate) Studied by TGA/FT-IR Spectroscopy. *Macromol. Symp.* **2008**, *265*, 183–194. [CrossRef]
36. Abdelwahab, M.A.; Flynn, A.; Chiou, B.S.; Imam, S.; Orts, W.; Chiellini, E. Thermal, mechanical and morphological characterization of plasticized PLA-PHB blends. *Polym. Degrad. Stab.* **2012**, *97*, 1822–1828. [CrossRef]

37. Arrieta, M.P.; Samper, M.D.; López, J.; Jiménez, A. Combined Effect of Poly(hydroxybutyrate) and Plasticizers on Polylactic acid Properties for Film Intended for Food Packaging. *J. Polym. Environ.* **2014**, *22*, 460–470. [CrossRef]
38. Faruk, O.; Bledzki, A.K.; Fink, H.P.; Sain, M. Biocomposites reinforced with natural fibers: 2000–2010. *Prog. Polym. Sci.* **2012**, *37*, 1552–1596. [CrossRef]
39. *EN 310 Wood-Based Panels;* Determination of Modulus of Elasticity in Bending and of Bending Strength. European Committee for Standardization: Brussels, Belgium, 1993.
40. *EN 319 Particleboards and Fibreboards;* Determination of Tensile Strength Perpendicular to the Plane of the Board. European Committee for Standardization: Brussels, Belgium, 1993.
41. Gumowska, A.; Robles, E.; Kowaluk, G. Evaluation of functional features of lignocellulosic particle composites containing biopolymer binders. *Materials* **2021**, *14*, 7718. [CrossRef]
42. Reis, K.C.; Pereira, L.; Melo, I.C.N.A.; Marconcini, J.M.; Trugilho, P.F.; Tonoli, G.H.D. Particles of coffee wastes as reinforcement in polyhydroxybutyrate (PHB) based composites. *Mater. Res.* **2015**, *18*, 546–552. [CrossRef]
43. Dasan, Y.K.K.; Bhat, A.H.H.; Faiz, A.; Ahmad, F.; Faiz, A. Polymer blend of PLA/PHBV based bionanocomposites reinforced with nanocrystalline cellulose for potential application as packaging material. *Carbohydr. Polym.* **2017**, *157*, 1323–1332. [CrossRef]
44. Panaitescu, D.M.; Nicolae, C.A.; Frone, A.N.; Chiulan, I.; Stanescu, P.O.; Draghici, C.; Iorga, M.; Mihailescu, M. Plasticized poly(3-hydroxybutyrate) with improved melt processing and balanced properties. *J. Appl. Polym. Sci.* **2017**, *134*, 44810. [CrossRef]
45. Auriga, R.; Gumowska, A.; Szymanowski, K.; Wronka, A.; Robles, E.; Ocipka, P.; Kowaluk, G. Performance properties of plywood composites reinforced with carbon fibers. *Compos. Struct.* **2020**, *248*, 112533. [CrossRef]
46. Jorda, J.; Cesprini, E.; Barbu, M.-C.; Tondi, G.; Zanetti, M.; Král, P. Quebracho Tannin Bio-Based Adhesives for Plywood. *Polymers* **2022**, *14*, 2257. [CrossRef]
47. Martin, O.; Avérous, L. Poly(lactic acid): Plasticization and properties of biodegradable multiphase systems. *Polymer* **2001**, *42*, 6209–6219. [CrossRef]
48. Soto-Valdez, H.; Auras, R.; Peralta, E. Fabrication of Poly(lactic acid) Films with Resveratroland the Diffusion of Resveratrol into Ethanol. *J. Appl. Polym. Sci.* **2011**, *121*, 970–978. [CrossRef]

polymers

MDPI

Article

Physical Properties of Fast-Growing Wood-Polymer Nano Composite Synthesized through TiO_2 Nanoparticle Impregnation

Istie Rahayu [1,*], Wayan Darmawan [1], Deded Sarip Nawawi [1], Esti Prihatini [1], Rohmat Ismail [2] and Gilang Dwi Laksono [1]

1 Department of Forest Products, Faculty of Forestry and Environment, IPB University, Kampus IPB Dramaga, Bogor 16680, Indonesia
2 Department of Chemistry, Faculty of Mathematics and Natural Sciences, IPB University, Kampus IPB Dramaga, Bogor 16680, Indonesia
* Correspondence: istiesr@apps.ipb.ac.id

Abstract: Mangium (*Acacia mangium* Willd.) is a fast-growing wood that is widely grown in Indonesia. The impregnation method is needed to improve the qualities of the wood. In this study, TiO_2 nanoparticle (79.17 nm) was produced using the hydrothermal method. The purpose of this study was to analyze the effect of TiO_2 nanoparticle impregnation on the density and dimensional stability of mangium and the effectiveness of the presence of TiO_2 nanoparticle in wood in degrading pollutants. The mangium samples (2 cm $\times$ 2 cm $\times$ 2 cm) were placed inside impregnation tube. The impregnation solutions included water (untreated), 1% TiO_2 nanoparticle, and 5% TiO_2 nanoparticles. The samples were analyzed for density, weight percent gain (WPG) dan bulking effect (BE). Samples were also analyzed by X-ray diffraction (XRD) and Fourier-transform infrared spectroscopy (FTIR). TiO_2 nanoparticle resulted in an increase in density, WPG, and BE-treated mangium. Based on XRD and FTIR results, TiO_2 nanoparticle was successfully impregnated into mangium wood. Scanning electron microscopy–energy-dispersive X-ray spectroscopy analysis indicated that TiO_2 nanoparticle covered the surface of the wood cells. The TiO_2-impregnated mangium wood has a higher photocatalyst activity than untreated, indicating better protection from UV radiation and pollutants.

Keywords: impregnation; mangium; TiO_2 nanoparticle; photocatalyst; physical properties

Citation: Rahayu, I.; Darmawan, W.; Nawawi, D.S.; Prihatini, E.; Ismail, R.; Laksono, G.D. Physical Properties of Fast-Growing Wood-Polymer Nano Composite Synthesized through TiO_2 Nanoparticle Impregnation. *Polymers* **2022**, *14*, 4463. https://doi.org/10.3390/polym14204463

Academic Editor: Antonios N. Papadopoulos

Received: 20 September 2022
Accepted: 14 October 2022
Published: 21 October 2022

1. Introduction

The need for quality wood that is strong, and durable remains very high in Indonesia and elsewhere. Wood is irreplaceable raw material for both inside (interior) and outside (exterior) applications. Based on data [1], sawn wood production in Indonesia reached 2.6 million m^3 in 2020. The demand for industrial wood under the Indonesian Forest Concession Association (APHI) [2] reached 2 million m^3 in 2019. Mangium (*Acacia mangium* Willd.) wood is widely used for making door and window frames and as raw material for furniture in Indonesia. Mangium wood has a specific gravity value of 0.61 (0.43–0.66) and belongs to the strong class II–III [3]. This fast-growing wood has several disadvantages that hinder its use, such as high-water content as well as hygroscopic properties that enable it to absorb water easily. The water content in wood can affect dimensional stability and mechanical properties [4]. According to Chu et al. [5], reducing the hygroscopic properties of wood is important because it can improve low-quality wood by increasing its dimensional stability. Several methods can be used to improve wood quality, including wood modification, drying, and preservation.

In this study, we use a wood impregnation method. This treatment inserts chemicals into the pores, lumens, and cell walls of wood, where they precipitate and stored without damaging the wood. In addition, the impregnation process involves chemical reaction between the impregnation compound and the most reactive groups in the wood cell wall [6,7]. Many studies have been conducted on impregnation of fast-growing wood to improve the quality, using chemicals such as methyl methacrylate [8–10], phenol formaldehyde [11,12], monoethylene glycol [13], melamine formaldehyde and furfuryl alcohol [14], and nano Fe_3O_4 [15].

The impregnation method using TiO_2 nanoparticle (NP) can also increase the wood's resistance to damage caused by weathering and UV radiation. In particular, the use of TiO_2 nanoparticle offers an alternative formulation for better protection, and Fuva and Hovde [16] reported that surface modification using TiO_2 nanoparticle can increase the wood's UV resistance. The resistance is because nanoparticle have a higher surface area. Wang et al. [17] showed that TiO_2 nanoparticle anatase is more suitable as a photocatalyst, while rutile TiO_2 performs better in blocking UV irradiation. TiO_2 serves as a highly functional materials, and it has been widely used in paints [18] because of its superior chemical stability and non-toxicity [18,19]. TiO_2 nanoparticle has also been used in wood to increase dimensional stability [20], fire resistance [21], rot protection [22], and prevent weathering [23–27].

In some previous studies, TiO_2 nanoparticle was mostly purchased from commercial sources. However, in this study we synthesized TiO_2 nanoparticle via hydrothermal method. This hydrothermal method involves conducting the reaction with water as the solvent in a closed system at a certain temperature and pressure [28]. Liu et al. [29] highlighted that following the reaction in a closed container, the contents can be recovered and reused after cooling to room temperature, which makes this method environmentally friendly. Because the hydrothermal method with water solvent enables controlling the morphology, thermal stability, and crystalline phase, it has been often used to synthesize nano-sized materials including TiO_2 photocatalyst [30,31]. The hydrothermal process has several advantages: such as the use of simple equipment, catalyst-free growth, low cost, large uniform production area, low process temperature, and ease of controlling particle size. Further, particle properties such as morphology and size can be controlled by adjusting the reaction temperature and time and concentration of precursors [32]. Based on the results of previous studies, information was obtained that TiO_2 nanoparticle have an increase in reactivity as photocatalysts when forming composites with other compounds such as toluene degradation reactions as organic volatile pollutants with Co_3O_4/TiO_2 nanocomposite [33,34], hydrocarbon compounds with RuO_2/TiO_2 nanocomposite [35], and ethanol with Pt/TiO_2 nanocomposite [36]. The purpose of this study is to analyze the effect of impregnation of TiO_2 nanoparticle on the density and dimensional stability of mangium and to investigate the effectiveness of the TiO_2 nanoparticle presence in wood as a polymer nanocomposite for degrading pollutants.

2. Materials and Methods

Ten-year-old mangium wood samples were obtained from community forests in the Bogor area, West Java, Indonesia. The mangium wood is then cut into 100 cm long pieces. All samples came from the same tree and were cut at the same time to obtain uniform wood samples. Chemicals used in this study included bulk TiO_2 anatase (Pure®), ethanol (Mallinckrodt chemical®), and deionized water.

2.1. Sample Preparation

Mangium wood was cut using a chain saw and table saw without distinguishing between sapwood and heartwood. The test samples each measured 2 cm × 2 cm × 2 cm [37], and consisted of 30 samples, with 10 replications at each level. The sample were used in testing the weight polymer gain (WPG), leachability (L), anti-swelling efficiency (ASE), water uptake (WU), bulking effect (BE), and density.

2.2. Synthesis of TiO_2 Anatase Nanoparticle Using the Hydrothermal Method

The preparation of TiO_2 nanoparticle solution was carried out using the hydrothermal method. Seven grams of pure TiO_2 powder was placed in a beaker, and then mixed with 75 mL of distilled water. The solution was then reacted in a Teflon stainless steel autoclave at 75 °C for 4 h, after which it was gradually cooled to room temperature over 24 h. The resulting precipitate was separated onto a porcelain dish and then washed with ethanol and calcined for 3 h at 500 °C [38].

2.3. Impregnation Method

The wood samples were oven dried before the impregnation process and then weighed and their dimensions were measured. Furthermore, the impregnation process is carried out using an impregnation tube. Wood samples were put into each container and an impregnation solution was poured which consisted of untreated, 1% (*w*/*v*) TiO_2 nanoparticle, and 5% (*w*/*v*) TiO_2 nanoparticle. The samples were under vacuum (−0.7 bar) for 30 min and then under pressure (1 bar) for 2 h. The test sample was removed from the tube and dried in the oven at 65 °C for 12 h and then left in the oven at 103 °C until the weight was constant. The impregnated samples were then used in density and dimensional stability test and the effectiveness of the presence of TiO_2 nanoparticle in wood in degrading pollutants was analyzed.

2.4. Physical Properties

Samples were weighed and measured to obtain their weight and volume both before and after the impregnation process. The weights and measurements were used to calculate the following values: weight percent gain (WPG), bulking effect (BE), density (ρ), anti-swelling efficiency (ASE), water uptake (WU), and leachability (L)

The WPG was calculated using Equation (1):

$$\text{WPG}\ (\%) = [(W_1 - W_0)/W_0] \times 100 \tag{1}$$

where W_0 is the initial oven-dried weight of the sample before impregnation, and W_1 is the oven-dried weight of the sample after impregnation.

ASE testing was carried out by repeated water soaking [4]. ASE was calculated with Equation (2):

$$\text{ASE}\ (\%) = [(S_u - S_t)/S_u] \times 100 \tag{2}$$

where S_u is the volume shrinkage of untreated wood sample, and S_t is the volume shrinkage of the treated wood.

WU testing on samples was carried out after immersion in water for 24 h. WU was calculated using Equation (3):

$$\text{WU}\ (\%) = [(W_2 - W_1)/W_1] \times 100 \tag{3}$$

where W_2 is the sample weight after immersion in water for 24 h.

The BE test was calculated using Equation (4):

$$\text{BE}\ (\%) = (V_1 - V_0)/V_0 \times 100 \tag{4}$$

where V_0 is the initial oven-dried volume of a wood sample before impregnation and V_1 is the oven-dried volume of wood sample after impregnation.

Oven-dried density (ρ) was calculated using Equation (5):

$$\rho\ (\text{g/cm}^3) = \frac{B}{V} \times 100 \tag{5}$$

where B is the weight of the sample before or after impregnation treatment, and V is the volume of the sample before or after impregnation treatment.

2.5. Photocatalyst Activity Test

Total of 0.1 g of the test sample was put into 100 mL of 10 ppm methylene blue solution and then stirred until homogeneous. The mixed solution was irradiated under UV light for 60 min, and every 10 min, the solutions were taken to measure the absorbance at a wavelength of 666 nm by UV-VIS Spectrophotometer [39].

2.6. Testing the Effect of UV-Vis Radiation

The test sample was irradiated for 6 h continuously with UV-Vis radiation in a closed room. After that, the effect of radiation on wood was analyzed by FTIR [24].

2.7. Characterization of Synthesized TiO_2 Nanoparticle

The synthesized TiO_2 nanoparticle were characterized using UV-Vis spectrophotometer, FTIR, and XRD [40]. Data processing is performed with Origin 8.5 software (Northampton, MA, USA) for spectrum and diffractogram data. Inorganic molecular elucidation was carried out with QualX software (Roma, Italy) and Mercury software (Cambridge, UK).

Characterization of optical properties using UV Vis spectrophotometer (Shimadzu 1800) was carried out by dissolving TiO_2 nanoparticle anatase at a concentration of 0.1 M in demineralized water and then measured its absorption in the UV wavelength range [41]. The results of this measurement are also used for band gap energy analysis using the Tauc method.

Characterization using FTIR (Perkin-Elmer Spectrum One) and XRD (PANAnalytical Empyrean) was carried out by filtering TiO_2 nanoparticle anatase in powder form using a 100 mesh sieve followed by FTIR analysis of pellet method with KBr and measured at wave numbers 400–4000 cm^{-1} and XRD analysis with powder measurement method at an angle range of 2θ which is 0–90°

2.8. Characterization of Impregnated Mangium Wood

2.8.1. Scanning Electron Microscopy and Energy-Dispersive X-Ray Spectroscopy

The penetration ability and distribution of TiO_2 nanoparticle into wood cell walls were analyzed using SEM (ZEISS EVO'50). Untreated and impregnated wood samples were cut to a size of 0.5 cm × 0.5 cm × 0.5 cm in a tangential plane, placed on a conductor adhesive, coated with gold, and observed under SEM at a voltage of 15 kV. Wood samples were also analyzed using energy dispersive X-ray analysis (EDX) to analyze the chemical content of the wood.

2.8.2. Fourier Transform Infrared Spectrometry

Untreated and impregnated samples were ground to a particle size of 200 mesh and embedded in potassium bromide (KBr) pellets in a ratio of 1:100. The pellets were analyzed by FT-IR (Perkin-Elmer Spectrum One) with a scanning range of 4000 to 400 cm^{-1} at a resolution of 4 cm^{-1} for 32 scans.

2.8.3. X-ray Diffraction Analysis

Wood samples were slashed using a 2 mm thick cutter in a tangential plane. The degree of crystallinity of the wood sample (incision) was analyzed by XRD-PANAnalytical Empyrean type with a 1D PIXcel detector. The parameters used in the device are: Cu Kα radiation with a graphite monochromator, a voltage of 40 Kv, a current of 30 mA, and a scan range of 2θ between 0 and 90° with a scanning speed of 2°/min.

2.9. Data Analysis

This study used a completely randomized design and data were evaluated using analysis of variance (ANOVA), followed by Duncan's test at a significance level of α = 5%. Statistical testing was performed using IBM SPSS Statistics (Statistical Package for Service Solutions) version 25.0 program Stanford, California, CA, USA.

3. Results

3.1. Synthesized TiO_2 Nanoparticle

TiO_2 powder forms a coarse spherical morphology with granules measuring approximately less than 100 nm. The magnification of the diffraction peak was confirmed by the small size of TiO_2 nanoparticle (see details in XRD discussion). These nanoparticle form agglomerations with approximately 500 nm (Figure 1), which Zanatta [42] previously attributed to van der Waals forces. The agglomeration of nanoparticle originates from the induced dipoles in the synthesis method. The short-range interactions arising from induced dipoles can be easily broken by the stresses that occur during impregnation.

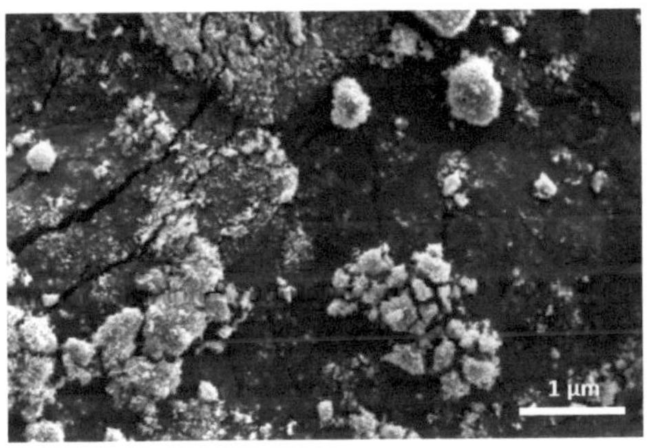

Figure 1. SEM images of TiO_2 nanoparticle synthesized by the hydrothermal method.

3.1.1. FTIR Result

The analysis of the FTIR spectrum of TiO_2 nanoparticle (Figure 2) led to the identification of the functional groups of Ti-O at a wave number of 545 cm^{-1} and Ti-O-Ti at a wave number of 802 cm^{-1}, which are bonds formed in the framework of TiO_2 compounds. The TiO_2 spectrum showed intense peaks at 3655 and 3860 cm^{-1} due to the stretching of the OH group of the H_2O molecule bonded to the TiO_2 compound [43]. This finding indicates that the TiO_2 compound has been successfully synthesized by the hydrothermal method.

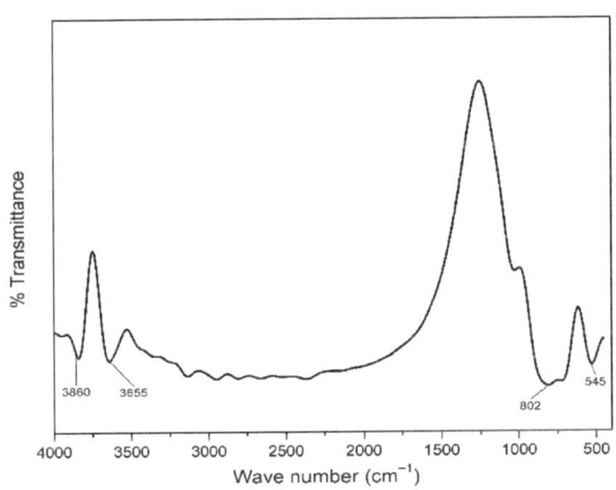

Figure 2. FTIR spectrum of TiO_2 nanoparticle.

3.1.2. XRD Result

The XRD analysis were intended to determine the phase, degree of crystallinity, and crystal size of TiO_2 synthesized by hydrothermal method. The diffractogram of the TiO_2 nanoparticle sample (Figure 3) shows peaks at the following 2θ values: 25.26, 36.92, 37.76, 38.63, 48.00, 53.86, 55.03, 62.66, 68.74, 70.46, 75.03, 82.65. The calculation results show that the synthesized TiO_2 nanoparticle have a high degree of crystallinity (99.86%). The crystal size of the TiO_2 nanoparticle is determined from the diffractogram data using the Scherrer equation, where λ = X-ray wavelength used; θ = diffraction angle; and K = is a constant whose magnitude depends on the crystal form factor, diffraction field (hkl), and β is the full width at half maximum (FWHM) or integral breadth of the peak [44].

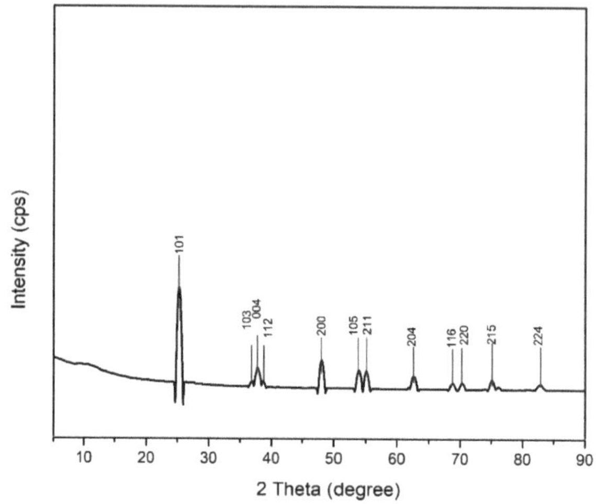

Figure 3. TiO_2 nanoparticle diffractogram.

The crystal size of TiO_2 nanoparticle calculated by the Scherrer equation (Equation (6)) is 79.17 nm. This proves that the synthesized TiO_2 is a nanoparticle because it has a particle size in the range of 1 to 100 nm based on crystal size [45].

$$D = \frac{K\lambda}{\beta cos\theta} \tag{6}$$

The UV spectrum analysis of TiO_2 nanoparticle (Figure 4) indicated wavelength absorbance value of 362 nm. In TiO_2, the valence band comes from the 2p orbital of the oxygen atom and the conduction band comes from the 3d orbital of the titanium atom [46].

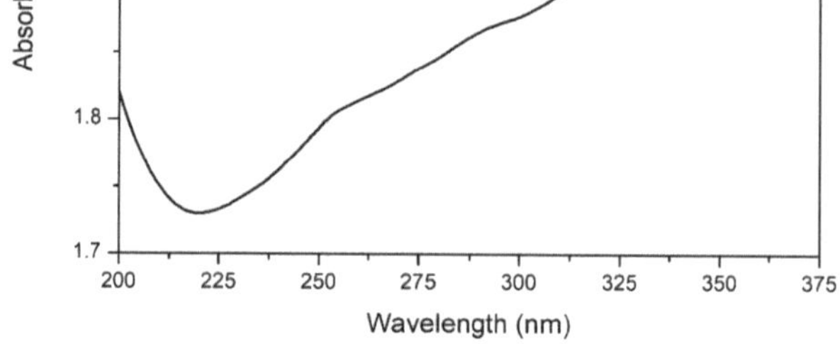

Figure 4. TiO_2 nanoparticle absorption spectrum.

The Tauc method was used for the band gap energy analysis (Figure 5) based on the assumption that the energy affected by the absorbance coefficient (α) can be determined using Equation (7):

$$(\alpha \cdot hv)^{\gamma} = B(hv - Eg) \tag{7}$$

where h is Planck's constant, v is the frequency of the proton, Eg is the band gap energy, and B is the constant. The factor depends on the natural electron transition value which is equivalent to 1/2 or 2 for indirect and direct transitions in the band gap energy [47]. The analysis of the energy band gap based on the synthesized TiO_2 nanoparticle yielded a value of 3.40 eV. This value is greater than the band gap value of bulk phase anatase TiO_2 nanoparticle which is 3.22 eV. These results confirm that when the size of the nanoparticle is smaller, the band gap will be larger [40].

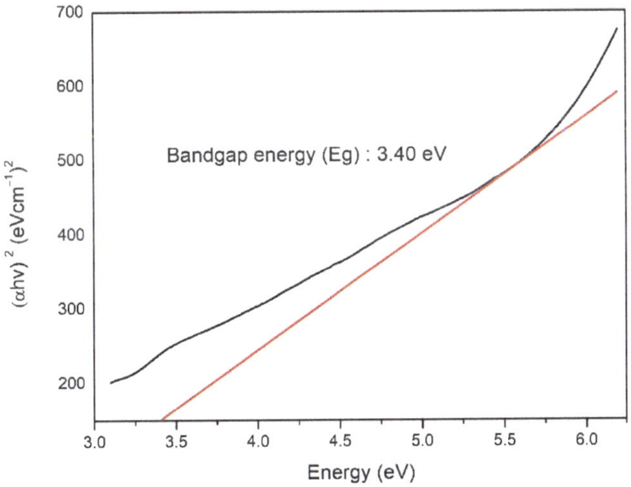

Figure 5. TiO_2 nanoparticle band gap analysis curve with Tauc method.

The crystal structure of TiO_2 nanoparticle was determination using QualX software developed by [48] and computerized with Mercury software. Based on the results of the analysis of the crystal structure, a tetragonal crystal form was obtained (Figure 6).

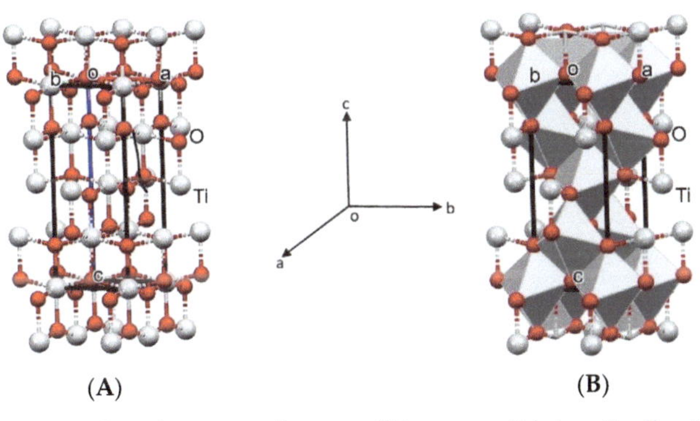

Figure 6. Crystal structure of anatases TiO_2 nanoparticle in unit cells with (**A**) ball stick and (**B**) polyhedron patterns. a, b, c and o is the notation of the three-dimensional coordinate system contains an origin (o) and formed by three mutually perpendicular coordinate axes: the x-axis (a), y-axis (b), and the z-axis (c).

3.2. Physical Properties of Mangium Wood

Table 1 presents the dimensional stability and density values of untreated and treated mangium wood. The density, WPG, BE, and ASE values showed an increasing trend, while the WU values showed a decreasing trend. The leaching value increased when 1% TiO_2 nanoparticle was used, decreased with 5% TiO_2 nanoparticle concentration. The same trend was previously shown for the parameters of dimensional stability and density [15,49].

Table 1. Value of dimensional stability and density of mangium wood.

Sample	Density	WPG	BE	WU	ASE	L
Untreated	0.54 ± 0.02^{a}	0.21 ± 0.14^{a}	1.04 ± 0.69^{a}	50.29 ± 4.45^{a}	0.00^{a}	0.44 ± 0.19^{a}
1% TiO_2 nanoparticle	0.55 ± 0.02^{ab}	0.53 ± 0.22^{a}	1.14 ± 0.47^{a}	49.08 ± 7.80^{a}	15.46 ± 9.52^{b}	0.57 ± 0.21^{a}
5% TiO_2 nanoparticle	0.57 ± 0.03^{b}	2.76 ± 0.72^{b}	2.10 ± 1.03^{b}	46.66 ± 6.67^{a}	21.76 ± 7.97^{b}	0.43 ± 0.15^{a}

[a,b] Values followed by different letters are significantly different based on the Duncan test.

The highest WPG value was obtained with 5% TiO_2 nanoparticle treatment (2.76 ± 0.72)%, which also yielded the highest density value (0.57 ± 0.03)%. In addition to the increasing WPG and density values, physical properties such as BE and ASE also had an increasing trend. The highest BE and ASE values were associated with 5% TiO_2 nanoparticle treatment (2.10 ± 1.03% and 21.76 ± 7.97%, respectively). A different trend occurred for WU value, which was lowest with the 5% TiO_2 nanoparticle treatment (46.66 ± 6.67)%.

3.3. Characteristics of Mangium Wood Impregnated with TiO_2 Nanoparticle

3.3.1. SEM-EDX Analysis

Figure 7a–f shows the differences between the untreated mangium wood and samples that were treated with TiO_2 nanoparticle. In both samples, TiO_2 nanoparticle appeared to cover the cell walls, with deposits also being present in the intercellular cavities (yellow color in Figure 7b,c,e,f).

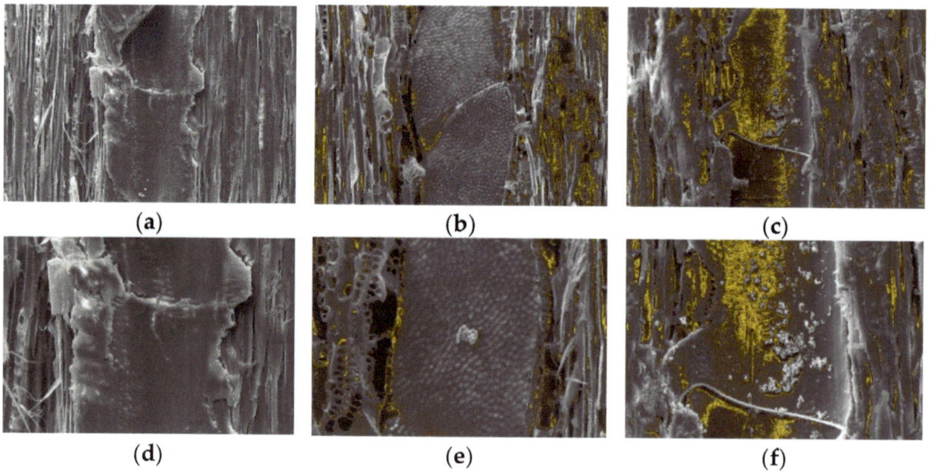

Figure 7. SEM images of untreated mangium wood 500× (**a**), mangium wood impregnated with 1% TiO_2 nanoparticle 500× (**b**) and mangium wood impregnated with 5% TiO_2 nanoparticle 500× (**c**), untreated mangium wood 1000× (**d**), mangium wood impregnated with 1% TiO_2 nanoparticle 1000×, (**e**) and mangium wood impregnated with 5% TiO_2 nanoparticle 1000× (**f**).

Based on the results of quantitative analysis using SEM EDX (Table 2), the TiO_2 content in mangium wood impregnated with 5% TiO_2 nanoparticle had a higher Ti content (4.69%) than that impregnated with 1% TiO_2 nanoparticle concentration (0.35%).

Table 2. SEM-EDX test results.

Sample	Ti (%Wt)
Untreated	-
1% TiO_2 nanoparticle	0.35
5% TiO_2 nanoparticle	4.69

3.3.2. Photocatalyst Activity Test

The concentration of methylene blue is directly proportional to its absorption value on a visible light spectrophotometer. Figure 8a shows that the presence of TiO_2 nanoparticle in wood significantly affected absorbance. Wood that was impregnated with TiO_2 nanoparticle had a higher photocatalytic activity than untreated, with the C/C_o ratio declining as the UV radiation treatment time increased.

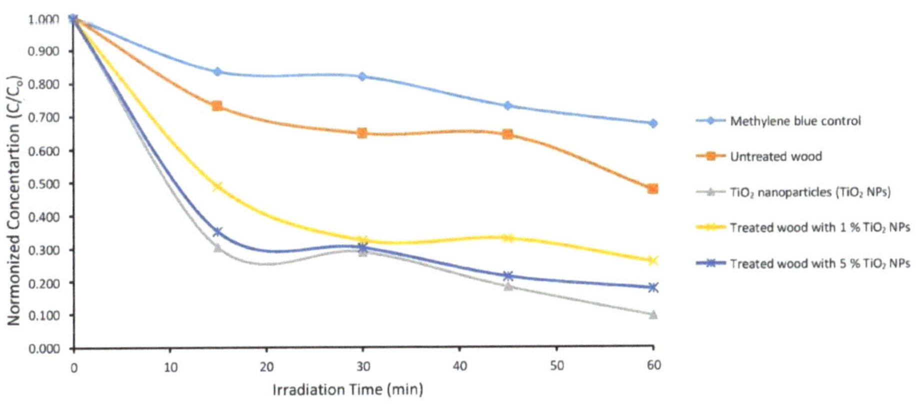

(**a**)

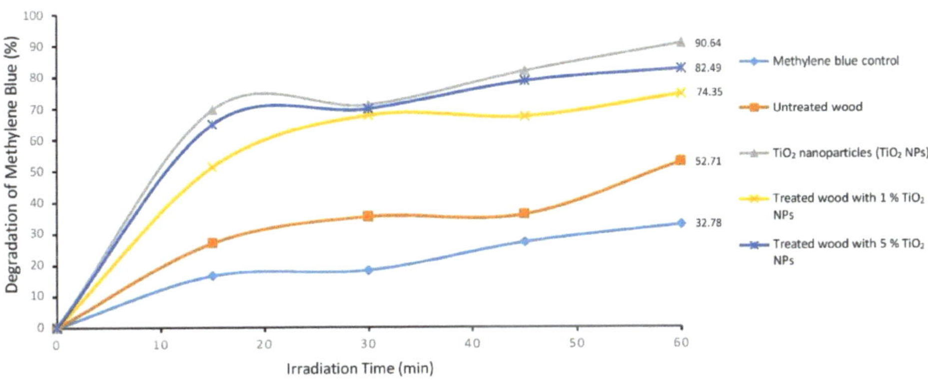

(**b**)

Figure 8. Curves (**a**) relative concentration of C/C_o and (**b**) percentage of degradation of methylene blue compounds.

Figure 8b shows that the role of TiO_2 nanoparticle can be seen significantly with the percentage value of methylene blue degradation increasing with time of UV radiation until it reaches 90.64%. The 1% TiO_2 nanoparticle (82.49%) and 5% TiO_2 nanoparticle (74.35%)—impregnated wood had a higher photocatalytic activity than the untreated. The end result of the photocatalytic reaction is carbon dioxide gas which can be released into nature and can be utilized by plants for the photosynthesis process [50].

3.3.3. Evaluation of TiO_2 Nanoparticle Anatase Efficiency on the Degradation of Methylene Blue

Electrical energy consumption is one of the important criteria in the photochemical process because it is related to the determination of the operating cost of the treatment. The figure-of-merit is the electrical energy per order, defined as the number of kWh of electrical energy required for reducing the concentration of a pollutant by 1 order of magnitude (i.e., 90% degradation), in 1 m^3 of the contaminated sample [51]. The electric energy per order, E_{EO} (kWh/m^3/order), is commonly used, which can be estimated using an equation for a batch reactor [52]. Equation (8) was as follows:

$$E_{E0} = \frac{P \times t \times 1000}{V \times 60 \times log\,\frac{C_o}{C_f}} \quad (8)$$

where P is the electrical power of the UV lamp (kW), t is irradiation time (min), V is the volume of the treated wastewater (L), C_o is the initial concentration of the pollutant (mg L^{-1}), and C_f is the final concentration of the pollutant (mg L^{-1}). The results of the analysis (Figure 9) showed that the highest energy consumption occurred in the untreated blue methylene with a value of 6955.58 kWh/m^3 and the lowest occurred in the anatase TiO_2 nanoparticle sample with a value of 1166.68 kWh/m^3. Energy consumption in untreated wood is 3679.12 kWh/m^3, this value is higher than treated wood with TiO_2 nanoparticle which is 2030.65 kWh/m^3 (1% TiO_2 nanoparticle) and 1585.59 kWh/m^3 (5% TiO_2 nanoparticle).

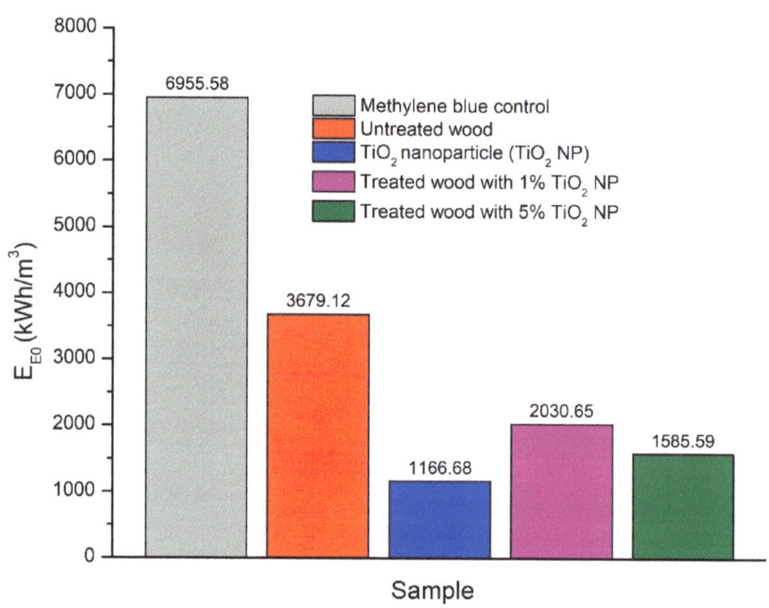

Figure 9. Energy consumption for the obtained samples.

To estimate the kinetic rate of the removal methylene blue reaction, applying the pseudo-first-order (n = 1) kinetic model using the following Equation (9) [53]:

$$-In\,\frac{C}{C_0} = K_{obs}\,t \tag{9}$$

where C_0 is the initial concentration of the pollutant, C is the concentration at time t, and K_{obs} and t are the observed rate constant and the irradiation time, respectively. According to Equation (9), K_{obs} can be determined as the slope of the plot of $-In\,\frac{C}{C_0}$ versus the reaction time (Figure 10). The linearity of lines (Figure 10) demonstrates that the photocatalytic kinetics for methylene blue obeyed the pseudo-first-order kinetic model. The lines on Figure 10, had high values of the correlation coefficients ($R^2 > 0.93$). This finding confirmed the validity of the pseudo-first order model [54]. The calculated values for K_{obs} were 0.0406 L/min (anatase TiO_2 NP), 0.0062 L/min (untreated methylene blue), 0.0334 L/min (untreated wood), 0.0262 L/min (1% TiO_2 nanoparticle), and 0.0123 L/min (5% TiO_2 nanoparticle).

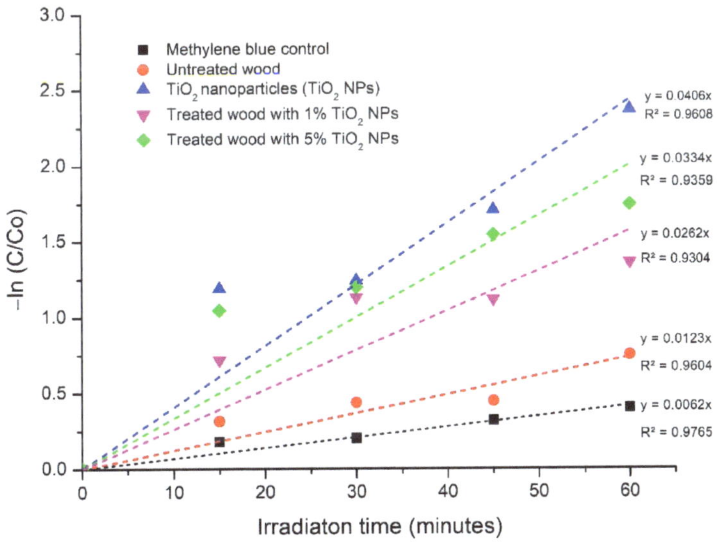

Figure 10. Reaction kinetics profile for the degradation of methylene blue in the presence of anatase TiO_2 nanoparticle. Experimental conditions photocatalyst dose = 1 g/L; methylene blue = 10 ppm.

3.3.4. UV-Vis Radiation Analysis

Mangium wood that has been impregnated with TiO_2 nanoparticle is treated with UV radiation at a wavelength of 366 nm and a power of 120 watts and visible light radiation from an incandescent lamp with a power of 150 watts for 6 h in a closed system. The radiation distance used is 15 cm. This test was conducted to determine the level of protection of TiO_2 nanoparticle against UV-Vis radiation.

The results of the FTIR spectrum analysis (Figure 11) showed a decrease in the intensity of the C-H aromatic and C=O functional groups which were part of the functional group of lignin compounds in mangium wood that were not impregnated with TiO_2 nanoparticle by UV-Vis radiation treatment. This indicates that there is a degradation initiation reaction in the main structure of the wood starting from the adhesive compound between wood, namely lignin. The same thing did not happen to mangium wood impregnated with TiO_2 nanoparticle which had the same spectrum pattern as untreated wood without UV-Vis radiation treatment.

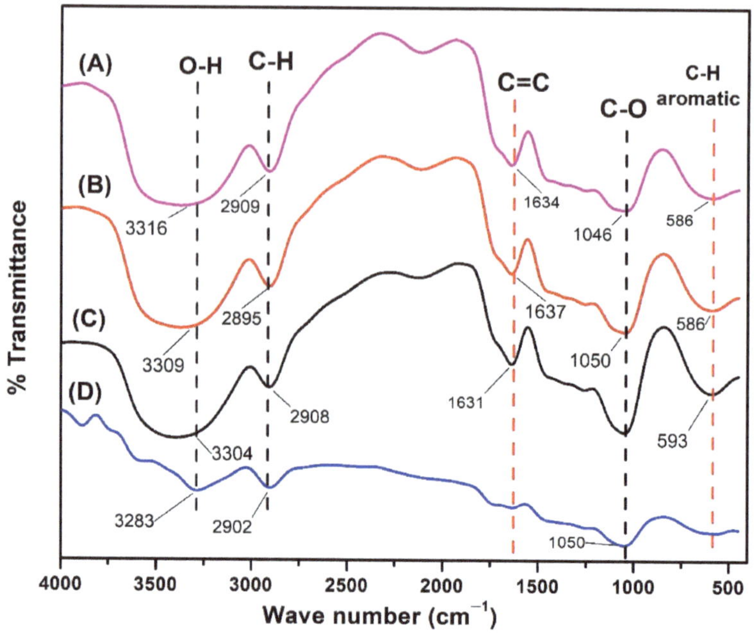

Figure 11. FTIR spectrum after UV-Vis radiation treatment on (**A**) impregnated mangium wood with 5% TiO_2 nanoparticle, (**B**) impregnated mangium wood impregnated with 1% TiO_2 nanoparticle, (**C**) untreated mangium wood without radiation, (**D**) untreated mangium wood with radiation.

3.3.5. FTIR

Analysis using FTIR (Figure 12) was carried out to identify the functional groups present in untreated and treated mangium wood. The FTIR spectrum for untreated revealed the C-H functional group from the framework of the aromatic compounds at a wave number of 593 cm^{-1}, the C-O functional group at a wave number of 1055 cm^{-1}, the C=C functional group at a wave number of 1631 cm^{-1}, the C-H functional group at a wave number of 2914 cm^{-1}, and the O-H functional group at wave number 3284 and 3887 cm^{-1} [55]. All functional groups identified in untreated mangium were also identified in treated mangium with 1% and 5% TiO_2 nanoparticle, but in treated mangium wood, functional groups were identified from Ti-O at wave numbers 523 and 533 cm^{-1}, and functional groups Ti-O-Tiat wave numbers 696 and 719 cm^{-1}, indicating the bonds formed in the framework of TiO_2 compounds [43].

3.3.6. XRD

XRD (Figure 13) was used on mangium wood samples impregnated with TiO_2 nanoparticle to identify its characteristics through phase analysis of cellulose and TiO_2 and to determine TiO_2 crystal size in mangium wood. The diffractogram of mangium wood impregnated with 1% TiO_2 nanoparticle showed peaks at a value of 2θ namely 25.25, 36.62, 38.09, 39.04, 47.83, 53.93, 55.03, 62.78, 69.13, 70.42, 75.38, and 83.05. In the diffractogram of mangium wood impregnated with 5% TiO_2 nanoparticle, there were peaks at 2θ values, namely 25.34, 36.54, 37.72, 39.09, 47.73, 53.66, 55.32, 62.72, 68.95, 70.29, 74.88, and 82.68. In the diffractogram of untreated mangium wood, peaks for 2θ occurred at 15.29, 21.975, and 34.31. In the diffractogram of mangium wood impregnated with 1% TiO_2 nanoparticle, there were peaks at 2θ values, namely 15.77, 22.20, and 34.39. In the diffractogram of mangium impregnated with 5% TiO_2 nanoparticle, there were peaks at 2θ values, namely 15.35, 21.97, and 34.47.

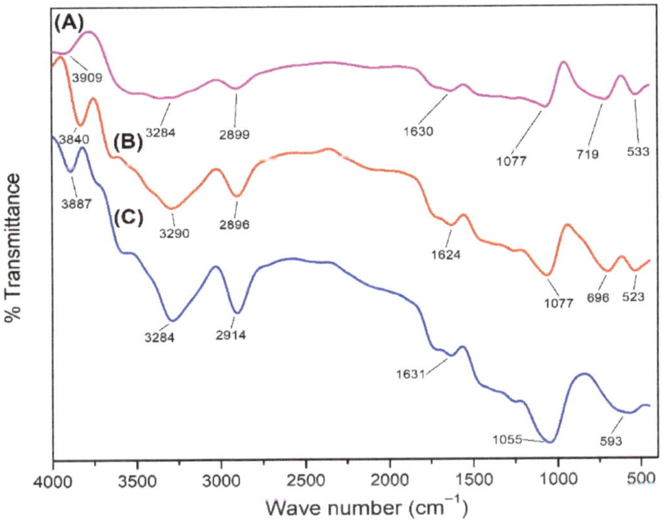

Figure 12. FTIR spectrum on (**A**) impregnated mangium wood with 5% TiO_2 nanoparticle, (**B**) impregnated mangium wood with 1% TiO_2 nanoparticle, (**C**) untreated mangium wood.

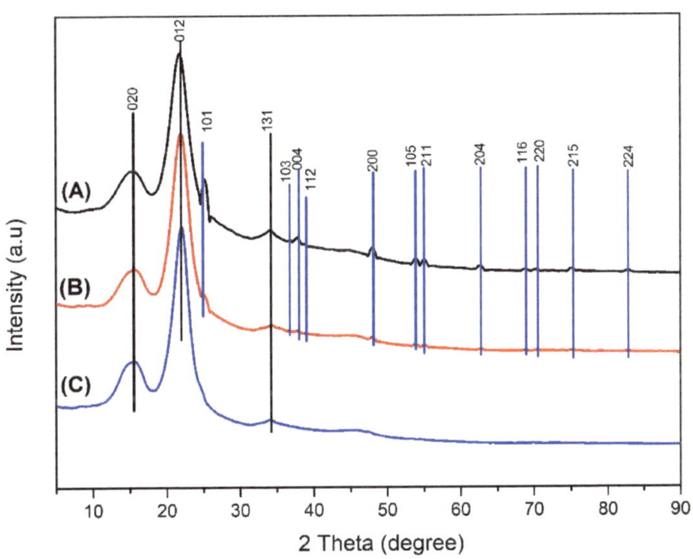

Figure 13. Diffractogram (**A**) impregnated mangium wood with 5% TiO_2 nanoparticle (**B**) impregnated mangium wood with 1% TiO_2 nanoparticle (**C**) untreated mangium wood.

The XRD analysis indicated that mangium wood impregnated with 1% TiO_2 nanoparticle experienced an increase in the value of the degree of crystallinity to 74.17% while mangium wood impregnated with TiO_2 nanoparticle 5% experienced a decrease in the degree of crystallinity to 65.13% (Table 3).

Table 3. Data on crystallinity test results for mangium wood impregnated with TiO_2 nanoparticle.

Sample	Degree of Crystallinity (%)
Untreated	71.98
1% TiO_2 nanoparticle	74.17
5% TiO_2 nanoparticle	65.13

4. Discussion

4.1. Synthesized TiO_2 Nanoparticle

XRD Analysis

A comparative analysis with the standard anatase TiO_2 diffractogram JCPDS card number 78-2486 [40] confirmed anatase TiO_2 compounds were successfully synthesized by the hydrothermal method. The degree of crystallinity of TiO_2 nanoparticle was determined by comparing the crystalline lattice with the total number of lattices in the diffractogram (amorphous and crystalline). The degree of crystallinity is very important because it is linearly related to the reactivity and optical properties of a compound [56]. These results in Figure 4 are in accordance with the research of Li et al. [57], showing that TiO_2 nanoparticle synthesized by the hydrothermal method have a maximum wavelength value in the UVA range of 320–400 nm. Maximum absorbance occurs when electrons are excited from the valence band to the conduction band [46].

The optical properties of a material can be determined through its interaction with electromagnetic radiation fields consisting of transmission, absorption, emission, reflection, refraction, diffraction, or scattering effects. In the special case of semiconductors such as TiO_2, this optical property can be tested in the ultraviolet, visible, and infrared wavelength ranges. This ability is related to the characteristics of the compound's band gap energy [58]. A smaller particle size is associated with a larger band gap, as fewer molecular orbitals are added to the possible ground state of the particle's energy. Therefore, absorption will occur at higher energies, so a shift toward shorter wavelengths will be seen [40]. Our findings (Figure 6) are in agreement with the study of Opoku et al. [59], who reported that among three TiO_2 phases namely rutile, anatase, and brookite, anatase was the phase that had the best photocatalytic activity with a body-centered tetragonal crystal structure, lattice parameters a = b = 3.782 and c = 9.502, and the space group I 41/amd.

4.2. Physical Properties of Mangium Wood

The observed increase in the WPG value indicated that TiO_2 nanoparticle was successfully dispersed in the wood, filling its cavities and cell walls of wood caused the weight of mangium wood to increase from untreated to treated as the TiO_2 concentration increased. Based on the report by Chu et al. [5], an increased WPG value indicates the presence of TiO_2 nanoparticle that is deposited in the cell cavities and fills them. The WPG value increased along with the increase in the density value. The increased density value was probably caused by the TiO_2 nanoparticle solution that was successfully dispersed into the wood, which caused the cavities and cell walls of the wood to become denser and thicker compared to untreated. As previously reported, the higher the density value of the treated wood, the more content in the wood cavities and the thicker the wood cell walls [60].

The increase in the ASE value is directly proportional to the BE value. This shows that the addition of TiO_2 nanoparticle made the wood cell walls expand and increased bulking, resulting in improved dimensional stability with treatment. Hill [6] showed that the higher the BE value, the more polymer filled in the wood cell wall which could increase the dimensional stability of the wood. This is consistent with a report by Devi and Maji [61] showing that the ASE value increases when TiO_2 nanoparticle was added. This result was attributed to TiO_2 nanoparticle being deposited into the empty space of the wood and making the cell walls thicker. This also causes the density value to increase and the shrinkage and swelling to be reduced. These changes therefore improved the dimensional stability of the wood.

Chu et al. [5] reported that the increased WPG value caused the WU value to decrease when wood was impregnated with TiO_2 nanoparticle. Based on Rathnam et al. [62], the value of WU WPC (wood polymer composite) decreased with the addition of nanoclay and TiO_2 nanoparticle. Like nanoclay, TiO_2 nanoparticle also provided barrier properties to prevent the wood from absorbing water. In addition, impregnated wood with 5% TiO_2 nanoparticle had a better leaching value than that with 1% TiO_2 nanoparticle, but between the two concentrations it shows a small possibility for leaching because the leaching value produced is quite low. Pouya and Younes [63] reported that at the end of a 6-day test period, almost no leaching of TiO_2 nanoparticle at any concentration occurred, compared with preservatives such as zinc sulfate and boron which are difficult to form and easy to leach and remove from wood.

4.3. Characteristics of Mangium Wood Impregnated with TiO_2 Nanoparticle

4.3.1. SEM-EDX Analysis

According to Liu et al. [64], TiO_2 nanoparticle are deposited evenly on the cell walls of the wood instead of accumulating in the pores. This nanoparticle interaction may be due to physical interactions or hydrogen bonds that occur between the OH present in the cell wall polymer and the hydrolyzed TiO_2 [17,65]. Mangium wood impregnated with 5% TiO_2 nanoparticle appeared to have more of the surface of the wood cells covered. This is in accordance with the physical properties test that the increase in WPG was directly proportional to the increase in the concentration of TiO_2 nanoparticle impregnated into mangium wood.

4.3.2. Photocatalyst Activity Test

Pollutants can increase the degradation rate of biopolymer materials [66]. Pollutants consist of two types, namely organic and inorganic pollutants [67]. The use of titanium oxide compounds can protect biopolymer materials such as wood from UV radiation due to the strong ability to absorb ultraviolet (UV) rays [68] while utilizing them to degrade organic pollutants through photocatalyst reactions [69]. TiO_2 compounds with particle sizes in the nano range have higher photocatalytic activity than the bulk phase. This can be attributed to the presence of a wider reactive surface area and an increase in the band gap energy through the mechanism of optical and electronic properties that depend on size; these properties are known as quantized particles (Q-particles) or quantum dots [70]. Photocatalytic activity can be analyzed using methylene blue in the wood samples. This substance is difficult to degrade, and its levels can be measured using a visible light spectrophotometer at a wavelength of 666 nm [71].

Titanium dioxide was studied extensively in the last two decades due to its strong activity in photocatalytic reactions. The photocatalytic activity of methylene blue substances can also be expressed as a percentage of their degradation [66]. The decrease in the concentration of methylene blue in the untreated mangium wood also occurred through adsorption, which can happen because the cellulose and lignin biomass in wood can behave as effective adsorbents. Adsorbents containing a high amount of cellulose can absorb basic dyes irreversibly [72]. The decrease in concentration of methylene blue in untreated sample also occurred because of heat from the UV radiation during the testing process [73].

4.3.3. Evaluation of TiO_2 Anatase Nanoparticle Efficiency on The Degradation of Methylene Blue

The calculation results of the E_{EO} (Figure 9) show that the availability of TiO_2 nanoparticle can reduce the E_{EO} value. This is in line with the results of research by Ghavi et al. [51] that TiO_2 can reduce the E_{EO} value because TiO_2 nanoparticle can reduce the activation energy for the reaction process of pollutant degradation. This result indicated that the use of TiO_2 nanoparticle and its application to mangium wood can reduce the cost of using electrical energy. The energy consumption of this research with TiO_2 nanoparticle is 1166.68 kWh/m^3, compared to the study of Varma et al. [74], who reported energy consumption with a value of 3000 kWh/m^3. It turned out that our finding is smaller. This was attributed to the utilization of nano-sized TiO_2, which had larger surface area and higher photocatalyst activity [75].

The decreased concentration of the methylene blue was linear in each treatment. This finding indicated that both on photocatalytic activity by anatase TiO_2 nanoparticle and impregnation processes applied the pseudo-first order kinetic of dye removal kinetic. The treated mangium wood had higher gradient slopes than untreated (Figure 10). It proved that treated mangium wood had faster photocatalytic reaction compared with untreated at the same conditions. E_{Eo} and kinetic profile showed that anatase TiO_2 nanoparticle is a reliable substance to degrade methylene blue as organic pollutant and catalyst to be used in photocatalytic reaction.

4.3.4. UV-Vis Radiation Analysis

Protection from UV radiation is important for the performance of the wood substrate and the coating that protects it. Ultraviolet light causes degradation of wood substrates and coating polymers [76]. This result is in line with research conducted by Rassam et al. [24] indicating that the initiation reaction of wood degradation due to UV-Vis radiation and radiation heat transfer is a decrease in the intensity of the C-H aromatic and C=C functional groups in lignin compounds. This result proves that the synthesized anatase TiO_2 nanoparticle can protect mangium wood from the degradation process due to UV-vis radiation even though it is still in the initiation reaction stage.

4.3.5. FTIR

The functional groups were identified based on the C-H stretching from cellulose and hemicellulose, the aromatic C-H strain and C=C strain in the lignin framework, the O-H strain from cellulose, and the C-O strain from hemicellulose [77]. Based on the FTIR spectrum, mangium wood impregnated with 1% and 5% TiO_2 nanoparticle (Figure 11) did not have the C-H functional group in the aromatic framework and had a decrease in the intensity of the C-H functional group. This may have happened because the TiO_2 nanoparticle covered the surface of cellulose, lignin, and hemicellulose, which are polymers [78] that are the main structural building blocks of wood [79], and blocked infrared radiation. In a previous study, FTIR analysis of jabon wood impregnated with magnetite nanoparticle showed that nano magnetite caused a bulking effect when filling the wood pores; new functional groups were not identified, and new peaks from the infrared spectrum did not appear [80].

4.3.6. XRD

Based on a comparative analysis of the standard JCPDS anatase TiO_2 diffractogram card number 78-2486 [40], the peaks are anatase TiO_2 compounds that successfully impregnated into mangium wood. In addition, XRD analysis was also used to identify the crystalline phase of wood after the impregnation with TiO_2 nanoparticle. Based on the comparative analysis with the standard JCPDS cellulose diffractogram No. 03-0226 [81], these peaks were identified as cellulose, which still had the same crystal lattice as the untreated. This finding indicates that the impregnation treatment did not affect the crystalline structure of the cellulose; however, the degree of crystallinity was changed in comparison with the untreated wood, which was 71.98%.

This increase in crystallinity value was likely caused by the rearrangement of the cellulose structure in the amorphous phase owing to its interaction with the impregnant solution using water as a solvent and the effect of bulking inorganic particles along the cellulose surface [82]. Inorganic particles playing a role included the TiO_2 nanoparticle. At higher TiO_2 concentrations, the degree of wood crystallinity decreased owing to the amount of hydrophobic TiO_2 [83] covering the surface of cellulose and consequently preventing rearrangement of cellulose in the amorphous phase. The results of the calculation of the crystal size of TiO_2 using the Scherrer equation [44] were 61.64 nm for 1% TiO_2 nanoparticle and 64.94 nm for 5% TiO_2 nanoparticle. This value indicates that the particle size of TiO_2 impregnated into the wood was smaller than the particle size prior to the impregnation process. This can have occurred because of the sonication treatment of the impregnated solution before the impregnation process, which can reduce the TiO_2 particle size [84].

5. Conclusions

1. Anatase TiO_2 nanoparticle has been successfully synthesized by hydrothermal method at low temperatures based on the results of characterization with spectrophotometer Uv-Vis, XRD, and FTIR. Anatase TiO_2 nanoparticle has maximum wavelength of 362 nm and band gap energy of 3.40 eV.
2. The values of E_{Eo} and kinetic profile proved that mangium wood impregnated with TiO_2 nanoparticle has the ability to degrade methylene blue as the organic pollutant.
3. Density and dimensional stability of treated mangium wood increased compared with untreated. Five percent TiO_2 nanoparticle treatment obtained higher WPG, BE, ASE, and leachability values; however, lower WU values than that of untreated and 1% TiO_2 nanoparticle.
4. Based on XRD and FTIR analysis, TiO_2 nanoparticle was successfully impregnated into mangium wood. Scanning electron microscopy–energy-dispersive X-ray spectroscopy analysis indicated that TiO_2 nanoparticle covered the surface of the wood cells. The TiO_2-impregnated mangium wood has a higher photocatalyst activity than untreated, indicating better protection from UV radiation and pollutants.

Author Contributions: Conceptualization, I.R.; methodology, I.R., E.P., R.I. and G.D.L.; software, R.I. and G.D.L.; validation, I.R., E.P., R.I., W.D. and D.S.N.; formal analysis, E.P., R.I. and G.D.L.; investigation, I.R., W.D., D.S.N., E.P., R.I. and G.D.L.; resources, I.R.; data curation, E.P., R.I. and G.D.L.; writing—original draft preparation, I.R., E.P., R.I. and G.D.L.; writing—review and editing, I.R., W.D., D.S.N.; visualization, E.P., R.I. and G.D.L.; supervision, I.R., W.D. and D.S.N.; project administration, I.R.; funding acquisition, I.R. All authors have read and agreed to the published version of the manuscript.

Funding: This research was funded by Ministry of Education, Culture and Research and Technology of Indonesia (Grant No. 0267/E5/AK.04/2022 and Grant No. 082/E5/PG.02.00.PT/2022).

Institutional Review Board Statement: Not applicable.

Informed Consent Statement: Not applicable.

Data Availability Statement: The data presented in this study are available on request from the corresponding author.

Acknowledgments: The authors are grateful for the support of the Ministry of Education, Culture and Research and Technology of Indonesia (Grant No. 0267/E5/AK.04/2022 and Grant No. 082/E5/PG.02.00.PT/2022).

Conflicts of Interest: The authors declare no conflict of interest.

References

1. Central Bureau of Statistics [BPS]. Central Bureau of Statistics. Available online: http://www.bps.go.id (accessed on 2 February 2022).
2. Indonesian Information Portal. Available online: https://www.indonesia.go.id/narasi/indonesia-dalam-angka/ekonomi/usaha-berbasis-kayu-tetap-jalan-hutan-pun-tetap-lestari (accessed on 2 February 2022).
3. Pandit, I.K.N.; Kurniawan, D. *Wood Anatomy: Wood Structure, Wood as Raw Material and Diagnostic Characteristics of Indonesian Trade Timber*; Bogor Agricultural University: Bogor, Indonesia, 2008.
4. Rowell, R.M.; Ellis, W.D. Determination of dimensional stabilization of wood using the water-soak method. *Wood Fiber Sci.* **1978**, *10*, 104–111.
5. Van Chu, T.; Van Chuong, P.; Tuong, V.M. Wettability of Wood Pressure-treated with TiO_2 Gel under Hydrothermal Conditions. *BioResources* **2014**, *9*, 2396–2404. [CrossRef]
6. Hill, C.A.S. *Wood Modification: Chemical, Thermal and Other Processes*; John Wiley and Sons: Hoboken, NJ, USA, 2006; ISBN 9780470021729.

7. Bryne, L.E.; Wålinder, M.E.P. Ageing of modified wood. Part 1: Wetting properties of acetylated, furfurylated, and thermally modified wood. *Holzforschung* **2010**, *64*, 295–304. [CrossRef]
8. Wardani, L.; Risnasari, I.; Yasni, H.Y. Resistance of jabon timber modified with styrene and methyl methacrylate against drywood termites and subterranean termites. In Proceedings of the 9th Pacific Rim Termite Research Group Conference, Hanoi, Vietnam, 27–28 February 2012; Science and Technics Publishing House: Hanoi, Vietnam, 2012; pp. 73–78.
9. Hadi, Y.S.; Rahayu, I.S.; Danu, S. Physical and mechanical properties of methyl methacrylate impregnated jabon wood. *J. Indian Acad. Wood Sci.* **2013**, *10*, 77–80. [CrossRef]
10. Hadi, Y.S.; Rahayu, I.S.; Danu, S. Termite resistance of jabon wood impregnated with methyl methacrylate. *J. Trop. For. Sci.* **2015**, *27*, 25–29.
11. Gabrielli, C.P.; Kamke, F.A. Phenol-formaldehyde impregnation of densified wood for improved dimensional stability. *Wood Sci. Technol.* **2010**, *44*, 95–104. [CrossRef]
12. Fukuta, S.; Watanabe, A.; Akahori, Y.; Makita, A.; Imamura, Y.; Sasaki, Y. Bending properties of compressed wood impregnated with phenolic resin through drilled holes. *Eur. J. Wood Wood Prod.* **2011**, *69*, 633–639. [CrossRef]
13. Rahayu, I.; Darmawan, W.; Zaini, L.H.; Prihatini, E. Characteristics of fast-growing wood impregnated with nanoparticles. *J. For. Res.* **2020**, *31*, 677–685. [CrossRef]
14. Prihatini, E.; Maddu, A.; Rahayu, I.S.; Kurniati, M.; Darmawan, W. Improvement of physical properties of jabon (*Anthocephalus cadamba*) through the impregnation of nano-SiO_2 and melamin formaldehyde furfuril alcohol copolymer. In *IOP Conference Series: Materials Science and Engineering*; IOP Publishing: Bristol, UK, 2020; Volume 935. [CrossRef]
15. Rahayu, I.; Prihatini, E.; Ismail, R.; Darmawan, W.; Karlinasari, L.; Laksono, G.D. Fast-Growing Magnetic Wood Synthesis by an In-Situ Method. *Polymers* **2022**, *14*, 2137. [CrossRef]
16. Fuva, S.M.; Hovde, P.J. Nano-based modifications of wood and their environmental impact: A review. In Proceedings of the World Conference on Timber Engineering, River Del Garda Trentino, Italy, 20–24 June 2010.
17. Wang, B.; Feng, M.; Zhan, H. Improvement of wood properties by impregnation with TiO_2 via ultrasonic-assisted sol-gel process. *RSC Adv.* **2014**, *4*, 56355–56360. [CrossRef]
18. Kim, T.K.; Lee, M.N.; Lee, S.H.; Park, Y.C.; Jung, C.K.; Boo, J.H. Development of surface coating technology of TiO_2 powder and improvement of photocatalytic activity by surface modification. *Thin Solid Films* **2005**, *475*, 171–177. [CrossRef]
19. Chen, X.D.; Wang, Z.; Liao, Z.F.; Mai, Y.L.; Zhang, M.Q. Roles of anatase and rutile TiO_2 nanoparticles in photooxidation of polyurethane. *Polym. Test.* **2007**, *26*, 202–208. [CrossRef]
20. Sun, Q.; Yu, H.; Liu, Y.; Li, J.; Lu, Y.; Hunt, J.F. Improvement of water resistance and dimensional stability of wood through titanium dioxide coating. *Holzforschung* **2010**, *64*, 757–761. [CrossRef]
21. Sun, Q.; Yu, H.; Liu, Y.; Li, J.; Cui, Y.; Lu, Y. Prolonging the combustion duration of wood by TiO_2 coating synthesized using cosolvent-controlled hydrothermal method. *J. Mater. Sci.* **2010**, *45*, 6661–6667. [CrossRef]
22. Shabir Mahr, M.; Hübert, T.; Stephan, I.; Militz, H. Decay protection of wood against brown-rot fungi by titanium alkoxide impregnations. *Int. Biodeterior. Biodegrad.* **2013**, *77*, 56–62. [CrossRef]
23. Mahltig, B.; Swaboda, C.; Roessler, A.; Böttcher, H. Functionalising wood by nanosol application. *J. Mater. Chem.* **2008**, *18*, 3180–3192. [CrossRef]
24. Rassam, G.; Abdi, Y.; Abdi, A. Deposition of TiO_2 nano-particles on wood surfaces for UV and moisture protection. *J. Exp. Nanosci.* **2012**, *7*, 468–476. [CrossRef]
25. Schmalzl, K.J.; Evans, P.D. Wood surface protection with some titanium, zirconium and manganese compounds. *Polym. Degrad. Stab.* **2003**, *82*, 409–419. [CrossRef]
26. Sun, Q.; Lu, Y.; Zhang, H.; Zhao, H.; Yu, H.; Xu, J.; Fu, Y.; Yang, D.; Liu, Y. Hydrothermal fabrication of rutile TiO_2 submicrospheres on wood surface: An efficient method to prepare UV-protective wood. *Mater. Chem. Phys.* **2012**, *133*, 253–258. [CrossRef]
27. Tshabalala, M.A.; Libert, R.; Schaller, C.M. Photostability and moisture uptake properties of wood veneers coated with a combination of thin sol-gel films and light stabilizers. *Holzforschung* **2011**, *65*, 215–220. [CrossRef]
28. Meng, L.Y.; Wang, B.; Ma, M.G.; Lin, K.L. The progress of microwave-assisted hydrothermal method in the synthesis of functional nanomaterials. *Mater. Today Chem.* **2016**, *1–2*, 63–83. [CrossRef]
29. Liu, S.X.; Chen, X.Y.; Chen, X. A TiO_2/AC composite photocatalyst with high activity and easy separation prepared by a hydrothermal method. *J. Hazard. Mater.* **2007**, *143*, 257–263. [CrossRef]
30. Kontos, A.I.; Arabatzis, I.M.; Tsoukleris, D.S.; Kontos, A.G.; Bernard, M.C.; Petrakis, D.E.; Falaras, P. Efficient photocatalysts by hydrothermal treatment of TiO_2. *Catal. Today* **2005**, *101*, 275–281. [CrossRef]
31. Zhu, J.; Zheng, W.; He, B.; Zhang, J.; Anpo, M. Characterization of Fe-TiO_2 photocatalysts synthesized by hydrothermal method and their photocatalytic reactivity for photodegradation of XRG dye diluted in water. *J. Mol. Catal. A Chem.* **2004**, *216*, 35–43. [CrossRef]
32. Aneesh, P.M.; Vanaja, K.A.; Jayaraj, M.K. Synthesis of ZnO nanoparticles by hydrothermal method. In Proceedings of the Nanophotonic Materials IV, San Diego, CA, USA, 26–30 August 2007; SPIE: Bellingham, WC, USA, 2007; Volume 6639, p. 66390J. [CrossRef]
33. Yang, Y.; Zhao, S.; Bi, F.; Chen, J.; Wang, Y.; Cui, L.; Xu, J.; Zhang, X. Highly efficient photothermal catalysis of toluene over Co_3O_4/TiO_2 p-n heterojunction: The crucial roles of interface defects and band structure. *Appl. Catal. B Environ.* **2022**, *315*, 121550. [CrossRef]

34. Yang, Y.; Zhao, S.; Bi, F.; Chen, J.; Li, Y.; Cui, L.; Xu, J.; Zhang, X. Oxygen-vacancy-induced O_2 activation and electron-hole migration enhance photothermal catalytic toluene oxidation. *Cell Rep. Phys. Sci.* **2022**, *3*, 101011. [CrossRef]
35. Morais, E.; O'Modhrain, C.; Thampi, K.R.; Sullivan, J.A. RuO_2/TiO_2 photocatalysts prepared via a hydrothermal route: Influence of the presence of TiO_2 on the reactivity of RuO_2 in the artificial photosynthesis reaction. *J. Catal.* **2021**, *401*, 288–296. [CrossRef]
36. Yin, Z.; Song, T.; Zhou, W.; Wang, Z.; Ma, Y. Highly isolated Pt NPs embedded in porous TiO_2 derived from MIL-125 with enhanced photocatalytic hydrogen production activity. *J. Catal.* **2021**, *402*, 289–299. [CrossRef]
37. *BS 373*; Methods of Testing Small Clear Specimens of Timber. British Standard: London, UK, 1957.
38. Shahat, A.M.; El-Hossary, F.M.; Ghitas, A.; Abd El-Rahman, A.M.; Ebnalwaled, A.A. Low-temperature Hydrothermal Synthesis of Titanium Dioxide Nanoparticles for Photocatalytic Applications. In *IOP Conference Series: Materials Science and Engineering*; IOP Publishing: Bristol, UK, 2021; Volume 1171, p. 012008. [CrossRef]
39. Krishnan, T. Photocatalytic Degradation of Dyes by TiO_2 Process in Batch Photoreactor. *Lett. Appl. NanoBioSci.* **2020**, *9*, 1502–1512.
40. Bagheri, S.; Shameli, K.; Abd Hamid, S.B. Synthesis and characterization of anatase titanium dioxide nanoparticles using egg white solution via Sol-Gel method. *J. Chem.* **2013**, *2013*, 6–11. [CrossRef]
41. Hong, S.; Lee, S.; Jung, H.J.; Yu, Y.; Shin, J.; Kwon, K.-Y.; Choi, M. Simple Preparation of Anatase TiO_2 Nanoparticles via Pulsed Laser Ablation in Liquid. *Bull. Korean Chem. Soc.* **2013**, *34*, 279–282. [CrossRef]
42. Zanatta, P.; Lazarotto, M.; de Cademartori, P.H.G.; da Silva Cava, S.; Moreira, M.L.; Gatto, D.A. The effect of titanium dioxide nanoparticles obtained by microwave-assisted hydrothermal method on the color and decay resistance of pinewood. *Maderas: Cienc. Tecnol.* **2017**, *19*, 495–506. [CrossRef]
43. Sharfudeen, B.F.J.M.; Latheef, A.F.A.; Ambrose, R.V. Synthesis and characterization of TiO_2 nanoparticles and investigation of antimicrobial activities against human pathogens. *J. Pharm. Sci. Res.* **2017**, *9*, 1604–1608.
44. Hargreaves, J.S.J. Some considerations related to the use of the Scherrer equation in powder X-ray diffraction as applied to heterogeneous catalysts. *Catal. Struct. React.* **2016**, *2*, 33–37. [CrossRef]
45. Khan, I.; Saeed, K.; Khan, I. Nanoparticles: Properties, applications and toxicities. *Arab. J. Chem.* **2019**, *12*, 908–931. [CrossRef]
46. Rathod, P.B.; Waghuley, S.A. Synthesis and UV-Vis spectroscopic study of TiO_2 nanoparticles. *Int. J. Nanomanufacturing* **2015**, *11*, 185–193. [CrossRef]
47. Makuła, P.; Pacia, M.; Macyk, W. How to Correctly Determine the Band Gap Energy of Modified Semiconductor Photocatalysts Based on UV-Vis Spectra. *J. Phys. Chem. Lett.* **2018**, *9*, 6814–6817. [CrossRef]
48. Altomare, A.; Corriero, N.; Cuocci, C.; Falcicchio, A.; Moliterni, A.; Rizzi, R. QUALX2.0: A qualitative phase analysis software using the freely available database POW-COD. *J. Appl. Crystallogr.* **2015**, *48*, 598–603. [CrossRef]
49. Dong, Y.; Yan, Y.; Zhang, Y.; Zhang, S.; Li, J. Combined treatment for conversion of fast-growing poplar wood to magnetic wood with high dimensional stability. *Wood Sci. Technol.* **2016**, *50*, 503–517. [CrossRef]
50. Le, S.T.T.; Khanitchaidecha, W.; Nakaruk, A. Photocatalytic reactor for organic compound removal using photocatalytic mechanism. *Bull. Mater. Sci.* **2016**, *39*, 569–572. [CrossRef]
51. Ghavi, A.; Bagherian, G.; Vahidian, H. Degradation of paraquat herbicide using hybrid AOP process: Statistical optimization, kinetic study, and estimation of electrical energy consumption. *Environ. Sci. Eur.* **2021**, *33*, 177. [CrossRef]
52. Makhatova, A.; Ulykbanova, G.; Sadyk, S.; Sarsenbay, K.; Atabaev, T.S.; Inglezakis, V.J.; Poulopoulos, S.G. Degradation and mineralization of 4-tert-butylphenol in water using Fe-doped TiO_2 catalysts. *Sci. Rep.* **2019**, *9*, 1–15. [CrossRef]
53. Salehi, M.; Hashemipour, H.; Mirzaee, M. Experimental Study of Influencing Factors and Kinetics in Catalytic Removal of Methylene Blue with TiO_2 Nanopowder. *Am. J. Environ. Engin.* **2012**, *2*, 1–7. [CrossRef]
54. Sescu, A.M.; Favier, L.; Lutic, D.; Soto-Donoso, N.; Ciobanu, G.; Harja, M. TiO_2 Doped with Noble Metals as an Efficient Solution for the Photodegradation of Hazardous Organic Water Pollutants at Ambient Conditions. *Water* **2021**, *13*, 19. [CrossRef]
55. Coates, J. *Interpretation of Infrared Spectra, A Practical Approach*; Encyclopedia of Analytical Chemistry, John Wiley & Sons, Ltd.: Chichester, UK, 2006. [CrossRef]
56. Hartland, G.V. Nanoparticle crystallinity: Is perfect better? *Nat. Mater.* **2007**, *6*, 716–718. [CrossRef] [PubMed]
57. Li, Z.; Shen, W.; He, W.; Zu, X. Effect of Fe-doped TiO_2 nanoparticle derived from modified hydrothermal process on the photocatalytic degradation performance on methylene blue. *J. Hazard. Mater.* **2008**, *155*, 590–594. [CrossRef] [PubMed]
58. Zanatta, A.R. Revisiting the optical bandgap of semiconductors and the proposal of a unified methodology to its determination. *Sci. Rep.* **2019**, *9*, 1–12. [CrossRef]
59. Opoku, F.; Govender, K.K.; van Sittert, C.G.C.E.; Govender, P.P. Understanding the synergistic effects, optical and electronic properties of ternary Fe/C/S-doped TiO_2 anatase within the DFT + U approach. *Int. J. Quantum Chem.* **2018**, *118*, e25505. [CrossRef]
60. Bowyer, J.L.; Schmulsky, R.; Haygreen, J.G. *Forest Products and Wood Science: An Introduction*, 5th ed.; Iowa State Press: Ames, IA, USA, 2007.
61. Devi, R.R.; Maji, T.K. Effect of nanofillers on flame retardancy, chemical resistance, antibacterial properties and biodegradation of wood/styrene acrylonitrile co-polymer composites. *Wood Sci. Technol.* **2013**, *47*, 1135–1152. [CrossRef]
62. Rathnam, V.; Kichu, A.; Dutta, N.; Maji, T.K.; Devi, N. Influence of organically modified nanoclay and TiO_2 nanopowder on the properties of Azadirachta indica wood flour-reinforced high-density polyethylene, low-density polyethylene, polypropylene, and polyvinyl chloride nanocomposite. *J. Thermoplast. Compos. Mater.* **2020**, *35*, 1–20. [CrossRef]

63. Marzbani, P.; Mohammadnia-Afrouzi, Y. Investigation on leaching and decay resistance of wood treated with nano- titanium dioxide. *Adv. Environ. Biol.* **2014**, *8*, 974–978.
64. Liu, Y.; Zhu, X.; Yuan, D.; Wang, W.; Gao, L. Preparation and characterization of TiO_2 based on wood templates. *Sci. Rep.* **2020**, *10*, 1–10. [CrossRef]
65. Pori, P.; Vilčnik, A.; Petrič, M.; Sever Škapin, A.; Mihelčič, M.; Šurca Vuk, A.; Novak, U.; Orel, B. Structural studies of TiO_2 /wood coatings prepared by hydrothermal deposition of rutile particles from $TiCl_4$ aqueous solutions on spruce (Picea Abies) wood. *Appl. Surf. Sci.* **2016**, *372*, 125–138. [CrossRef]
66. Tsang, C.H.A.; Li, K.; Zeng, Y.; Zhao, W.; Zhang, T.; Zhan, Y.; Xie, R.; Leung, D.Y.C.; Huang, H. Titanium oxide based photocatalytic materials development and their role of in the air pollutants degradation: Overview and forecast. *Environ. Int.* **2019**, *125*, 200–228. [CrossRef]
67. Kermani, M.; Jonidi Jafari, A.; Gholami, M.; Shahsavani, A.; Goodarzi, B.; Fanaei, F. Extraction and determination of organic/inorganic pollutants in the ambient air of two cities located in metropolis of Tehran. *Environ. Monitor. Assess.* **2022**, *194*, 1–20. [CrossRef]
68. Guo, H.; Klose, D.; Hou, Y.; Jeschke, G.; Burgert, I. Highly Efficient UV Protection of the Biomaterial Wood by A Transparent TiO_2/Ce Xerogel. *ACS App. Mater. Inter.* **2017**, *9*, 39040–39047. [CrossRef]
69. Han, L.; Gao, Y.; Gao, X.; Gong, W.; Wang, H. Degradation of Organic Pollutants by Titanium Dioxide Matrix Composites. In *E3S Web of Conferences*; EDP Sciences: Ulis, France, 2021; Volume 245, p. 2039. [CrossRef]
70. Jain, A.; Vaya, D. Photocatalytic Activity of Titanium Dioxide Nanomaterial. *J. Chil. Chem. Soc.* **2017**, *4*, 3683–3690. [CrossRef]
71. Baeissa, E. Photocatalytic degradation of methylene blue dye under visible light irradiation using In/ZnO nanocomposite. *Front. Nanosci. Nanotechnol.* **2016**, *2*, 1–5. [CrossRef]
72. Luminita Udrea, M.; Ion, R.-M. Modelling of Methylene Blue Dye Adsorption on Beech and Fir Wood Sawdust As Adsorbent Support Materials. *J. Sci. Arts Year* **2019**, *19*, 675–686.
73. Raheb, I.; Manlla, M.S. Kinetic and thermodynamic studies of the degradation of methylene blue by photo-Fenton reaction. *Heliyon* **2021**, *7*, e07427. [CrossRef]
74. Varma, K.S.; Tayade, R.J.; Shah, K.J.; Joshi, P.A.; Shukla, A.D.; Gandhi, V.G. Photocatalytic degradation of pharmaceutical and pesticide compounds (PPCs) using doped TiO_2 nanomaterials: A review. *Water-Energy Nexus* **2020**, *3*, 46–61. [CrossRef]
75. Azeez, F.; Al-Hetlani, E.; Arafa, M.; Abdelmonem, Y.; Abdel Nazeer, A.; Amin, M.; Madkour, M. The effect of surface charge on photocatalytic degradation of methylene blue dye using chargeable titania nanoparticles. *Sci. Rep.* **2018**, *8*, 1–9. [CrossRef]
76. Pandey, K.K.; Ramakantha, V.; Chauhan, S.S.; Arun Kumar, A.N. Wood is good: Current trends and future prospects in wood utilization. In *Wood is Good*; Springer: New York, NY, USA, 2017; pp. 1–480. [CrossRef]
77. Hazarika, A.; Maji, T.K. Synergistic effect of nano-TiO_2 and nanoclay on the ultraviolet degradation and physical properties of wood polymer nanocomposites. *Ind. Eng. Chem. Res.* **2013**, *52*, 13536–13546. [CrossRef]
78. Mayerhöfer, T.G.; Popp, J. The electric field standing wave effect in infrared transflection spectroscopy. *Spectrochim. Acta-Part A Mol. Biomol. Spectrosc.* **2018**, *191*, 283–289. [CrossRef]
79. Pereira, H.; Graça, J.; Rodrigues, J. Wood Chemistry in Relation to Quality. *Cheminform* **2004**, *35*, 53–86. [CrossRef]
80. Bi, W.; Li, H.; Hui, D.; Gaff, M.; Lorenzo, R.; Corbi, I.; Corbi, O.; Ashraf, M. Effects of chemical modification and nanotechnology on wood properties. *Nanotechnol. Rev.* **2021**, *10*, 978–1008. [CrossRef]
81. Osman, A.I.; Blewitt, J.; Abu-Dahrieh, J.K.; Farrell, C.; Al-Muhtaseb, A.H.; Harrison, J.; Rooney, D.W. Production and characterisation of activated carbon and carbon nanotubes from potato peel waste and their application in heavy metal removal. *Environ. Sci. Pollut. Res.* **2019**, *26*, 37228–37241. [CrossRef]
82. Dong, Y.; Yan, Y.; Zhang, S.; Li, J. Wood/polymer nanocomposites prepared by impregnation with furfuryl alcohol and Nano-SiO_2. *BioResources* **2014**, *9*, 6028–6040. [CrossRef]
83. Wang, M.; Wang, W.; He, B.L.; Sun, M.L.; Bao, Y.S.; Yin, Y.S.; Liu, L.; Zou, W.Y.; Xu, X.F. Corrosion behavior of hydrophobic titanium oxide film pre-treated in hydrogen peroxide solution. *Mater. Corros.* **2011**, *62*, 320–325. [CrossRef]
84. Sompech, S.; Srion, A.; Nuntiya, A. The effect of ultrasonic treatment on the particle size and specific surface area of $LaCoO_3$. *Procedia Eng.* **2012**, *32*, 1012–1018. [CrossRef]

polymers

MDPI

Article

Evaluation of Selected Cellulose Macromolecular Properties after Its Chemical Treatment Using Size Exclusion Chromatography

Tereza Jurczyková [1], František Kačík [2], Iveta Čabalová [2,*] and Kateřina Hájková [1]

1 Department of Wood Processing and Biomaterials, Faculty of Forestry and Wood Sciences, Czech University of Life Sciences Prague, Kamýcka 129, 16000 Prague, Czech Republic
2 Department of Chemistry and Chemical Technologies, Faculty of Wood Sciences and Technology, Technical University in Zvolen, T. G. Masaryka 24, 96001 Zvolen, Slovakia
* Correspondence: cabalova@tuzvo.sk; Tel.: +421-455-206-375

Abstract: This work evaluates the effect of using selected inorganic chemicals as the main components of waterborne wood preservative systems on the degradation of the cellulose constituent in wood from model samples. The polymeric properties of cellulose and the homogeneity of the degradation process primarily reflect very well the degree of cellulose deterioration. Whatman papers, as pure cellulose model samples, were impregnated with 10 different 5 wt% solutions of inorganic salts and distilled water and consequently subjected to wet-thermal accelerated aging (T = 85 °C, RH = 65%, for 30 days). The samples were then derivatized to cellulose tricarbanilates (CTCs) through two different procedures (by precipitation in a methanol–water mixture/by evaporation of pyridine from the reaction mixture) and finally analyzed using size exclusion chromatography (SEC). Chemically treated and aged cellulose samples showed different changes in the degree of polymerization (DP) and polydispersity (PD) in terms of untreated non-aged standard caused by different ongoing degradation reactions, such as dehydration, hydrolysis, oxidation, and crosslinking. In general, the lowest degradation rate after treatment by chemicals and after accelerated aging was observed in samples treated by borates, NaCl, and $ZnSO_4 \cdot 7H_2O$. The greatest depolymerization after treatment and after accelerated aging was caused by sulphates containing NH^{4+}, Cu^{2+}, and Fe^{3+} cations, with aging by NH_4Cl and $(NH_4)_2HPO_4$-treated samples also leading to significant depolymerization. The higher DP values are linked to the precipitated method of CTC preparation, though not for chlorides and phosphates. PD is also generally higher in precipitated and aged samples and is heavily influenced by the presence of low molecular weight products. This paper brings new insights regarding the complex evaluation of the polymeric properties of degraded cellulose by considering all important factors affecting the sample and the analysis itself through the use of statistics. From the statistical point of view, the influences of all factors (solution, aging, method) and their interactions (except aging*method) on DP are statistically significant. The influence of the sample processing method used for analysis of the desired results becomes important mainly in practice. This work recommends the evaporation method for more accurate description of more degraded cellulose.

Keywords: cellulose; wood preservatives; degradation; degree of polymerization; cellulose tricarbanilates; size exclusion chromatography

Citation: Jurczyková, T.; Kačík, F.; Čabalová, I.; Hájková, K. Evaluation of Selected Cellulose Macromolecular Properties after Its Chemical Treatment Using Size Exclusion Chromatography. *Polymers* **2023**, *15*, 573. https://doi.org/10.3390/polym15030573

Academic Editor: Jalel Labidi

Received: 20 December 2022
Revised: 3 January 2023
Accepted: 17 January 2023
Published: 22 January 2023

1. Introduction

Although wood is one of the most widespread and essential natural materials, it also has disadvantages. Its most well-known disadvantage is its natural flammability and high ability to generate smoke. For this reason, wood is treated with protective substances. Inorganic salts (borates, phosphates, sulphates, chlorides) are active components of waterborne wood preservatives. These inorganic chemicals induce changes in the molecular structure of wood polymers depending on their concentration, pH value, exposure time, dissociation constant, and other conditions [1–3]. Meanwhile, the effect of these compounds when used for wood protection has mostly been observed only in terms of the application process, fixation, leaching, hygroscopicity, corrosivity to metal fasteners, their thermal stability, and dimensional changes in wood mass caused by crystallizing salts from preservative solutions after drying [4–8]. Most researchers have focused on the effect of biocides and flame retardants on the strength properties of the wood under study [1,9–11]. Surprisingly little attention has been given to the chemical aspects of this problem, and few controlled experimental studies have been made on the subject [12–14].

This work focuses on the evaluation of changes in the macromolecular structure of chemically treated cellulose (a main wood component) with solutions of selected inorganic salts in terms of long-term exposure. However, cellulose isolated from wood always shows a certain degree of degradation caused by agents used in the isolation process, and pure cellulose model samples in the form of Whatman paper were used for better qualitative and quantitative description of changes in the investigated average degree of polymerization (DP) [12,15–17].

The extent of cellulose deterioration, i.e., the magnitude of changes in its structure and properties, depends on the extent of damage to its chemical, microscopic, and macroscopic structure. The cellulose synthesized in nature has a certain degree of polydispersity which influences its physical properties. Viscosity and/or the molecular weight distribution (MWD) of cellulose are important parameters for all of its end uses. Therefore, based on determination of MWD and DP, it is possible to accurately evaluate the degradation of cellulose and predict its physical and mechanical properties and ongoing chemical reactions. The result of cellulose degradation is an increased proportion of low molecular weight fractions and a decrease in average DP and polydispersity (PD) [18–21]. Moreover, DP, as well as crystallinity and crystallite size, have an influence on the thermal stability of cellulose [22,23].

The molecular weight of macromolecules such as cellulose can be determined by absolute and relative methods. Therefore, size exclusion chromatography (SEC) has gained wide acceptance as a preferred method. This method allows for the characterization of MWD and the determination of more of the parameters of the molecular weights. SEC also provides information on degradation fractions and thus helps clarify the mechanism of degradation in cellulose [24,25].

Formulation of Individual Scientific Goals

To evaluate the influence of the individual inorganic salts, which are active components of fire retardants and biocides, on the degradation of cellulose through the determination of DP via the SEC method.

To compare the DP of cellulose samples degraded by various chemicals among themselves, both with the reference sample (treated only by distilled water) and the standard sample (untreated Whatman paper).

To compare the changes in DP within each series of samples before and after aging.

To assess the changes in DP for each sample series caused by two different methods of CTC sample preparation (evaporation/precipitation in methanol–water mixture) and accordingly identify the samples of treated cellulose with the largest proportion of low molecular weight fractions due to degradation.

To determine whether the solution, aging, method, or combined effect of their interaction (solution*aging, solution*method, method*aging, solution*aging*method) statistically significantly affects the value of the calculated DP.

2. Materials and Methods

2.1. Material

Whatman filtration papers (Grade No. 1, α-cellulose content >98 wt%, basis weight 87 g/m^2, air flow rate 10.5 s/100 mL/in^2) with dimensions of 15 × 100 mm were firstly conditioned to a constant weight at a temperature of 23 ± 1 °C and with a relative air humidity of 50 ± 2% according to the ISO 187 standard [26]. For the treatment of

10 sample series, aqueous solutions of selected chemicals at a 5 wt% concentration level were used, i.e., $Na_2B_4O_7 \cdot 10H_2O$, H_3BO_3, $CuSO_4 \cdot 5H_2O$, $ZnSO_4 \cdot 7H_2O$, $Fe_2(SO_4)_3$, NaCl, NH_4Cl, $(NH_4)_2SO_4$, $(NH_4)_2HPO_4$, and $NH_4H_2PO_4$. Moreover, the standard non-treated unaged samples and reference unaged and aged samples impregnated by distilled water were also included in this experiment. The impregnation of individual series of samples was performed through immersion in a laboratory beaker with a 5 wt% salt solution or distilled water under atmospheric pressure for 30 min. Some pieces of polyethylene mesh were used as interlayers to avoid sticking the individual samples together. The wet-thermal accelerated aging technique according to ISO 5630-3 [27], i.e., at 80 °C and 65% RH for a duration of 30 days, was chosen for aging of one half of each sample series based on the knowledge of previously published studies and existing standards. (Samples releasing ammonia during accelerated aging were separately subjected to elevated temperature and humidity in a second climatic chamber with the intention of preventing the other samples from being contaminated with ammonia.) Prior to testing, all specimens were re-equilibrated to a constant weight [26].

2.2. Methods

2.2.1. Derivatization of Cellulose into the Form of Cellulose Tricarbanilates (CTCs)

Accurate determination of molar mass distribution for cellulose samples is a challenging task and a universal method for accurate molar mass distribution analysis of cellulose samples still does not exist [28].

In our work, cellulose tricarbanilates (CTCs) were prepared in two different ways. The first was the modified method of precipitation [29,30]. Briefly, the cellulose samples were dried over silica gel for several days. Anhydrous pyridine (8.0 mL), cellulose (50 mg), and phenyl isocyanate (1.0 mL) were sealed in a 50 mL dropping flask. The flasks were immersed in an oil bath at 70 °C for 72 h. At the end of the reaction, methanol (2.0 mL) was added to the mixture to eliminate excess phenyl isocyanate. The yellow solutions were then added drop wise into a rapidly magnetic stirring 7:3 methanol/water mixture (150 mL). The solids were collected via filtration and washed with 7:3 methanol/water mixtures (1 × 50 mL) and water (2 × 50 mL) to a neutral reaction. The CTC was air-dried overnight and subsequently under vacuum at 60 °C to remove any traces of pyridine. The CTC was then redissolved in 10 mL of THF and filtered through a glass filter (Membrane Solutions, Auburn, WA, USA) with a pore size of 0.7 μm before then being analyzed by SEC.

In the second method, the pyridine was evaporated from the reaction mixture instead of precipitated, thereby producing a sample containing all the non-volatile products of the derivatization procedure [31]. Subsequently, 2.0 mL of the pyridine reaction mixture was transferred to a 25 mL round bottomed flask and the pyridine was evaporated under vacuum at 40 °C. The syrupy liquid which remained was dissolved in approximately 15 mL of acetone. Following this, 2.0 mL aliquots of the acetone solution were transferred to a 10 mL flask and evaporated with a stream of nitrogen. The residual material was dried overnight under vacuum at 60 °C to remove any traces of pyridine and then redissolved in 10 mL of THF and filtered as above.

2.2.2. Size Exclusion Chromatography (SEC)

The SEC analysis was carried out on an Agilent 1200 HPLC (Agilent Technologies, Santa Clara, CA, USA) system consisting of an autosampler, quaternary pump, degasser, column thermostat, diode array detector (DAD) working at 240 nm, and differential refractive index detector (DRI). The separation was performed at 35 °C with THF at a flow rate of 1 mL min^{-1} on two PLgel 10 μm (7.5 × 300 mm) MIXED B columns proceeded by a PLgel 10 μm (7.5 × 50 mm) GUARD column (Agilent Technologies, Santa Clara, CA, USA). Data acquisition was carried out with Chemstation software (Agilent Technologies, Santa Clara, CA, USA) and calculations were performed with the Clarity GPC module (DataApex, Prague, Czech Republic). The system was calibrated with polystyrene standards with an average MW in the range of 500–6,035,000 (Polymer Laboratories, Shropshire, UK; Tosoh

Corporation, Tokyo, Japan). A universal calibration for determination of molecular weights was used with the constants $K = 1.095 \times 10^{-3}$ and $\alpha = 0.955$ [32]. All SEC results represent the mean of two different samples, and each CTC was chromatographed twice (total of four runs for each sample).

2.2.3. Calculations and Statistical Evaluation

The values of molecular weights, the degree of polymerization (DP), and polydispersity indexes (PDI) were calculated after data conversion in Clarity 15.7.1 (DataApex, Praha, Czech Republic). All SEC results represent the mean of two different samples, each CTC was chromatographed twice, and five signals from two detectors ($1 \times$ DRI, $4 \times$ DAD) were obtained from each measurement (but only one was used for calculation). The results from chromatographic analysis of CTCs were statistically evaluated by using analysis of variance (ANOVA), and the multiple comparison Duncan's test was used for the determination of statistically significant differences in DP mean values. The statistical evaluation took 23 situations of variously modified samples into account, as well as the influence of 3 factors (solution, aging, method) and 4 interactions (solution*aging, solution*method, aging*method, solution*aging*method), including 4 repeated measurements of DP for each situation (Figure 1).

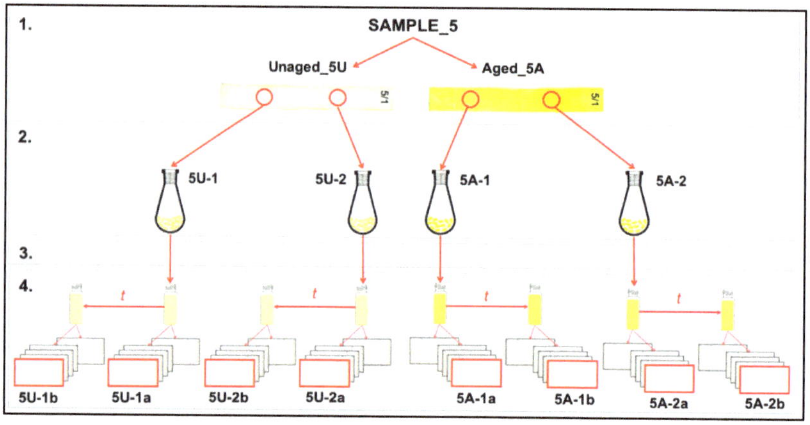

Figure 1. Illustrative overview of three influencing factors (solution, aging, method) on the DP of examined cellulose samples.

The layout of experimental runs and conditions (example for sample 5: treated by $Fe_2(SO_4)_3$) was as follows: 1. Two pieces were cut from each sample of chemically treated cellulose before and after aging. 2. These pieces of cellulose were converted into the form of CTCs. 3. Two vials with dissolved CTCs in THF were prepared from each CTC sample. 4. One vial was measured twice over time by SEC. This scheme describes the procedure using only one method of CTC preparation. In our case, using two methods, double sampling and measurement were performed for each sample.

3. Results and Discussion

Aging and degradation of cellulose due to chemical treatment, elevated temperature, and increased humidity results in a drop in DP and an increase in the proportion of low molecular weight fractions. The adverse consequences of the degradation of cellulose are also reflected in the deterioration of mechanical and optical properties [33,34].

3.1. Degree of Polymerization (DP)

It has been previously reported that some chemicals tend to reduce cellulose's molecular weight (MW) and degree of polymerization (DP) [30,35–37]. It is apparent from our experiments (Table 1) that there was a greater or a lesser reduction in DP as a consequence of the degradation process caused by presence and action of inorganic salts in the cellulose fiber structure, which was especially evident due to accelerated aging in conditions of elevated temperature and increased relative humidity.

Table 1. The degree of polymerization (DP) data (mean + SD) and the accumulated DP loss in chemically treated and aged cellulose samples.

Samples	DP of Treated Samples		DP of Treated and Aged Samples		Accumulated DP Loss (ω_{DP})	
	Evaporated	Precipitated	Evaporated	Precipitated	Evaporated	Precipitated
Standard	1444 ± 3	1516 ± 17	—	—	—	—
Reference	1673 ± 6	1754 ± 5	875 ± 4	912 ± 4	0.48	0.48
$Na_2B_4O_7.10H_2O$	1308 ± 106	1261 ± 114	1002 ± 53	1139 ± 49	0.23	0.10
H_3BO_3	913 ± 68	1029 ± 37	885 ± 1	925 ± 9	0.03	0.10
$CuSO_4.5H_2O$	681 ± 5	1145 ± 25	37 ± 1	630 ± 47	0.95	0.45
$ZnSO_4.7H_2O$	877 ± 39	1244 ± 88	847 ± 30	1463 ± 40	0.03	−0.18
$Fe_2(SO_4)_3$	689 ± 5	1689 ± 21	275 ± 8	520 ± 9	0.60	0.31
NaCl	1675 ± 7	1721 ± 26	1285 ± 23	1212 ± 4	0.23	0.30
NH_4Cl	1344 ± 53	1326 ± 12	273 ± 1	271 ± 5	0.20	0.80
$(NH_4)_2SO_4$	638 ± 26	1409 ± 16	12 ± 0	376 ± 3	0.98	0.73
$(NH_4)_2HPO_4$	972 ± 67	490 ± 20	34 ± 3	176 ± 13	0.97	0.64
$NH_4H_2PO_4$	1090 ± 71	296 ± 19	796 ± 67	116 ± 6	0.27	0.61

The non-treated unaged standard samples had an average DP of 1444 (when the evaporation method for CTC preparation was used) and 1516 (for the precipitation method). The reference samples, which were only treated with distilled water free of salt cations and anions, recorded a slight increase in DP of about 15.9–15.7% depending on the method of CTC preparation with regard to the standard. This increase in DP, as well as an increase in the crystallinity index caused by preferred degradation of cellulose in its amorphous part, was also observed previously [38–41]. This anomalous behavior may be also attributed to the formation of covalent intermolecular bonds, i.e., cross-links. Dehydration reactions leading to the formation of predominantly ether bonds have been known to occur during subsequent thermal treatment of cellulose [42].

The highest values for DP before accelerated aging, which were higher than the standard samples, were observed in samples treated with NaCl (about 10.3–16.0%) for both methods of CTC preparation. Moreover, in the group of samples where CTCs were prepared via the precipitation method, treatment by $Fe_2(SO_4)_3$ provided a relatively high DP (about 11.4% higher than standard samples). However, it is evident that this chemical causes a drastic reduction in the DP of cellulose, as it was also determined by the evaporation method. The drop in DP caused by $Fe_2(SO_4)_3$ was measured at approx. 52.3% in comparison to the evaporated standard, which is a decrease of about 59.2% in comparison to the simultaneously precipitated standard sample. A DP higher than 1000 was observed in celluloses treated by $Na_2B_4O7{\cdot}10H_2O$ and NH_4Cl and was confirmed by both methods. That means that these two chemicals should be considered the most gentle salt solutions in the preservation of cellulose, at least in this first impregnation step. The precipitation method provided a DP higher than 1000 for almost all samples besides ammonium phosphates, and the decrease in DP was in the range of 67.7–80.5%. On the other hand, cellulose treated by $NH_4H_2PO_4$ from precipitated samples provided the lowest DP value from this group, although the group of evaporated samples provided DP values of up to 1000. The highest drop in DP in the group of unaged evaporated samples was observed in celluloses degraded by $(NH_4)_2SO_4$, $CuSO_4{\cdot}5H_2O$, and $Fe_2(SO_4)_3$.

The lowest deterioration in chemically treated cellulose after aging, described by high DP values, was caused by NaCl, $Na_2B_4O_7{\cdot}10H_2O$, and H_3BO_3. Even higher DP values than those observed for reference samples were found, with values being in the range of 1.1% for H_3BO_3 to 46.9% for NaCl. These findings are confirmed by both the precipitation and evaporation method of CTC preparation. Treatment with $ZnSO_4{\cdot}7H_2O$ also results in higher DP values when the precipitation method is used (even the highest value from aged precipitated samples was about 60.4 % higher than the reference sample). However, there is a slight decrease (about 3.2%) compared to the reference sample when the evaporation method is used. This chemical, as well as borates and NaCl, can be considered as an acceptable chemical for maintaining a high DP even after aging. The most negative influence on cellulose degradation was observed with $(NH_4)_2SO_4$, $(NH_4)_2HPO_4$, and $Fe_2(SO_4)_3$ when the evaporation method was used (the drop in DP was in the range of 95.8–98.6% compared to the evaporated reference sample), and ammonium phosphates and NH_4Cl demonstrated the greatest negative influence when the precipitation method was used (the drop in DP was in the range of 70.3–87.3% compared to the precipitated reference sample).

3.2. Accumulated Degree of Polymerization Loss (ωDP)

Degradation of the cellulose can also be accurately characterized as a decrease in DP, which is calculated by $\omega_{DP} = 1 - DP_\tau/DP_0$, where ω_{DP} is the accumulated DP loss of cellulose, DP_0 is the degree of polymerization at an initial time, and DP_τ is the real degree of polymerization at the τ time of accelerated aging [43,44]. The same accumulated DP loss values were obtained with both methods of CTC preparation in reference samples. Values corresponding to the samples treated by borates, $ZnSO_4{\cdot}7H_2O$, and NaCl, i.e., samples with a lower degree of depolymerization, are also in good agreement. On the other hand, a greater than two-fold difference between values obtained from the two different methods of CTC preparation was observed in samples treated by $CuSO_4{\cdot}5H_2O$ and $Fe_2(SO_4)_3$, where higher values were reported when the evaporation method was used. For samples impregnated by NH_4Cl and $NH_4H_2PO_4$, conversely higher values were observed when the precipitation method was used. Based on these results, we can conclude that the lowest difference in DP between treated and aged samples, i.e., the lowest depolymerization caused by accelerated aging in the presence of a certain chemical, can be observed when using $ZnSO_4{\cdot}7H_2O$ and H_3BO_3 for treatment. $Na_2B_4O_7{\cdot}10H_2O$ and NaCl are, from this point of view, also acceptable. The precipitation method provides the greatest differences in DP between aged and unaged cellulose samples treated by all salts containing ammonia, while the evaporation method confirms that only $(NH_4)_2SO_4$ and $(NH_4)_2HPO_4$ from the group of chemicals based on ammonia have a negative effect on treated celluloses in terms of degradation during accelerated aging.

3.3. Polydispersity Index (PDI)

It was reported that the artificial aging of cellulose has an influence on PDI. Emsley et al. [45] found a slight PDI increase in cotton linters aged at 120, 140, and 160 °C. On the other hand, a decrease in PDI was observed in irradiated and chemically or enzymatically treated cellulose [46].

Changes in PDI (Table 2) were found for reference samples before aging in comparison to standard samples. The PDI of the reference evaporated sample was about 8.9% lower and the PDI of the reference precipitated sample was up to 20.3% lower than the PDI of the corresponding standard samples. PDI generally differs more in the group of precipitated samples and, of course, after aging.

In the series of unaged samples, the PDI of most samples remained fairly constant, i.e., values of between 3.15 and 4.04 for the evaporated series and values of between 2.12 and 3.87 for the precipitated series of samples. In these samples, depolymerization proceeds randomly without preferential breakdown of the longest cellulosic chains. When the hydrolysis of the glycosidic bonds is dominant, the ratio increases. Higher PDI values were observed in the group of unaged samples when using ammonium sulphate and phosphates for treatment. The range of PDI values for these samples is from 4.12 to 4.58 for the evaporation method and from 5.36 to 6.52 for the precipitation method. The precipitation method provides slightly higher values than the evaporation method for samples treated by phosphates and all sulphates. On the other hand, the PDI values for the standard and reference samples and borate- and chloride-treated celluloses are lower in the case of precipitation method (up to 40.4% for $Na_2B_4O_7 \cdot 10H_2O$). $Fe_2(SO_4)_3$ treatment leads to significantly higher PDI values that are approx. 3.3 times higher in the case of the evaporation method and approx. 6.5 times higher in the case of the precipitation method than the corresponding standard samples. This chemical leads, in all cases (independent of method or aging), to an extreme increase in PDI compared to other investigated samples. This means that a great amount of the products of low molecular weight degradation are still present, which thus increases PDI; however, they have no influence on MWD. Additionally, they are usually lost during the precipitation step involving methanol or ethanol when CTCs are prepared [47]. When the precipitation method was used, all three celluloses treated by sulphates containing transition metal cations provided higher PDI values compared to the other samples.

Table 2. Polydispersity index (PDI = Mw/Mn) for chemically treated and aged cellulose samples (mean ± SD).

Samples	PDI of Treated Samples		PDI of Treated and Aged Samples	
	Evaporated	Precipitated	Evaporated	Precipitated
Standard	3.71 ± 0.07	2.66 ± 0.03	—	—
Reference	3.38 ± 0.03	2.12 ± 0.01	2.94 ± 0.05	2.43 ± 0.03
$Na_2B_4O_7.10H_2O$	3.81 ± 0.27	2.27 ± 0.18	3.38 ± 0.16	2.84 ± 0.08
H_3BO_3	3.74 ± 0.21	3.15 ± 0.21	3.02 ± 0.01	2.72 ± 0.06
$CuSO_4.5H_2O$	3.69 ± 0.07	8.45 ± 0.26	2.14 ± 0.07	8.53 ± 0.29
$ZnSO_4.7H_2O$	4.04 ± 0.03	11.00 ± 0.70	3.22 ± 0.09	5.83 ± 0.06
$Fe_2(SO_4)_3$	12.32 ± 0.44	17.18 ± 0.23	10.56 ± 0.67	8.01 ± 0.03
NaCl	3.31 ± 0.10	2.37 ± 0.02	5.45 ± 0.28	3.48 ± 0.05
NH_4Cl	3.24 ± 0.10	2.98 ± 0.04	2.42 ± 0.01	2.53 ± 0.02
$(NH_4)_2SO_4$	4.42 ± 0.09	5.73 ± 0.12	1.34 ± 0.03	11.28 ± 0.44
$(NH_4)_2HPO_4$	4.12 ± 0.03	6.52 ± 0.12	2.24 ± 0.14	6.33 ± 0.41
$NH_4H_2PO_4$	4.58 ± 0.19	5.36 ± 0.28	31.36 ± 1.78	5.07 ± 0.20

In the series of aged samples, the calculated PDI values differed quite a lot depending on the method of CTC preparation. The highest PDI values for aged samples obtained via the evaporation method were observed in celluloses treated with $NH_4H_2PO_4$ (PDI of 31.36, which is also the highest PDI value among all samples) and $Fe_2(SO_4)_3$ (10.56). High DPI values were also reported for aged precipitated samples impregnated by all sulphates (8.01–11.28). The comparatively higher PDI values observed with precipitation methods were also caused by treatment using phosphates (5.07–6.33). The highest difference in PDI between both methods of CTC preparation was caused when the cellulose was impregnated by ammonium salts (except of NH_4Cl) and $CuSO_4 \cdot 5H_2O$. For reference samples, borates, $Fe_2(SO_4)_3$, NaCl, and $NH_4H_2PO_4$, the PDI values were lower when the precipitation method was used.

3.4. Statistical Evaluation of the Experiment

The results of determined DP obtained via SEC were statistically evaluated by the three-way analysis of variance (ANOVA) method.

The final interpretation of ANOVA results with multiple screenings consists of an evaluation of the mutual effect of main factors and the interactions involved in the experiment. To the resulting ANOVA table (Table 3), it should be noted that each mean square is just the sum of squares divided by its degrees of freedom, and the F value is the ratio of the mean squares. Moreover, for determination of the statistically significant difference in mean DP values, the multiple comparison Duncan's test was used. Statistical significance among mean DP values is demonstrated by a calculated p-value lower than $\alpha = 0.05$.

Table 3. ANOVA table for three-way analysis of the average degree of polymerization (DP) for the examined cellulose samples influenced by three factors (solution, aging, method) at 15 levels (11 different salt solutions, two types of aging, i.e., unaged and aged samples, two methods of CTC preparation) and their four combined interactions.

Source of Variation	Sum of Squares	Degrees of Freedom	Mean Square (Variance)	F-Test	*p*-Significance Level
Total mean	140,282,339	1	140,282,339	86,771.49	0.000
Solution (a)	18,125,070	10	1,812,507	1121.12	0.000
Aging (b)	11,334,045	1	11,334,045	7010.66	0.000
Method (c)	775,295	1	775,295	479.56	0.000
Solution*Aging (a*b)	5,319,434	10	531,943	329.03	0.000
Solution *Method (a*c)	6,479,499	10	647,950	400.79	0.000
Aging*Method (b*c)	663	1	663	0.41	0.523
Solution*Aging*Method (a*b*c)	1,272,813	10	127,281	78.73	0.000
Random effect	213,403	132	1617	—	—

Based on the basic statistical characteristics listed in Table 3, it is possible to suggest that the influences of solution, aging, and method and the interactions of solution*aging, solution*method, and solution*aging*method on DP are statistically significant. Conversely, the influence of aging*method interaction does not provide statistically significant results ($p > 0.05$). From the F-test that tests the differences between two variances, it is apparent that the most important influence on DP from the three factors (solution, aging, method) was aging, while the factor of method was the least significant. From all influencing interactions, the solution*method interaction was the most powerful. The aging*method interaction did not provide statistically significant results because the null hypothesis, in which these two variances are equal at the 0.05 significance level, was rejected. Graphs of cellulose DP showing the influence of various interactions within the three-way ANOVA are shown in Figures 2–4.

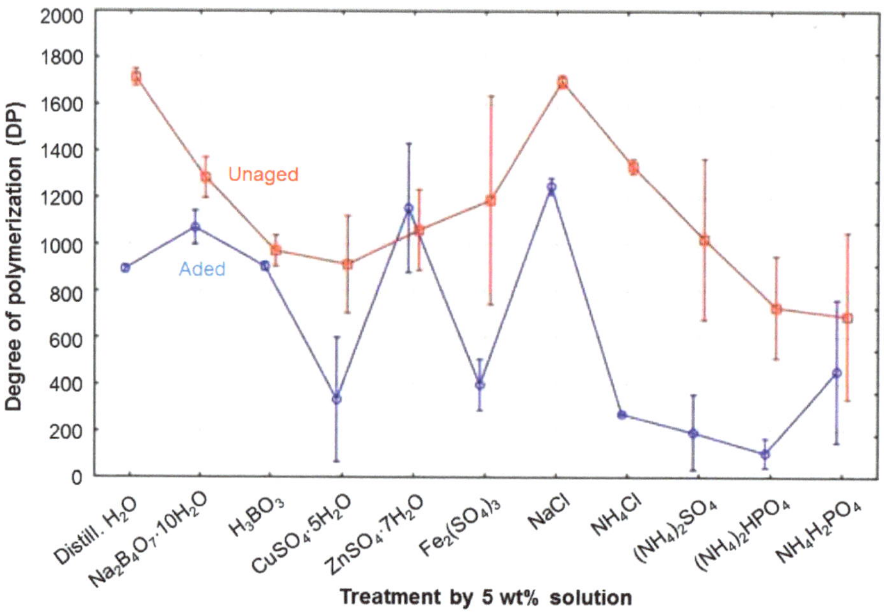

Figure 2. Influence of solution*aging on the degree of polymerization in cellulose.

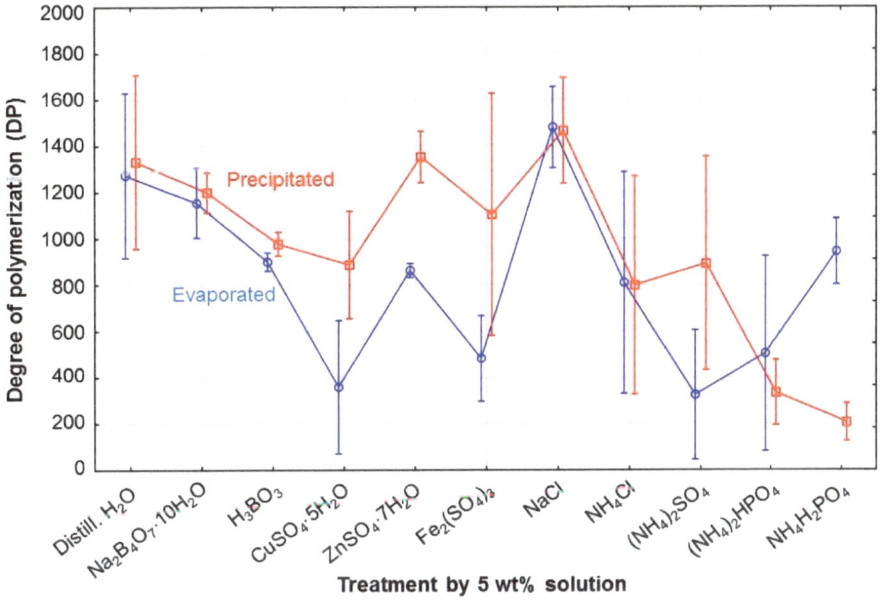

Figure 3. Influence of solution*method on the degree of polymerization in cellulose.

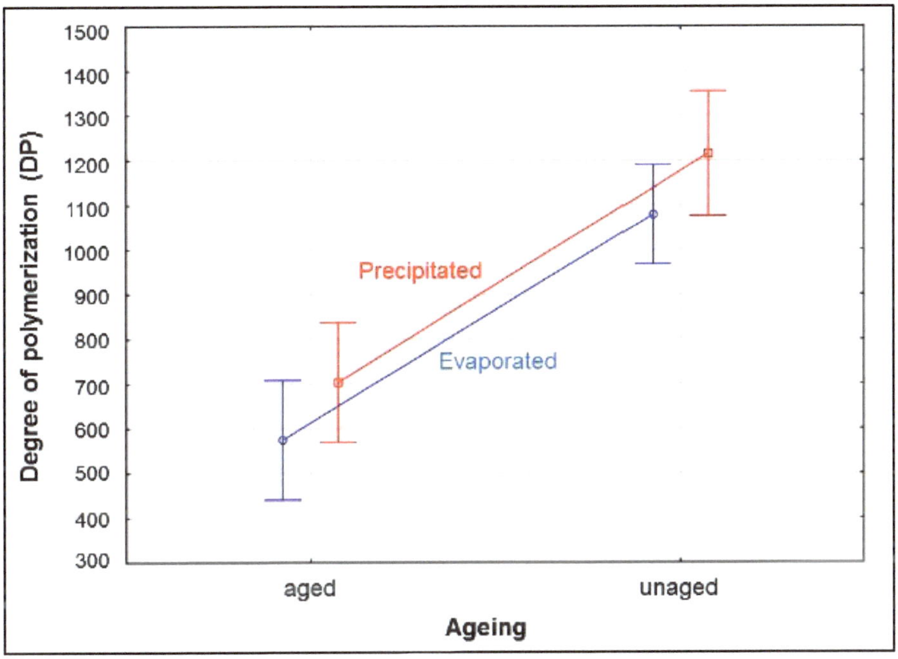

Figure 4. Influence of aging*method on the degree of polymerization in cellulose.

In general, the precipitation method from the perspective of solution*method interaction provides a higher DP value than the evaporation method (Figure 3). However, when the cellulose was treated by chlorides, the DP values obtained from the evaporation method were slightly higher, and even higher values were observed after treatment with phosphates. Reference samples and samples treated by NaCl, $Na_2B_4O_7{\cdot}10H_2O$, and $ZnSO_4{\cdot}7H_2O$ (only when CTCs were prepared by precipitation in the case of $ZnSO_4{\cdot}7H_2O$) showed the highest DP values. Low DP values depended more on the method used: $(NH_4)_2SO_4$, $CuSO_4{\cdot}5H_2O$, and $Fe_2(SO_4)_3$ caused significant depolymerization when the evaporation method was used, and $NH_4H_2PO_4$, $(NH_4)_2HPO_4$, and NH_4Cl had a negative effect on DP when the precipitation method was used. Very similar DP values that fell within the tolerance of both performed methods of CTC preparation were provided by samples impregnated by H_2O (i.e., the reference sample), $Na_2B_4O_7{\cdot}10H_2O$, NaCl, and NH_4Cl. Similar DP values were also observed in cellulose treated by H_3BO_3 and $(NH_4)_2HPO_4$. This means that MWD is quite uniform without extremely low molecular weight fractions. These low molecular fractions are probably broadly present in samples with sulphate content (NH_4^+, Cu^{2+}, Zn^{2+}, Fe^{3+}), as shown by the great difference between DP values for both methods. The evaporation method provided lower DP values due to the presence of low molecular weight fractions that are more reflected by this method. A very similar difference between DP values for the two methods was also observed in the sample treated with $NH_4H_2PO_4$. However, in this case, a higher DP value is reported for the evaporation method.

The DP values for samples precipitated in the methanol–water mixture presented a very similar statistical trend to the DP values of evaporated samples (Figure 4), although the evaporated samples generally provided lower DP values. For aged evaporated samples, the average DP value for all samples was approx. 17% lower than the corresponding precipitated samples, while values for unaged evaporated samples were approx. 11% lower. This fact indicates that precipitation of the CTC samples in a methanol–water mixture still results in a partial loss of low molecular weight fractions.

The MWD of precipitated CTC samples may not be representative of the MWD of the original cellulose if the cellulose contains a low molecular weight fraction [31]. This is the reason why evaporated CTC samples were also prepared, as they contain (according to assumptions) all the nonvolatile products from a derivatization procedure. Thus, using the evaporation method seems to be suitable for more accurate description of more degraded cellulose treated with various sulphates (differences between the evaporated and precipitated samples are shown in Figure 4), and it may also be useful in general for aged samples with a higher amount of low molecular weight fractions. Similarly, Pitkänen and Sixta [28] showed that size exclusion chromatography of non-derivatized cellulose coupled with multi-angle light scattering (MALS) and differential refractive index (DRI) detectors suffers from low sensitivity in the low molar mass range. They suggest using a combination of two calibration strategies: MALS/DRI for the polymeric region of the cellulose sample and conventional calibration for the oligomeric region.

Comparison of the polymeric properties of treated or aged samples in certain experimental series should always be conducted within one of these methods. For this study, which was focused on cellulose degradation with different chemicals in conditions of wet-thermal accelerated aging, application of the evaporation method is considered to be more accurate. Another reason for using this method, apart from preserving the products of low molecular weight degradation, is the fact that the observed trends in DP decline in the series of evaporated samples correspond more, compared to the precipitation method, with other investigated trends among individual samples within the scope of our previous research works on this topic, e.g., a decrease in cellulose yields, a drop in the degree of cellulose crystallinity [12], the generation of new chromophoric structures, etc. [48].

The results in Figure 5 show that different methods of sample preparation for SEC analysis provide different absolute mean DP values. However, some of the main effects of individual factors on DP have already been found using analysis of variance, though which of these factors have the greatest influence it is still unknown. Therefore, it was necessary to conduct post hoc comparisons between pairs of treatments.

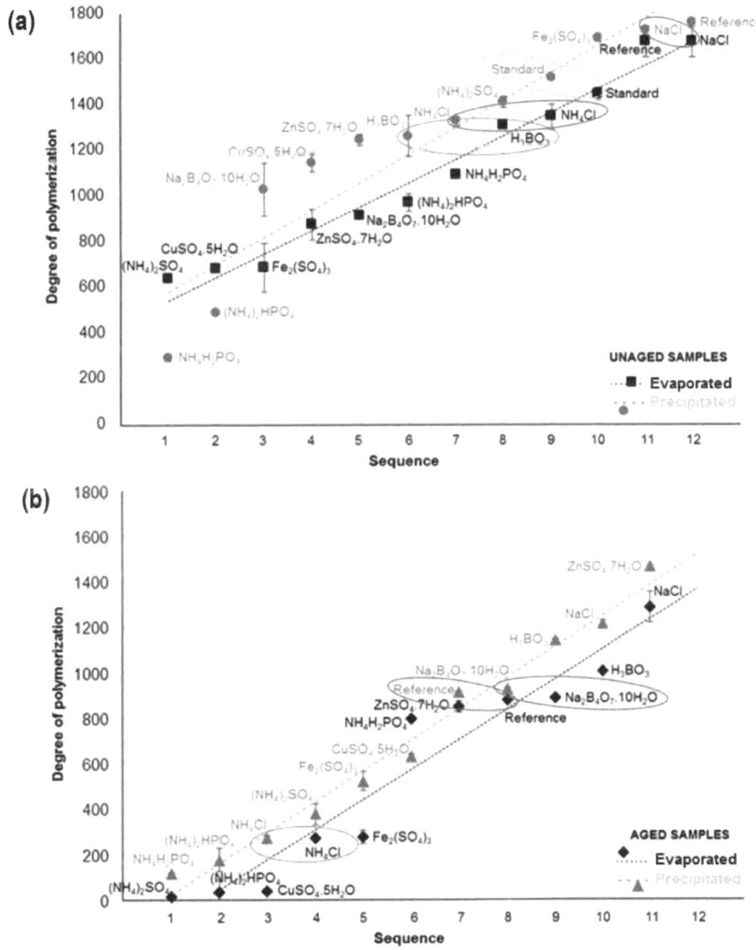

Figure 5. Graphical comparison of average degrees of polymerization for chemically treated cellulose when determination was performed using precipitated and evaporated samples within the group of (**a**) unaged samples and (**b**) samples after accelerated aging.

Duncan's multiple range test is based on the comparison between the range of a subset of sample means and a confidence level mean with a calculated least significant range. This least significant range increases with the number of sample means in the subset. If the range of the subset exceeds the least significant range, then the population means can be considered to represent differences with statistical significance ($p < 0.01$) [49]. If $0.01 < p < 0.05$, then the parameter demonstrates a moderately statistically significant difference, which is this study can be seen between DP means for unaged precipitated and evaporated standard samples and also between aged NaCl-treated precipitated and evaporated samples. There are also defined pairs of samples between which a statistically significant difference ($p > 0.05$) does not exist. This similarity is present within individual groups of unaged and aged chemically treated celluloses, and moreover between some samples from unaged and aged groups as well. Pairs of samples (precipitated and evaporated) without any significantly different DP values are the H_3BO_3-treated, NaCl-treated, and NH_4Cl-treated celluloses in the unaged group and the reference samples, $Na_2B_4O_7 \cdot 10H_2O$-treated samples, and NH_4Cl-treated celluloses in the aged group. The pairs of samples across unaged and aged groups that do not have any significant differences in DP means are the $CuSO_4{\cdot}5H_2O$-treated unaged evaporated and aged precipitated samples, the $ZnSO_4{\cdot}7H_2O$-treated unaged evaporated and aged evaporated samples, and both (precipitated and evaporated) $Na_2B_4O_7{\cdot}10H_2O$-treated aged and unaged samples.

It is also interesting to highlight the similarities between some of the chemically treated celluloses with standard (non-treated unaged) and reference (distilled water-treated unaged and aged) samples that are evident from Duncan's test. There was no significant difference between evaporated and precipitated standard samples and precipitated aged $ZnSO_4{\cdot}7H_2O$-treated samples, nor was there a significant difference between evaporated standard and precipitated unaged $(NH4)_2SO_4$-treated samples. Pairs of samples without any significantly different DP values included the following: unaged evaporated reference sample and unaged precipitated $Fe_2(SO_4)_3$-treated cellulose/unaged precipitated as well as evaporated NaCl-treated cellulose; unaged precipitated reference sample and unaged precipitated NaCl-treated cellulose; aged evaporated reference sample and aged evaporated as well as precipitated $Na_2B_4O_7{\cdot}10H_2O$-treated cellulose/unaged evaporated $Na_2B_4O_7{\cdot}10H_2O$-treated cellulose/unaged evaporated $ZnSO_4{\cdot}7H_2O$-treated cellulose. Aged precipitated reference samples had the same relations as aged evaporated reference samples, and there was no significant difference between aged precipitated reference samples and unaged evaporated $(NH_4)_2HPO_4$-treated cellulose.

4. Conclusions

Cellulose degrades due to the effect of inorganic salts, which are active ingredients in biocides and flame retardants. The rate of degradation increases with time and exposure to elevated temperature and humidity. DP, as well as PDI, are reduced during the degradation of cellulose.

Our results show that depolymerization of cellulose was present to a greater or lesser extent in all samples treated with various chemicals, and was also present in the sample treated with distilled water. Acid hydrolysis (catalyzed by metal cations) of glycosidic bonds, oxidation of glucopyranose rings, dehydration, and crosslinking occurred in cellulose degraded by most of these chemicals.

The highest DP values after impregnation were observed in samples treated with distilled water, NaCl, NH_4Cl, and $Na_2B_4O_7{\cdot}10H_2O$. After aging, the highest DP values were again observed in samples treated by NaCl, $Na_2B_4O_7{\cdot}10H_2O$, and H_3BO_3. The greatest depolymerization after treatment and accelerated aging was observed in sulphates containing NH_4^+, Cu^{2+}, and Fe^{3+} cations, as well as in groups of aged NH_4Cl and $(NH_4)_2HPO_4$-treated samples. The lowest accumulated DP loss throughout the process (from initial chemical treatment to the aged form after 30 days) was mainly connected with $ZnSO_4{\cdot}7H_2O$ and borates, and the biggest decline was associated with ammonium salts, especially NH_4Cl. PD was generally higher in evaporated and aged samples and was heavily influenced by the presence of low molecular weight products. Chemicals such as sulphates (mainly $Fe_2(SO_4)_3$) and ammonium phosphates (mainly $NH_4H_2PO_4$) lead, in all cases, to extreme increases in PD compared to other samples.

Therefore, wood preservatives containing ammonia that is bound mainly to phosphates or sulphates and chlorides, as well as wood preservatives containing copper and iron sulphates, are not recommended for the long-term protection of timber due to the large-scale degradation of cellulose constituents. The most acceptable wood protective agents among those studied, in terms of preventing cellulose in wood from being damaged by depolymerization, are borates (primarily alkaline solution of $Na_2B_4O_7{\cdot}10H_2O$), NaCl, and $ZnSO_4{\cdot}7H_2O$. These assumptions were already confirmed by our previous studies [12,48] dealing with the effect of inorganic salts on wood.

Statistical evaluation of the experiment indicates that aging had the most important influence on DP from the three factors (solution, aging, method), with the factor of method being the least significant.

This study evaluated the appropriateness of using the evaporation method for more accurate description of cellulose samples degraded by sulphates or aging. Duncan's multiple-range test helped to define which pairs of samples did not have a statistically significant difference. These findings again confirm the close similarity between standard samples (or more precisely the reference sample in this case) and celluloses treated with $ZnSO_4 \cdot 7H_2O$, $Na_2B_4O_7 \cdot 10H_2O$, and NaCl, which are characterized by a low degree of deterioration.

In any case, SEC analysis proved to be suitable and highly recommended for analysis of chemically treated and aged celluloses. This measurement technique provides important information concerning degraded fractions, thus leading to insights into degradation mechanisms.

Author Contributions: T.J. and F.K. conceived and designed the experiments; T.J. and F.K. carried out the laboratory experiments; T.J., K.H., F.K. and I.Č. analyzed the data, interpreted the results, prepared figures, and wrote the manuscript. All authors have read and agreed to the published version of the manuscript.

Funding: This work was supported by the Slovak Research and Development Agency under contract No. APVV-16-0326 (50%) and by the VEGA Agency of the Ministry of Education, Science, Research and Sport of the Slovak Republic, No. 1/0117/22 (50%).

Institutional Review Board Statement: Not applicable.

Data Availability Statement: The data presented in this study are available on request from the corresponding author.

Acknowledgments: We would like to thank Vladimír Vacek (Department of Mathematics and Descriptive Geometry, Faculty of Wood Sciences and Technology, Technical University in Zvolen, T.G. Masaryka 24, 960 01 Zvolen, Slovakia) for help with statistical evaluation.

Conflicts of Interest: The authors declare no conflict of interest.

References

1. Winandy, J. Effects of waterborne preservative treatment on mechanical properties: A review. In Proceedings of the 91st Annual Meeting of the American Wood-Preservers' Association, New York, NY, USA, 21–24 May 1995; Volume 91, pp. 17–34.
2. Kögl, J.; Hartmann, P.; Beikircher, W. Performance of different fire retardant products applied on Norway spruce tested in a Cone calorimeter. *MATEC Web Conf.* **2013**, *9*, 02011. [CrossRef]
3. Tsapko, Y.; Lomaha, V.; Vasylyshyn, R.; Melnyk, O.; Balanyuk, V.; Tsapko, A.; Bondarenko, O.; Karpuk, A. Establishing regularities in the reduction of flammable properties of wood protected with two-component intumescent varnish. *East.-Eur. J. Enterp. Technol.* **2022**, *3*, 63–71. [CrossRef]
4. Lebow, S. *Leaching of Wood Preservative Components and Their Mobility in the Environment—Summary of Pertinent Literature*; General Technical Report FPL–GTR–93; Department of Agriculture, Forest Service, Forest Products Laboratory: Madison, WI, USA, 1996.
5. Cook, D.R. Making the connection; corrosive effects of new wood preservatives on steel fasteners and connections. *Struct. Eng.* **2003**, *5*, 28–31.
6. Zelinka, S.L.; Glass, S.V. Water Vapor Sorption Isotherms for Southern Pine Treated with Several Waterborne Preservatives. *J. Test. Eval.* **2010**, *38*, 1–5. [CrossRef]
7. Sonowal, J.; Gogoi, P.K. Dimensional Stability, Thermal Degradation and Termite Resistant Studies of Chemically Treated Wood. *Int. J. Chem.* **2010**, *2*, 218–225. [CrossRef]
8. Tsapko, Y.; Tsapko, A. Establishment of the mechanism and fireproof efficiency of wood treated with an impregnating solution and coatings. *East.-Eur. J. Enterp. Technol.* **2017**, *3*, 50–55. [CrossRef]
9. Ayrilmis, N. Effect of fire retardants on internal bond strength and bond durability of structural fiberboard. *Build. Environ.* **2007**, *42*, 1200–1206. [CrossRef]
10. Kartal, S.N.; Hwang, W.-J.; Imamura, Y. Combined effect of boron compounds and heat treatments on wood properties: Chemical and strength properties of wood. *J. Mater. Process. Technol.* **2008**, *198*, 234–240. [CrossRef]
11. Li, P.; Zhang, Y.; Zuo, Y.; Lu, J.; Yuan, G.; Wu, Y. Preparation and characterization of sodium silicate impregnated Chinese fir wood with high strength, water resistance, flame retardant and smoke suppression. *J. Mater. Res. Technol.* **2020**, *9*, 1043–1053. [CrossRef]
12. Tribulová, T.; Kačík, F.; Evtuguin, D.V.; Čabalová, I.; Ďurkovič, J. The effects of transition metal sulfates on cellulose crys-tallinity during accelerated ageing of silver fir wood. *Cellulose* **2019**, *26*, 2625–2638. [CrossRef]
13. Sun, L.; Han, J.; Wu, J.; Huang, W.; Li, Y.; Mao, Y.; Wang, L.; Wang, Y. Cellulose pretreatment with inorganic salt hydrate: Dissolution, regeneration, structure and morphology. *Ind. Crop. Prod.* **2022**, *180*, 114722. [CrossRef]
14. Malešič, J.; Marinšek, M.; Cigić, I.K. Evaluation of Bookkeeper mass deacidification based on historical book papers. *Cellulose* **2002**, *29*, 6889–6905. [CrossRef]

15. Zou, X.; Uesaka, T.; Gurnagul, N. Prediction of paper permanence by accelerated ageing. I. Kinetic analysis of the ageing process. *Cellulose* **1996**, *3*, 243–267. [CrossRef]
16. Dupont, A.-L.; Tetreault, J. Cellulose Degradation in an Acetic Acid Environment. *Stud. Conserv.* **2000**, *45*, 201–210. [CrossRef]
17. Eronen, P.; Österberg, M.; Jääskeläinen, A.-S. Effect of alkaline treatment on cellulose supramolecular structure studied with combined confocal Raman spectroscopy and atomic force microscopy. *Cellulose* **2009**, *16*, 167–178. [CrossRef]
18. Hon, D.N.; Shiraishi, N. *Wood and Cellulosic Chemistry*, 2nd ed.; CRC Press: New York, NY, USA, 2000; p. 928.
19. Han, J.; Seo, Y. Comparison of Cellulose Dissolution Behavior by Alkaline and Sulfuric Acid Solvents and Their Films' Physical Properties. *Materials* **2021**, *14*, 5273. [CrossRef]
20. Thango, B.A. Feedforward Artificial Neural Network (FFANN) Application in Solid Insulation Evaluation Methods for the Prediction of Loss of Life in Oil-Submerged Transformers. *Energies* **2022**, *15*, 8548. [CrossRef]
21. Ghoneim, S. Determination of Transformers' Insulating Paper State Based on Classification Techniques. *Processes* **2021**, *9*, 427. [CrossRef]
22. Eugenio, M.E.; Ruiz-Montoya, M.; Martín-Sampedro, R.; Ibarra, D.; Díaz, M.J. Influence of Cellulose Characteristics on Pyrolysis Suitability. *Processes* **2021**, *9*, 1584. [CrossRef]
23. Špérová, M.; Nasadil, P.; Průšová, A.; Kučerík, J. A hint on the correlation between cellulose fibers polymerization degree and their thermal and thermo-oxidative degradation. *J. Therm. Anal. Calorim.* **2012**, *110*, 71–76. [CrossRef]
24. Dupont, A.-L.; Mortha, G. Comparative evaluation of size-exclusion chromatography and viscometry for the characterisation of cellulose. *J. Chromatogr. A* **2004**, *1026*, 129–141. [CrossRef] [PubMed]
25. Malešič, J.; Kraševec, I.; Cigić, I.K. Determination of Cellulose Degree of Polymerization in Historical Papers with High Lignin Content. *Polymers* **2021**, *13*, 1990. [CrossRef]
26. *ISO 187*; Paper, Board and Pulps—Standard Atmosphere for Conditioning and Testing and Procedure for Monitoring the Atmosphere and Conditioning of Samples. ISO: Geneva, Switzerland, 2022.
27. *ISO 5630-3*; Paper and Board—Accelerated Ageing. Part 3: Moist Heat Treatment at 80 °C and 65% Relative Humidity. ISO: Geneva, Switzerland, 1996.
28. Pitkänen, L.; Sixta, H. Size-exclusion chromatography of cellulose: Observations on the low-molar-mass fraction. *Cellulose* **2020**, *27*, 9217–9225. [CrossRef]
29. Josefsson, T.; Lennholm, H.; Gellerstedt, G. Changes in cellulose supramolecular structure and molecular weight distribu-tion during steam explosion of aspen wood. *Cellulose* **2001**, *8*, 289–296. [CrossRef]
30. Hubbell, C.A.; Ragauskas, A.J. Effect of acid-chlorite delignification on cellulose degree of polymerization. *Bioresour. Technol.* **2010**, *101*, 7410–7415. [CrossRef] [PubMed]
31. Wood, B.F.; Conner, A.H.; Hill, C.G., Jr. The effect of precipitation on the molecular weight distribution of cellulose tricarbanilate. *J. Appl. Polym. Sci.* **1986**, *32*, 3703–3712. [CrossRef]
32. Kačík, F.; Podzimek, Š.; Vizárová, K.; Kačíková, D.; Čabalová, I. Characterization of cellulose degradation during accelerat-ed ageing by SEC-MALS, SEC-DAD, and A4F-MALS methods. *Cellulose* **2016**, *23*, 357–366. [CrossRef]
33. Bansa, H. Accelerated Ageing of Paper: Some Ideas on its Practical Benefit. *Restaurator* **2002**, *23*, 106–117. [CrossRef]
34. Havermans, J.B. The impact of European research on paper ageing and preventive conservation strategies. Protection and treatment of paper, leather and parchment. In Proceedings of the EC 5th Conference, Krakow, Poland, 16–18 May 2002; pp. 87–91.
35. Kumar, R.; Mago, G.; Balan, V.; Wyman, C.E. Physical and chemical characterizations of corn stover and poplar solids resulting from leading pretreatment technologies. *Bioresour. Technol.* **2009**, *100*, 3948–3962. [CrossRef]
36. Sweet, M.; Winandy, J. Influence of Degree of Polymerization of Cellulose and Hemicellulose on Strength Loss in Fire-Retardant-Treated Southern Pine. *Holzforschung* **1999**, *53*, 311–317. [CrossRef]
37. Berggren, R. Cellulose Degradation in Pulp Fibres Studied as Changes in Molar Mass Distributions. Ph.D. Thesis, Royal Institute of Technology, Stockholm, Sweden, 2003; 183p.
38. Inagaki, T.; Siesler, H.W.; Mitsui, K.; Tsuchikawa, S. Difference of the Crystal Structure of Cellulose in Wood after Hydro-thermal and Aging Degradation: A NIR Spectroscopy and XRD Study. *Biomacromolecules* **2010**, *11*, 2300–2305. [CrossRef] [PubMed]
39. Lionetto, F.; Del Sole, R.; Cannoletta, D.; Vasapollo, G.; Maffezzoli, A. Monitoring Wood Degradation during Weathering by Cellulose Crystallinity. *Materials* **2012**, *5*, 1910–1922. [CrossRef]
40. Jasiukaityté-Grojzdek, E.; Kunaver, M.; Poljanšek, I. Influence of cellulose polymerization degree and crystallinity on kinetics of cellulose degradation. *Bioresources* **2012**, *7*, 3008–3027.
41. Kačík, F.; Šmíra, P.; Kačíková, D.; Reinprecht, L.; Nasswettrová, A. Chemical changes in fir wood from old buildings due to ageing. *Cellul. Chem. Technol.* **2014**, *48*, 79–88.
42. Chamberlain, D.C.; Priest, D.J. Comparison of physical and chemical methods for analysing cellulose degradation following artificial ageing treatment. *Cellul. Chem. Technol.* **1998**, *32*, 35–41.
43. Ding, H.Z.; Wang, Z.D. On the degradation evolution equations of cellulose. *Cellulose* **2008**, *15*, 205–224. [CrossRef]
44. Jablonský, M.; Kazíková, J.; Botková, M.; Holúbková, S. Kinetic dependences for the decrease of polymerization of paper undergoing accelerated ageing. *Cellul. Chem. Technol.* **2012**, *46*, 625–630.
45. Emsley, A.; Ali, M.; Heywood, R. A size exclusion chromatography study of cellulose degradation. *Polymer* **2000**, *41*, 8513–8521. [CrossRef]

46. Strobin, G.; Wlochowicz, A.; Ciechanska, D.; Boryniec, S.; Struszczyk, H.; Sobczak, S. Molecular parameters of bacterial cellulose. Effect of temperature and pH biosynthesis medium. *Polimery* **2003**, *48*, 779–783. [CrossRef]
47. Dupont, A.L. Gelatin Sizing of Paper and Its Impact on the Degradation of Cellulose during Aging. Ph.D. Thesis, University of Amsterdam, Amsterdam, The Netherlands, 2003; 251p.
48. Tribulová, T.; Kačík, F.; Evtuguin, D.V.; Čabalová, I. Assessment of Chromophores in Chemically Treated and Aged Wood by UV-Vis Diffuse Reflectance Spectroscopy. *Cellul. Chem. Technol.* **2016**, *50*, 659–667.
49. Bewick, V.; Cheek, L.; Ball, J. Statistics review 9: One-way analysis of variance. *Crit. Care* **2004**, *8*, 130–136. [CrossRef] [PubMed]

polymers

MDPI

Article

Chromophores' Contribution to Color Changes of Thermally Modified Tropical Wood Species

Tereza Jurczyková [1,*], Ondřej Šárovec [1], František Kačík [2], Kateřina Hájková [1], Tomáš Jurczyk [3] and Richard Hrčka [4]

[1] Department of Wood Processing and Biomaterials, Faculty of Forestry and Wood Sciences, Czech University of Life Science Prague, Kamýcká 129, 165 21 Prague, Czech Republic; ondrejsarovec@seznam.cz (O.Š.); hajkovakaterina@fld.czu.cz (K.H.)

[2] Department of Chemistry and Chemical Technology, Faculty of Wood Sciences and Technology, Technical University in Zvolen, T. G. Masaryka 24, 960 01 Zvolen, Slovakia; kacik@tuzvo.sk

[3] TIBCO Software s.r.o., Klimentská 1216/46, 110 00 Prague, Czech Republic; tjurczyk@tibco.com

[4] Department of Wood Science, Faculty of Wood Sciences and Technology, Technical University in Zvolen, T. G. Masaryka 24, 960 01 Zvolen, Slovakia; hrcka@tuzvo.sk

* Correspondence: jurczykova@fld.czu.cz

Abstract: This work examines the effect of thermal modification temperature (180, 200, and 220 °C) in comparison with reference (untreated) samples on selected optical properties of six tropical wood species—Sp. cedar (*Cedrala odorata*), iroko (*Chlorophora excelsa*), merbau (*Intsia* spp.), meranti (*Shorea* spp.), padouk (*Pterocarpus soyauxii*), and teak (*Tectona grandis*). The main goal is to expand the existing knowledge in the field of wood thermal modification by understanding the related degradation mechanisms associated with the formation of chromophoric structures and, above all, to focus on the change in the content of extractive substances. For solid wood, the CIELAB color space parameters (L^*, a^*, b^*, and ΔE^*), yellowness (Y), ISO brightness, and UV-Vis diffuse reflectance spectra were obtained. Subsequently, these wood samples were extracted into three individual solvents (acetone, ethanol, and ethanol-toluene). The yields of the extracted compounds, their absorption spectra, and again L^*, a^*, b^*, ΔE^*, and Yi parameters were determined. With increasing temperatures, the samples lose brightness and darken, while their total color difference grows (except merbau). The highest yield of extractives (mainly phenolic compounds, glycosides, and dyes) from thermally modified samples was usually obtained using ethanol. New types of extractives (e.g., 2-furaldehyde, lactones, formic acid, some monomer derivatives of phenols, etc.) are already created around a temperature of 180 °C and may undergo condensation reactions at higher temperatures. For padouk, merbau, teak, and partially iroko modified at temperatures of 200 and 220 °C, there was a detected similarity in the intensities of their UV-Vis DR spectra at the wavelength regions corresponding to phenolic aldehydes, unsaturated ketones, quinones, stilbenes, and other conjugated carbonyl structures. Overall, a statistical assessment using PCA sorted the samples into five clusters. Cluster 3 consists of almost all samples modified at 200 and 220 °C, and in the other four, the reference and thermally modified samples at 180 °C were distributed. The yellowness of wood (Y) has a very high dependence (r = 0.972) on its brightness (L^*) and the yellowness index of the extractives in acetone Yi(Ac), whose relationship was described by the equation Y = −0.0951 × Y(Ac) + 23.3485.

Keywords: ThermoWood; chemical degradation; extractives; optical properties; UV-Vis diffuse reflectance; total color difference

check for updates

Citation: Jurczyková, T.; Šárovec, O.; Kačík, F.; Hájková, K.; Jurczyk, T.; Hrčka, R. Chromophores' Contribution to Color Changes of Thermally Modified Tropical Wood Species. *Polymers* **2023**, *15*, 4000. https://doi.org/10.3390/polym15194000

Academic Editor: Antonio Pizzi

Received: 30 July 2023
Revised: 29 September 2023
Accepted: 3 October 2023
Published: 5 October 2023

1. Introduction

Wood has multipurpose uses that are impossible to deny and it is difficult to note all of them in our daily lives. Wood is used commercially worldwide for construction and fencing, furniture, floors, household uses, shipbuilding, artworks, musical instruments, wooden toys, etc. Among the characteristics that are evaluated first is its color—a visual perception influenced by the spectral composition of the reflected light rays. The incident rays are partially absorbed by the wood chromophores, i.e., parts of the molecules capable of integrating with ultraviolet (UV) and visible (Vis) wavelengths of electromagnetic radiation [1]. As a result, chromophoric structures are responsible for the color of wood, being part of all its main components (lignin as well as polysaccharide parts) and extractive substances, more significantly represented in tropical wood species [2]. Their usually low concentration is increased by the depolymerization, condensation, or oxidation of individual wood components due to stressful situations [3–5].

One such situation is the exposure of wood to heat, which uses thermal modifications with the aim of improving its selected properties compared to unmodified wood and increasing its application. This process usually takes place in a thermal chamber in an environment of air, steam, nitrogen gas, hot oil, or vacuum, most often at a temperature between 160 and 240 °C. The structural arrangement of thermally treated wood occurs firstly at its molecular chemical level, mainly due to the degradation of hemicelluloses, the cleavage of hydroxyl groups from hemicelluloses and lignin, and the 3D spatial crosslinking of hemicelluloses with highly thermally stable lignin. At the same time, the essence of thermal modification is targeted intervention in its chemical composition and structure, which also affects the color and representation of chromophoric structures [6–9]. The discoloration reactions that occur during heat treatment are often explained by the formation of colored oxidation and degradation products concerning the main wood components and extractives. In lignin, there is a cleavage of β-O-4 linkages, resulting in a higher concentration of phenolic groups and a demethoxylation, leading to a more condensed structure via lignin autocondensation. Changes in phenolic extractives have been identified as another probable reason for the discoloration of wood. Temperature and oxygen cause polymerization of ellagitannins, while vacuum drying effectively prevents discoloration due to a lack of oxygen. The first part of the color change process is because of the formation of chromophoric groups such as carbonyl and carboxyl groups resulting mainly from the degradation of carbonyl, biphenyl, and ring-conjugated double bond structures in lignin. The release of acids and formation of quinoid compounds and carboxylic groups during heat treatment were also confirmed [10–13].

Thermally modified wood is not just an academic topic these days. Thermal wood treatment has already been penetrating the market for several years. The most common heat-modified woods in North America and Europe include red oak, Norway spruce, European and North American white ash, hemlock, radiata and Scots pine, beech, poplar, etc. The typical dark tones of these heat-treated wood species increase their economic value [14–18]. There is still some input into research and development. For example, the thermal modification of tropical wood species is a relatively young topic, but it is the subject of interest for many researchers: Kačíková et al. [19,20] dealt with chemical changes in lignin due to thermal modification in meranti, padouk, merbau, teak, and iroko wood. Gašparík et al. [21] and Gaff et al. [22] evaluated the impact of thermal modification on color, mechanical, and chemical changes in teak, meranti, padouk, merbau, iroko, and mahogany wood. Korkut [23] assessed the performance of three thermally treated tropical woods (sapele, limba, and iroko) commonly used in Turkey based on changes in density, weight loss, swelling, color difference, surface, and physical properties. Mburu et al. [24] characterized the physicochemical properties of heat-treated Southern silky oak (*Grevillea robusta*), a tropical wood species with a low natural durability, to improve its decay resistance. Lengowski et al. [25] compared the anatomical, chemical, physical, mechanical, colorimetric, and thermal stability properties of thermally treated teak (*Tectona grandis*) with the characteristics of the untreated teakwood to understand the effects of the thermal modification technique beyond color changes. Esteves et al. [26] investigated the changes in the content and composition of lignin, cellulose, hemicelluloses, and extractives in dichloromethane, ethanol, and water in thermally modified African teak (*Pericopsis elata*) and wild mahogany (*Tapirira guianensis*).

However, none of these or other works in the field of wood thermal modification specifically focus on the topic of color in connection with the chromophoric structures associated with it. At a constant heat treatment duration, an increase in treatment temperature determines the intensity of changes to a darker color, however, reported results provide ambiguous data as to whether and to what degree the extension of heat treatment duration at a constant temperature at the same time changes the color and properties of the wood [27]. An important carrier of color can be chromophores contained not only in the wood itself but in the case of tropical wood species to a greater extent in its extractive substances, too. Many works report correlations between color change and wood mechanical properties, but these dependencies are questionable in the case of wood species with a large content of different extractives, e.g., tropical wood species and black locust [2,21,22,28].

Goals

The objective of this work is to investigate the color properties of six thermally modified tropical woods by different spectroscopic techniques in order to obtain a modeling of the changes validated by rigorous statistical methods together with principal component analysis. These results may contribute to the understanding of the chemical degradation mechanism of thermally modified wood by monitoring the changes in the extractive content yield and the development of relevant chromophores. Because the aesthetic aspect is often the prevailing criterion of a decision made by the end user, the color of thermally treated wood should be controlled during the production process, too. The results of this experiment should show certain trends in color development in the production process of thermally modified tropical wood species, which can be optimized with this in mind.

This new approach was applied for evaluating samples of Spanish cedar (*Cedrala odorata*), iroko (*Chlorophora excelsa*), merbau (*Intsia* spp.), meranti (*Shorea* spp.), padouk (*Pterocarpus soyauxii*), and teak (*Tectona grandis*), thermally modified at temperatures of 180 °C, 200 °C, and 220 °C, including the reference unmodified samples.

Other partial goals of this research leading to the main objective are

1. Determination of the influence of the thermal modification temperature on individual parameters (L^*, a^*, and b^*) of the CIELAB color space, including total color difference (ΔE^*) and on UV-Vis diffuse reflectance parameters, i.e., k/s intensity, yellowness (Y), and ISO brightness, of selected wood species.
2. Determination of the effect of the thermal modification temperature on individual parameters (L^*, a^*, and b^*) of the CIELAB color space and yellowness index (Yi) of extractives isolated from these woods into acetone, ethanol, and ethanol-toluene and their mass yields (wt.%).
3. Quantitative and qualitative evaluation of extracted chromophoric compounds in the above-mentioned individual solvents using UV-Vis spectroscopy.
4. Statistical evaluation of the measured data of solid samples and extractives.

2. Materials and Methods

2.1. Materials

Six different tropical woods, which are summarized in Table 1, were used in this study [19,29,30].

2.2. Sample Preparation

2.2.1. Thermal Modification

Radially cut samples with dimensions of 20 × 20 × 300 mm (r × d × l) were divided into 4 groups for each type of wood. One group was left as a reference (denoted as 20 °C) and the other three were intended for thermal modification at temperatures of 180, 200, and 220 °C. All samples were pre-conditioned in a Memmert climate chamber at a temperature of 20 ± 1 °C and a relative air humidity of 65 ± 3% to achieve an equilibrium moisture content of 12 ± 1%.

Table 1. Characteristics of the analyzed wood species.

Trade Name (Samples Marking)	Latin Name	Occurrence	Color/Appearance
Spanish cedar (SC)	*Cedrela odorata*	Central and South America and the Caribbean; also, plantations	heartwood is light pink to reddish brown
Iroko (IR)	*Chlorophora excelsa*	Africa, especially in the Ivory Coast	heartwood is yellow to golden or medium brown, with color tending to darken over time; pale yellow sapwood is demarcated from the heartwood
Padouk (PA)	*Pterocarpus soyauxii*	Central and West Africa (e.g., Congo, Angola)	fresh heartwood is blood-red in color, changing to a dark purplish-brown with a red tinge over time
Meranti (ME)	*Shorea* spp.	Southeast Asia (e.g., Laos, Philippines)	core varies depending on the origin and species, ranging from brownish-pink to red to dark red; it darkens further in the light or, on the contrary, pales
Merbau (MB)	*Intsia* spp.	Southeast Asia (e.g., Burma, Indonesia), East Africa, and Australia (New Guinea)	fresh cut orangish-brown color, which ages to a darker reddish-brown; small yellow mineral deposits that can cause staining
Teak (TE)	*Tectona grandis*	Native to southern Asia; widely on plantations throughout tropical regions of Africa, Asia, and Latin America	heartwood tends to be a golden or medium brown, with color darkening with age; often with dark brown or black stripes (2–8 mm wide)

The thermal modification of the test specimens was carried out in a Katres thermal chamber according to the principle of the Thermowood® method (Thermo S for 180 °C and Thermo D for 200 and 220 °C) in three phases [31]:

1. Rapid heating to a temperature of 100 °C followed by gradual heating to a temperature of 130–140 °C for about 3–4 h (heating step 8–10 °C per hour) depending on the type of wood, its thickness, and moisture content. During this phase, the wood was dried to a nearly 0% moisture content.
2. After reaching the required temperature (180, 200, and 220 °C), the thermal modification itself takes place for 3 h. Sensors control the constant temperature level of the samples' cross-section. During heat treatment, water vapor was introduced to the chamber, which served as a protective medium against ignition and favorably influenced the ongoing chemical reactions.
3. The final stage is a cooling and regeneration process (moistening) lasting 5–15 h. By gradually watering the wood and dropping the temperature to 80–90 °C, the samples return to the desired moisture level of approximately 5–7%. The chamber was opened at a temperature of 40 °C to avoid thermal shock in the modified wood.

2.2.2. Preparation of Wood Samples for Analyses

From all reference and thermally modified woods, seven samples with dimensions of 3 × 20 × 30 mm were cut for color measurement of solid wood samples using a spectrophotometer device (Figure 1). After cutting, the samples were stored in darkness and conditioned at 65% relative humidity and a temperature of 21 °C to reach a 12% moisture content. The radial surface was used for the measurements.

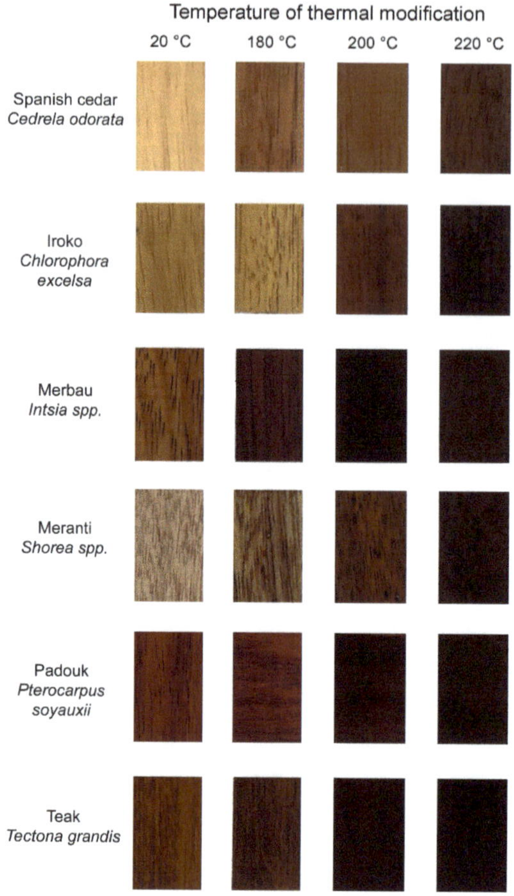

Figure 1. The appearance of reference and thermally modified samples at temperatures of 180, 200, and 220 °C.

Moreover, a part of each original tested series was also disintegrated using a knife mill (MF 10 BASIC, Germany) and then sieved through a standardized set of nets. To determine the extractives content, conditioned samples of the fraction 0.425–0.250 mm (40–60 mesh) were used. This wood sampling and preparation was carried out according to TAPPI T 257 cm-02 [32] and ASTM D 1107—96 [33] standard procedures. From each test series, 2 samples were prepared in this way, on which, the measurement took place.

2.3. Methods

All experiments were performed according to the scheme in Figure 2.

2.3.1. Measurement of Color Parameters Including UV-Vis Diffuse Reflectance Spectra of Solid Wood Samples

The characteristics describing the color of the sample (L^*, a^*, and b^*) as well as the characteristics of diffuse reflectance SCI (specular component included), which are k/s parameters for 360–740 nm, yellowness Y, and ISO brightness were measured using a benchtop-type spectrophotometer Konica Minolta CM-5 (Konica Minolta, Osaka, Japan). The parameters ΔL^*, Δa^*, Δb^*, and ΔE^*, and brightness reversion were subsequently calculated from the measured data.

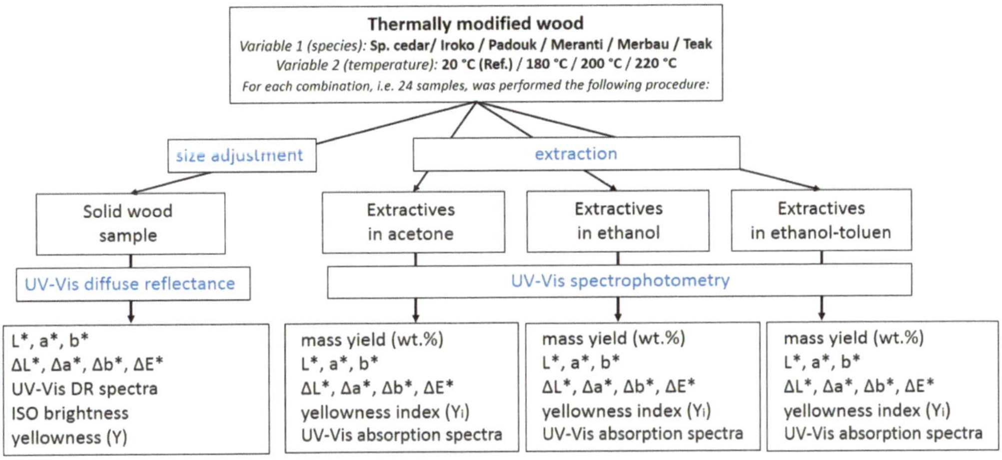

Figure 2. Design of an experiment for the analysis of color and chromophoric groups in thermally modified wood.

Before measurement, samples were conditioned at a room temperature of 23 ± 1 °C and a relative air humidity of 50 ± 2% according to standard ISO 187 [34].

The measurement of the color change took place according to the principle of spectral reflection with a wavelength resolution of 10 nm and a measurement diameter spot of 3 mm. The spectrophotometer was calibrated before each measurement according to the procedure recommended by the producer guided by a computer using the Color Data Software CM-S100W SpectraMagicTNNX (Konica Minolta, Osaka, Japan).

The color parameters according to ISO 11664-4 method [35] were determined for reference and thermally modified wood samples (for each series on seven replicates) by the CIELAB system, in which L^* denotes lightness (from 0 black to 100 white), a^* denotes color (−a green to +a red), and b^* denotes color (-b blue to +b yellow). ΔL^*, Δa^*, and Δb^* represent the differences in the individual coordinates of the color space before and after thermal modification. The total color differences ΔE^* for individual samples were then evaluated according to ISO 11664-6 [36] and following Equation (1):

$$\Delta E^* = \sqrt{(\Delta L^*)^2 + (\Delta a^*)^2 + (\Delta b^*)^2}, \quad (1)$$

The reflectance spectra were measured in the wavelength range of 360 to 740 nm under illuminant D65 with specular components included. The 2° standard observer was set before the computation of color coordinates. The reflectance spectra were converted into k/s spectra using the Kubelka–Munk equation according to Equation (2) [37]:

$$\frac{k}{s} = \frac{(1-R)^2}{2R}, \quad (2)$$

where R is the measured reflectance and k and s are the absorption and scattering coefficients, respectively. Four replicates were analyzed for each sample series and the resulting spectrum is an average of them. Yellowness and brightness values were provided directly by the instrument, Konica Minolta CM-5.

2.3.2. Extraction

For the determination of extractive substances, the following solvents that are capable of extracting various substances from wood were chosen: acetone (C_3H_6O), ethanol (C_2H_6O), and ethanol-toluene binary mixture in a volume ratio 7:3 (C_2H_6O–C_7H_8).

The complete extraction method with acetone was performed according to TAPPI T 280 pm-99, Acetone Extractives of Wood and Pulp [38], and methods with ethanol (95%) and ethanol–toluene (7:3) were performed according to TAPPI T 204 cm-97, Solvent Extractives of Wood and Pulp [39]. The moisture content of each sample was determined prior to extraction according to TAPPI T 264 cm-97: Preparation of Wood for Chemical Analysis [40].

2.3.3. UV-Vis Spectrophotometry Measurement of Extractives in Solvents

The analysis of samples of dissolved extractives in the liquid phase in individual solvents was carried out on a DR-6000 UV-VIS spectrophotometer (Hach, Colorado). The instrument was calibrated automatically on a cuvette with distilled water. First of all, the characteristics of used pure solvents were measured and then always subtracted from measured color parameters as well as spectra of the corresponding dissolved extract sample.

According to the CIELAB system, the characteristics describing the color (L^*, a^*, and b^*) were determined for all samples (in two repetitions) and subsequently, the parameters ΔL^*, Δa^*, Δb^*, and ΔE^* were calculated. Measured data were evaluated similarly to the solid samples. Thus, the color deviation ΔE^* was calculated using Equation (1).

Yellowness index (Yi) values were provided by the instrument DR-6000 at the wavelength of 720 nm.

The absorbance spectra of samples in vitro were recorded in the wavelength range of 190 to 900 nm. At specific wavelengths, the following chromophoric structures were observed and evaluated: phenols at 460 nm, tannins at 700 nm, organics at 254 nm, volatile organic acids at 495 nm, and suspended solids at 810 nm.

2.3.4. Statistical Evaluation

Statistical evaluation of the measured data of solid samples and extracts was performed using TIBCO Statistica software version 14 (TIBCO Software Inc., Palo Alto, CA, USA) for the indicated statistical methods: cluster analysis, principal component analysis (PCA), correlation (Pearson's correlation coefficient), and regression analysis.

3. Results and Discussion

3.1. Measurement of Color Parameters of Solid Wood Samples Using UV-Vis Diffuse Reflectance

The color parameters of a^* and b^* acquire only positive values for all samples and their colors therefore have shades of red (a^*) and yellow (b^*) as shown in Table 2. Padouk wood, namely PA20 ($a^* = 29.00$) and PA180 ($a^* = 26.53$), is the most red-colored of all the samples. The a^* value of the other reference tropical wood samples ranges from 9.81 to 14.77. Thermally modified samples IR220, PA220, TE220, and ME200 have the lowest values of parameter a^* in the range of 2.91–4.24. Regarding the redness parameter, we can observe that its greatest decrease was recorded between 200 and 220 °C for all samples, except for merbau. There is no similar trend for brightness and yellowness. The samples of padauk (PA20, PA180) together with TE20 and IR180 have the highest values of the b^* parameter (30.57–26.19). The average range of these values for the other reference samples is 17.91–24.93. The lowest values of the b^* parameter (4.17–5.24) are again, as with the a^* parameter, observed for samples IR220, PA220, TE220, and ME200. The SC20 sample (66.32) and also the MT20, IR20, and IR180 samples ($L^* > 50$) have the highest brightness (L^*). The brightness of the other reference samples is within a narrow range of 39–41. The lowest brightness values ($L^* < 20$) are shown by samples IR220, ME220, TE220, MB200, and TE200. L^* values clearly decrease with the increase in thermal treatment for all samples of tropical woods compared to the reference sample. This trend was also observed in the work of many other authors [41–44] investigating other wood species. Monaco et al. [41], who have studied the effect of thermal treatment on chestnut wood (*Castanea sativa* Mill.) at temperatures of 140, 170, and 200 °C, moreover observed that ΔE values, calculated for the three thermal treatments in respect to untreated samples, depend mainly on L^* variations in the case of their samples because the chromatic coordinates a* and b* did not change significantly except for an increase clearly observed at 170 °C and corresponding to an increase in both yellow and red components.

Table 2. Measured (L^*, a^*, b^*, ISO brightness, and yellowness) and calculated (ΔE^*) color values for samples of selected species of thermally modified tropical wood.

Wood Sample	Modification Temperature (°C)	L^*	a^*	b^*	ΔE^*	ISO Brightness	Yellowness (Y)
Sp. cedar (SC)	20 (ref.)	66.32 ± 0.34	12.24 ± 0.14	24.10 ± 0.14	-	20.06 ± 0.19	34.89 ± 0.24
	180	40.42 ± 0.68	14.67 ± 0.25	23.78 ± 0.46	26.02	7.99 ± 0.16	14.95 ± 0.27
	200	35.49 ± 0.79	13.33 ± 0.09	20.49 ± 0.14	31.06	6.74 ± 0.15	11.82 ± 0.24
	220	26.00 ± 0.60	8.30 ± 0.29	13.93 ± 0.47	41.77	5.10 ± 0.04	7.58 ± 0.14
Iroko (IR)	20 (ref.)	52.58 ± 1.26	9.81 ± 0.14	24.93 ± 1.23	-	10.99 ± 0.13	22.64 ± 0.29
	180	52.27 ± 0.61	10.86 ± 0.25	28.69 ± 0.30	3.92	11.88 ± 0.14	23.62 ± 0.26
	200	29.44 ± 0.79	10.58 ± 0.41	14.86 ± 1.16	25.25	5.67 ± 0.04	8.33 ± 0.25
	220	17.52 ± 1.07	2.91 ± 0.36	4.19 ± 0.50	41.32	4.63 ± 0.14	5.03 ± 0.13
Padouk (PA)	20 (ref.)	39.45 ± 0.81	29.00 ± 0.73	30.57 ± 1.35	-	5.32 ± 0.13	10.72 ± 0.37
	180	28.89 ± 0.87	26.53 ± 0.40	26.19 ± 0.62	11.70	4.52 ± 0.03	8.99 ± 0.12
	200	22.03 ± 0.41	9.44 ± 0.42	10.68 ± 0.74	32.90	4.72 ± 0.17	6.08 ± 0.23
	220	22.40 ± 0.51	3.92 ± 0.30	4.17 ± 0.28	40.21	5.51 ± 0.06	6.08 ± 0.09
Meranti (MT)	20 (ref.)	58.96 ± 0.31	10.09 ± 0.21	17.91 ± 0.24	-	18.66 ± 0.30	28.24 ± 0.40
	180	40.25 ± 1.17	9.26 ± 0.49	19.91 ± 1.06	18.83	8.64 ± 0.22	14.54 ± 0.58
	200	30.87 ± 1.12	10.57 ± 0.42	19.68 ± 0.84	28.15	5.75 ± 0.14	10.01 ± 0.36
	220	18.09 ± 1.11	6.13 ± 0.72	10.67 ± 1.79	41.70	4.27 ± 0.04	5.70 ± 0.11
Merbau (MB)	20 (ref.)	39.29 ± 3.26	13.54 ± 0.91	23.50 ± 4.55	-	7.27 ± 0.11	15.19 ± 0.21
	180	24.13 ± 0.29	11.70 ± 0.33	12.11 ± 0.39	19.06	5.40 ± 0.05	7.42 ± 0.09
	200	17.43 ± 0.91	4.24 ± 0.51	5.24 ± 0.13	29.97	4.65 ± 0.03	5.24 ± 0.05
	220	22.52 ± 0.42	6.06 ± 0.31	7.94 ± 0.54	24.07	5.22 ± 0.06	6.54 ± 0.05
Teak (TE)	20 (ref.)	40.38 ± 1.17	14.77 ± 0.23	30.19 ± 0.39	-	6.19 ± 0.09	13.62 ± 0.21
	180	33.18 ± 1.08	11.32 ± 0.22	21.05 ± 0.15	12.14	6.26 ± 0.08	11.82 ± 0.16
	200	19.49 ± 0.92	8.87 ± 0.54	10.27 ± 0.69	29.47	4.33 ± 0.06	5.56 ± 0.13
	220	15.05 ± 0.44	4.17 ± 0.31	5.21 ± 0.41	37.13	4.03 ± 0.12	4.47 ± 0.11

In the next step, a cluster analysis (Figure 3) was performed for the parameters L^*, a^*, and b^*, which divided all samples into six clusters according to similarity. The first group includes six thermally modified samples, MT220, SC220, PA200, TE200, IR200, and MB180. The second cluster consists of samples modified at higher temperatures, PA220, TE220, IR220, MB220, and MB200. The third cluster includes MT200, MT180, SC200, SC180, TE180, and MB20. It is interesting that this group also includes the merbau reference sample, which has similar optical properties to thermally modified samples of other woods at temperatures of 180 °C and 200 °C. At the same time, there is twice the similarity of two samples of the same wood and at different modification temperatures (MT200 and MT180, and SC200 and SC180). The fourth cluster includes IR180, IR20, and TE20, and the fifth includes MT20 and SC20. The samples of both clusters differ from the others by a higher L^* value and from each other mainly by the values of a^* and b^*. The last cluster consists of samples PA20 and PA180 with a high redness value.

The graph in Figure 4 shows the development of the average values of the calculated total color difference (ΔE^*), i.e., the total color change compared to the respective reference samples. In all cases, except for ME220, there is an increase in ΔE^* along with an increasing thermal modification temperature. In the case of merbau, the highest ΔE^* is at 200 °C, and a decrease is noted at 220 °C. A temperature of 180 °C had the greatest impact on the overall color change for SP180 (ΔE^* = 26) and the least for IR180, which remained without a significant color change (ΔE^* = 4). At a temperature of 200 °C, the largest color change occurred at PA200 (ΔE^* = 33) and the smallest again at IR200 (ΔE^* = 25.5). It can be said that the ΔE^* caused by the temperature of 200 °C of most samples, except iroko, is very similar (32.90–28.15), and compared to the other temperatures, the smallest fluctuations in values are observed within this group. The highest modification temperature of 220 °C caused color changes at a similar level ($\Delta E^* \approx 40$–42) for SC220, IR220, PA220, and MT220. ME220 was relatively least affected by this temperature ($\Delta E^* = 24$). The most significant increase in ΔE^* occurred in the case of SC, MT, and MB during thermal modification from 20 °C to 180 °C. For IR, PA, and TE, this increase was the highest from 180 °C to 200 °C.

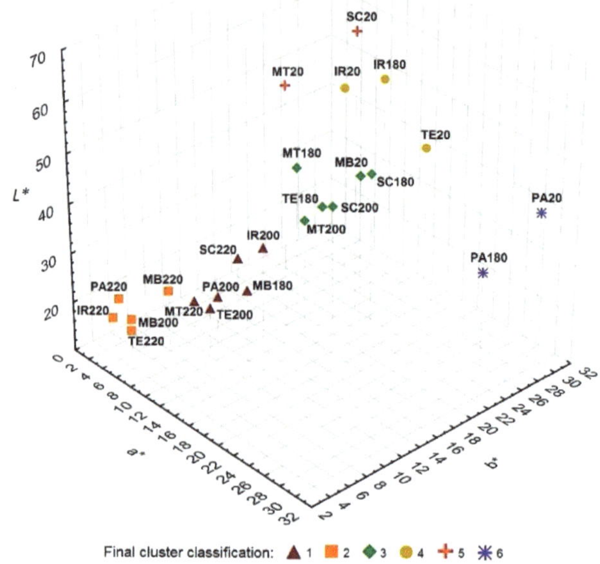

Figure 3. A 3D graph of individual L^*, a^*, *and* b^* values of the color space defining the color of selected types of thermally modified wood samples—cluster analysis.

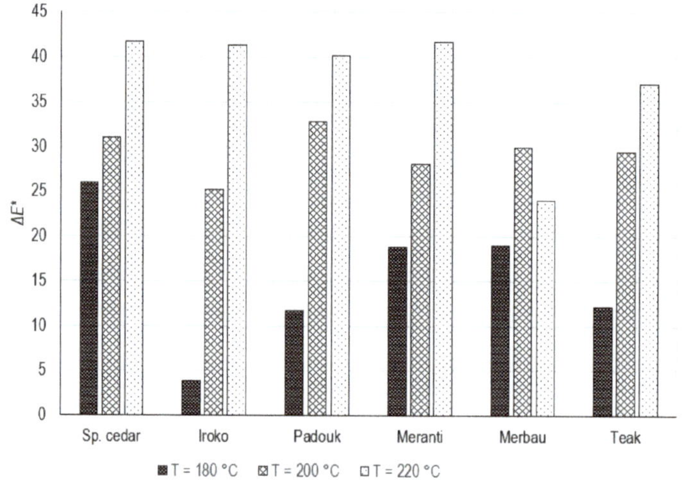

Figure 4. Total color difference (ΔE^*) of selected types of thermally modified tropical wood samples at temperatures of 180, 200, and 220 °C.

The measured yellowness (Y) of the samples is a number calculated from the spectrophotometric data and indicates the degree of deviation of the wood color from colorless or the preferred white towards yellow (ASTM E313 [45]). ISO brightness indicates the actual percentage of light reflected from the wood sample at 457 nm (Tappi North America, 2017). In general, within one series of samples, both parameters decrease with increasing modification temperature. Small deviations from this trend, i.e., a slight increase in these values compared to a lower modification temperature, are observed in the case of ISO brightness for samples IR180, TE180, PA200, PA220, and MB220, and in the case of yellowness for samples IR180 and MB220. Determined values of ISO brightness and yellowness listed in Table 2 were further included in the statistical evaluation using a principal component analysis (PCA).

The thermal modification temperature and wood species are the most important factors that affect the final color. When the temperature is higher, the wood color is darker [46]. Like for Baar and Gryc [47], it was also confirmed that lighter wood is more susceptible to changes in surface color and brightness, and dark wood shows the smallest deviations. In general, there is an increase in the parameter ΔE^* as the temperature of thermal modification increases, as in the studies published by Gaff et al. [22] for padouk merbau, mahogany, and iroko wood, Cuccui et al. [48] for teak wood, or Kačík et al. [28] for acacia. The strong correlation between the lightness difference value (ΔL^*) and the content of hemicelluloses was moreover published by Gaff et al. [22]. Such changes also increase ΔE^* [49,50] relative to untreated wood.

3.2. Measurement of UV-Vis DR Spectra and Quantitative Evaluation of Chromophoric Groups in Solid Wood Samples

UV-Vis diffuse reflectance spectra of individual tropical wood samples thermally modified at temperatures of 180, 200, and 220 °C, including a comparison with the reference sample (20 °C), are presented in Figure 5. k/s values, which in the range of this measurement (360–740 nm) indicate the most frequent occurrence of chromophoric groups [5], were monitored and used for their quantitative evaluation. The course of recorded spectra was also evaluated.

The absorption coefficient in the UV region at 360 nm corresponds to the chromophoric structures containing phenolic aldehydes (coniferaldehyde), unsaturated ketones, quinones, stilbenes, other conjugated carbonyl structures, and charge transfer complexes [51,52]. It is known that lignin has only a small contribution to absorption in the visible wavelength region but increases rapidly with decreasing wavelength into the ultraviolet regions [5].

The absorbance in the visible region of blue and violet color (380–520 nm) is recognized as a yellow color. In the visible region of measured spectra, there were no significant absorption bands (except PA180), which would indicate the presence of chromogens at the chromaticity centers. The absorption coefficient at 440 nm can be attributed to the formation of colored quinoid compounds originating not only from the degradation and oxidation of the aromatic hydroxyl groups of lignin but also from aromatic extractives upon heat treatment [53]. The absorption coefficient of all samples (except IR180, TE180, and PA220) after thermal modification at this wavelength is always higher than that for the corresponding reference sample. Equally, the brightness of all thermally modified wood (except the same samples mentioned above) is always lower in comparison with that of the reference wood (note the k/s values at 457 nm).

Similar to the case of Sp. cedar (Figure 5a), it can also be observed for meranti that the value of k/s increases with the increasing temperature of thermal modification and decreases with increasing wavelengths. In addition, the spectra of SC20 and MT20 have very similar courses and k/s values. At 360 nm, the k/s intensity of all thermally modified SC samples is 1.7–1.8× higher than that of SC20, which indicates an increase in the content of chromophoric groups.

For iroko (Figure 5b), the intensity of the k/s spectra does not always increase with temperature, such as for SC and MT. For IR180, there was a decrease in this intensity, even below the values of IR20 in the entire measurement area. The most significant increase in k/s intensity in this group was observed at 457 nm for sample IR200 (2.4× compared to IR180) and sample IR220 (another 1.2× compared to IR200). In the case of iroko, it can be observed that the k/s values of all four samples in the 360 nm region are very close (9.4–10.6). Similar clustering is also evident in the merbau (9.6–10.6) and teak (11.2–11.9) series.

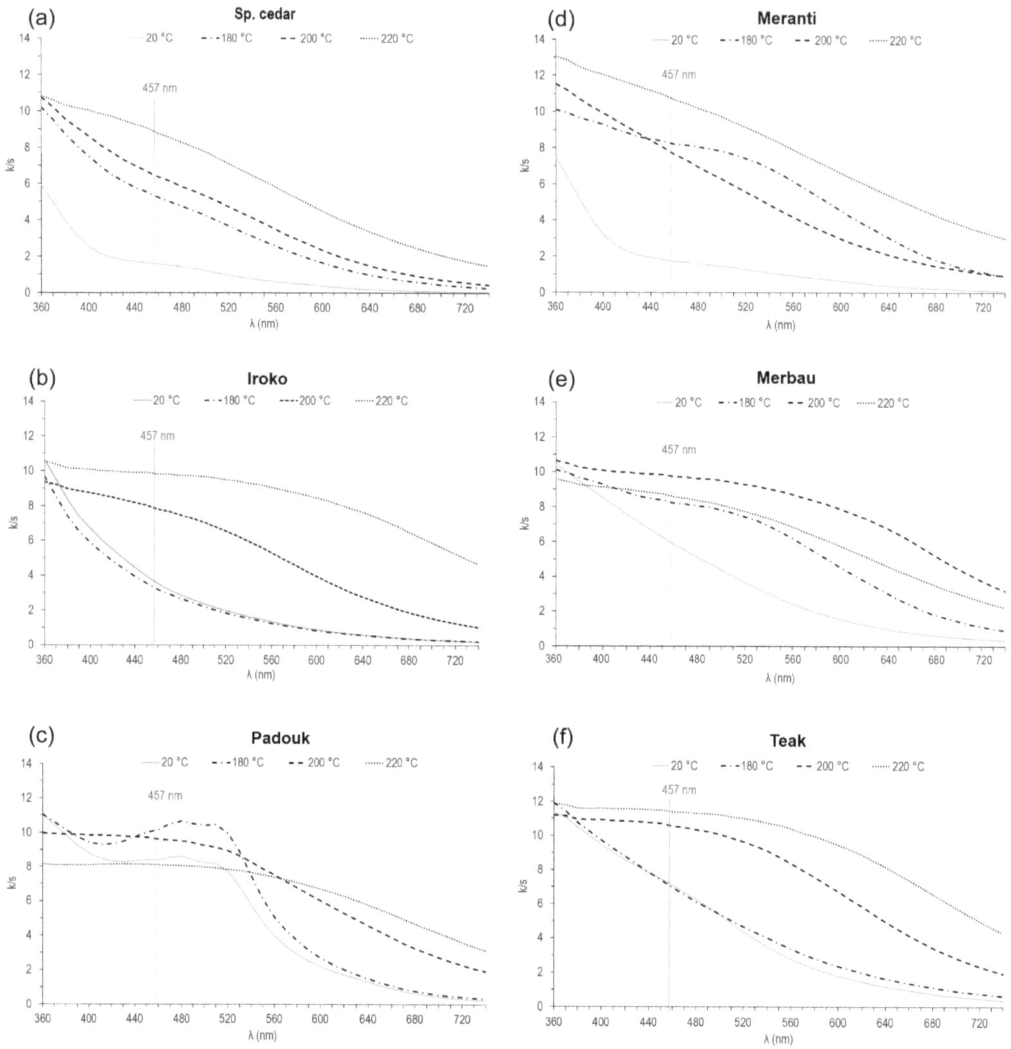

Figure 5. UV-Vis DR spectra of selected tropical wood samples. (**a**) Sp. cedar, (**b**) iroko, (**c**) padouk, (**d**) meranti, (**e**) merbau, and (**f**) teak thermally modified at temperatures of 180, 200, and 220 °C compared to reference unmodified samples.

In the case of padauk (Figure 5c), in the range of wavelengths from 360 to 580 nm, it is not possible to talk about a specific relationship between the intensity of the k/s spectra and the modification temperature, since there is an intermingling of the individual spectra. The course of the spectrum of PA220 (descending and slightly convex) is particularly interesting. Its intensity is the lowest at 360 nm (8.2), compared to the other measured spectra in this group, and remains at this intensity approximately until the wavelength of 520 nm, when the PA20 spectrum is below its level, and from 535 nm, when the PA180 spectrum is also below its level. It is also worth noting the non-standard deviation of the PA180 spectrum in the range of wavelengths 440–540 nm, when its intensity exceeds the spectra of samples modified at higher temperatures. A similar deviation in the same range of wavelengths, but to a lesser extent, is observed in the spectrum of PA20.

MT220 (Figure 5d) achieves the highest k/s values of all sample types at wavelengths of 360 nm (13.1) and 392 nm (12.2).

In the case of MB200 (Figure 5e), k/s values of the monitored wavelengths (360, 440, and 457 nm) are at a very similar level (ca. 10). The same behavior and a low spread of values around 10 is observed in the case of the IR220 and PA200 spectra, too. The spectrum of MB220 is abnormally located below the other spectra of samples from this given series up to a wavelength of approx. 385 nm. Again, a certain similarity with sample PA220 is obvious.

TE220 (Figure 5f) is unique with the highest value of k/s (11.4) at a wavelength of 457 nm. The highest increase in the intensity of k/s spectra at a wavelength of 457 nm is between samples TE180 and TE200 (by 3.5). For all samples in the teak series, very similar trends appeared for iroko samples, too.

The apparent absorption changes in spectra of investigated samples after thermal treatment suggested the generation of some types of chromophores in the degradation, condensation, and oxidation progress, which explains the samples yellowing and their loss of brightness after the action of elevated temperature. However, it is very difficult to differentiate structural differences between these chromophores, because the nature of color change is complex and their composition depends on the extraction method used [54]. Considering that tropical woods have a high content of extractive substances, it can be assumed that thermal oxidation processes play a significant role in this case. Bekhta and Niemz [55] have reported that the color changes caused due to elevated temperatures can be because of the formation of secondary condensation products and/or degradation products, quinoid substances, for instance, which are known to be intensely colored. Keating et al. [52], as well as Chen and Li [53], add that the reactive compounds may include degradation products from the cleavage of α- and β-aryl ether bonds in lignin and degradation products from hydrolyzed hemicelluloses [56,57], too.

3.3. Determination of the Extractive Content in Wood

Wood extractives are natural products extraneous to a lignocellulose cell wall. They are present within a cell wall but are not chemically attached to it. It is known that some tropical wood species have high extractive contents in the heartwood [58]. These extractives influence the color, odor, durability, and technological properties of the different wood species [59]. In general, the amount of lipophilic fatty acids in tropical wood species compared to softwood and hardwood is very low [60]. From a group of hydrophilic constituents, many phenolic compounds (i.e., phenolic acids, flavonoids, sterols, stilbenes, tannins, and lignans), terpenes and terpenoids, free sugars, and sugar alcohols can be detected [61,62]. Some of the most volatile compounds produced during the thermal modification of wood are released, e.g., fats, waxes, furfural, and hydroxymethylfurfural [63], and on the contrary, new extractive substances are created by the degradation of the main components of wood, e.g., dehydration and degradation products of polysaccharides and phenolic compounds like catechol, vanillin, vanillic acid, 3-vanillyl propanol, and coniferyl aldehyde, probably resulting from lignin or phenolic extractives [64].

These remaining substances in the wood can be extracted by several solvents. Polar solvents penetrate well into the cell wall, due to which the cell wall partially swells. Organic salts, oligosaccharides, and carbohydrates can be specifically extracted from wood by water (hot or cold). Ethanol removes the tannins, dyes, or glucosides. Acetone extracts mainly the fatty and resinous acids and sterols. Non-polar solvents (e.g., benzene, petroleum ether, ether, and toluene) extract the fats, fatty acids, and their esters, resins and resin acids, waxes, and sterols from wood. The advantage of using an ethanol–toluene mixture is its non-carcinogenicity compared to benzene. Using this mixture, it is possible to extract well from the wood most phenolic substances (lignin fraction, sterol, tannin, and plobaphen), some organic acids (resin acid, amino acid, vanillic acid, and syringic acid), and other substances such as pigments and syringaldehyde [65,66].

Table 3 lists the percentage mass representation w (wt.%) of individual extractive substances soluble in acetone, ethanol, and ethanol–toluene binary mixture in thermally modified and reference tropical wood samples, including parameters describing their optical properties (ΔE^*—total color difference; Yi—yellowness index) from the next chapter.

Table 3. Determined mass yields (w) and color characteristics (ΔE^*, Yi) for extractives in selected solvents obtained from investigated thermally modified tropical wood species.

Wood Sample	Modification Temperature (°C)	Extractives in Acetone			Extractives in Ethanol			Extractives in Ethanol-Toluene		
		w (wt.%)	ΔE^*	Yi	w (wt.%)	ΔE^*	Yi	w (wt.%)	ΔE^*	Yi
Sp. cedar (SC)	20 (ref.)	5.00 ± 0.00	-	40.70	5.13 ± 1.32	-	180.80	2.31 ± 0.60	-	49.40
	180	1.82 ± 0.13	15.81	62.40	2.95 ± 0.95	60.43	74.70	2.60 ± 0.24	33.85	97.00
	200	2.14 ± 0.06	29.73	85.00	4.37 ± 0.20	51.91	82.80	2.89 ± 0.65	30.99	97.40
	220	0.11 ± 0.00	49.90	107.20	3.30 ± 0.47	24.66	177.00	1.38 ± 0.04	76.42	164.90
Iroko (IR)	20 (ref.)	4.98 ± 0.00	-	28.90	2.44 ± 0.61	-	129.30	3.64 ± 0.82	-	103.40
	180	0.42 ± 0.04	9.93	35.90	1.40 ± 0.23	28.33	89.90	0.35 ± 0.05	36.00	149.60
	200	1.36 ± 0.11	95.50	89.50	3.59 ± 0.00	12.93	137.00	1.27 ± 0.16	24.90	135.10
	220	0.79 ± 0.06	72.33	137.70	2.26 ± 0.08	28.06	189.00	1.27 ± 0.01	63.78	228.10
Padouk (PA)	20 (ref.)	13.20 ± 0.00	-	210.80	4.81 ± 0.17	-	232.60	4.90 ± 0.00	-	271.70
	180	2.45 ± 0.36	16.53	199.40	5.05 ± 0.22	34.40	214.60	3.05 ± 0.03	10.53	251.80
	200	1.20 ± 0.08	35.96	215.70	0.70 ± 0.02	47.92	138.90	2.29 ± 0.26	10.68	242.30
	220	3.05 ± 0.08	28.04	159.00	1.69 ± 0.08	48.18	155.30	1.39 ± 0.07	38.87	189.00
Meranti (MT)	20 (ref.)	4.15 ± 0.00	-	15.00	1.23 ± 0.41	-	29.80	3.27 ± 0.40	-	17.80
	180	0.65 ± 0.08	38.02	54.10	3.57 ± 0.37	60.63	124.20	1.63 ± 0.37	56.62	111.60
	200	1.46 ± 0.12	83.58	125.00	1.57 ± 0.37	79.67	151.40	1.21 ± 0.06	62.49	105.40
	220	1.80 ± 0.02	66.87	112.90	1.70 ± 0.12	86.25	159.90	1.16 ± 0.17	64.50	111.20
Merbau (MB)	20 (ref.)	7.37 ± 0.57	-	116.70	4.86 ± 0.15	-	221.50	7.49 ± 0.02	-	202.80
	180	0.90 ± 0.03	26.44	99.00	4.56 ± 0.04	16.25	240.30	1.57 ± 0.22	10.71	209.60
	200	0.84 ± 0.03	31.94	169.40	4.21 ± 0.09	78.93	277.30	1.40 ± 0.05	15.07	169.80
	220	2.17 ± 0.13	47.37	189.10	1.69 ± 0.38	59.25	91.90	1.75 ± 0.23	36.04	134.20
Teak (TE)	20 (ref.)	4.6 ± 0.03	-	182.80	1.95 ± 0.09	-	153.90	1.27 ± 0.11	-	202.70
	180	2.57 ± 0.03	35.88	132.50	3.06 ± 0.18	27.61	125.20	1.38 ± 0.11	38.83	132.60
	200	1.75 ± 0.46	25.50	134.70	0.60 ± 0.03	35.93	85.40	1.33 ± 0.05	30.35	175.00
	220	0.83 ± 0.08	26.50	149.50	1.83 ± 0.01	22.49	163.70	3.05 ± 0.47	40.77	166.00

In the following graphs in Figure 6, we can see even better the average percentage by weight of the extracted substances from individual samples in the three selected solvents.

Upon closer analysis, it is evident that in the case of the six reference samples, four of them (IR20, PA20, MT20, and TE20) had the greatest gain in extractive substances when acetone was used. The greatest gain for samples thermally modified at 180 °C was, in all six cases, when ethanol was used. For samples modified at higher temperatures (200 and 220 °C), in most cases, ethanol was the most effective during extraction (except for padauk and teak, when acetone or a binary mixture of ethanol–toluene was more efficient).

Specifically, the greatest amount of substances extracted into acetone was obtained for PA20 (13.2 wt.%), even in comparison with other samples extracted in other solvents. The lowest amount of extractives across all samples and solvents (0.1 wt.%) can be observed in SC220 in the case of acetone, which is approximately a twenty-fold decrease compared to the SC200 sample. The amount of extractive substances in samples SC220, IR220, and TE220 is below the limit of 1 wt.%, and compared to the modification at a temperature of 200 °C, their representation decreases. On the contrary, for PA220, MT220, and MB220, an increase in the obtained extractives was observed when the temperature increased from 200 to 220 °C.

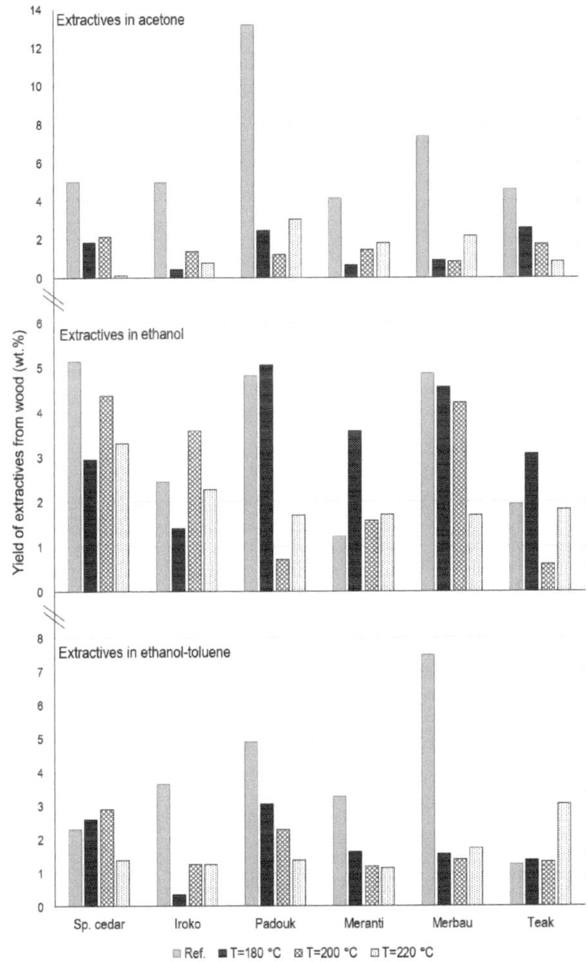

Figure 6. The average amount of extracted substances in acetone, ethanol, and ethanol–toluene from thermally modified and reference samples of selected tropical woods.

The greatest amount of substances soluble in ethanol was obtained for samples SC20 and PA180 (both 5.1 wt.%), MB20 (4.9 wt.%), and PA20 (4.8 wt.%), and the lowest for TE200 (0.6 wt.%) and PA200 (0.7 wt.%).

In the case of extraction into ethanol–toluene, MB20 shows the highest value of the representation of extractive substances and, in comparison with other types of wood and concerning the solvents used, it has the second highest yield. On the contrary, the lowest amount of extractives soluble in this solvent (0.35 wt.%) was obtained from the IR180 sample, when compared to IR20, there was a roughly ten-fold decrease. In addition, as can be seen from the graph in Figure 6c, an interesting group of 13 samples is observed during extraction into this binary mixture, from which a very similar amount of extractives of around 1.3 ± 0.5 wt.% was obtained.

The color stability of heat-treated wood during artificial weathering has long been the subject of much research, as wood treated in this way tends to lose color, especially on the exterior by the action of light, or just by the leaching of extractive substances and degradation products, which can be carriers of chromophoric structures [67–69]. Although many improvements have been made, this issue is still a challenge. In future research, it is therefore necessary to look for new technologies for color stability of heat-treated wood using new surface treatments (varnishes, metal oxides, nanoparticles, etc.) [41].

3.4. *In Vitro Color Measurement of Extractive Compounds Using UV-Vis Spectrophotometry*

The color characteristics (ΔE^*, respectively, L^*, a^*, b^*, and Yi) of the individual extractive samples are not further discussed separately, but they were used for statistical evaluation in the PCA analysis.

The absorbance spectra of all investigated samples in vitro were recorded in the wavelength range of 200 to 900 nm. The spectrum of the respective solvent was always subtracted from the spectra of the dissolved extractive samples. At specific wavelengths, depending on the extractive solvent, the following chromophoric structures could be observed: organic acids at 254 nm, phenols at 460 nm, volatile organic acids at 495 nm, tannins at 700 nm, and suspended solids at 810 nm. The course of the recorded spectrum was evaluated, too.

The most effective solvent in the case of thermally modified samples (mainly at a temperature of 180 °C), which extracted the highest percentage of compounds from wood, was ethanol. The high content of ethanol extractives is because it is a more polar solvent than acetone that can dissolve the most represented polyphenols [70]. For that reason, the absorption spectra of the extractives dissolved in ethanol were also analyzed and are presented in Figure 7.

In dissolved extractives, light is mainly absorbed by lignin below 500 nm and by phenolic extractives, such as tannins, flavonoids, stilbenes, and quinones, above 500 nm [71]. The discoloration reactions that occur during heat treatment are often justified by the formation of colored oxidation and degradation products involving the cell wall constituents and extractives located in cell vacuoles [50,72].

3.4.1. The Evaluation of the Measured Spectra in the Region below 500 nm

Thermally modified samples at the highest temperature of 220 °C showed the highest absorption in the respective series in the cases of SC220, IR220, and TE220. On the contrary, the lowest absorption associated with a decrease in volatile organic acids and phenols compared to the standard samples was observed in the cases of the PA220 and MB220 samples. For SC, IR, PA, and TE samples thermally modified at temperatures of 180 and 200 °C, the measured absorbance in these areas was lower than in the case of the respective reference samples. The spectrum of the reference sample PA20 is the only one that shows the highest intensity in the region below 500 nm compared to the other thermally modified samples from its series. On the contrary, the reference sample MT20 shows the lowest intensity in the entire measurement area.

For all samples, the highest absorption bands are observed in the wavelength range around 460 nm, where an increase in the phenolic hydroxyl group can be detected. As an auxochrome group, it could also contribute to the darkening of the thermally treated wood.

The increased absorption band at approx. 495 nm suggests the release of acidic substances, specifically volatile organic acids (acetic and formic acid), after the degradation of acetyl groups in polyose [73]. Acids can catalyze the condensation and degradation reactions in lignin structures [49] and extractives after heat treatment in the presence of water and contribute to the discoloration of the wood.

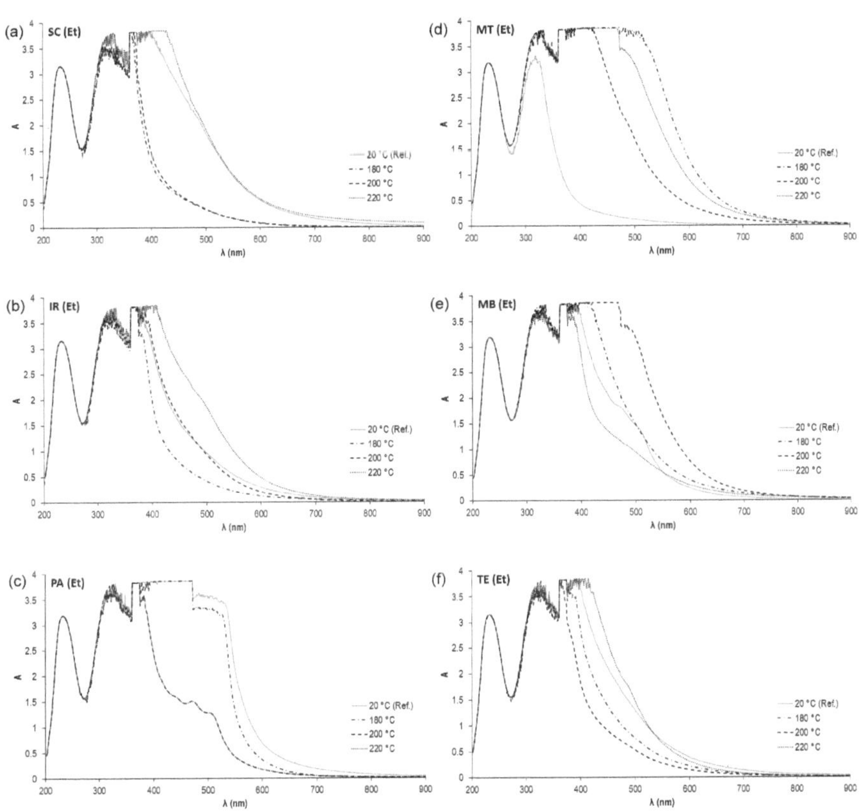

Figure 7. Absorption spectra of ethanol extractives from selected tropical wood samples. (**a**) Sp. cedar, (**b**) iroko, (**c**) padouk, (**d**) meranti, (**e**) merbau, and (**f**) teak thermally modified at temperatures of 180, 200, and 220 °C compared to reference unmodified samples.

3.4.2. The Evaluation of the Measured Spectra in the Region above 500 nm

It has been established that around wavelengths of 700 nm, a mixture of extractable components is detected, such as tannins and their derivates, e.g., ellagitannins due to their polymerization by oxygen and increased temperatures [74,75]. Absorbance in the region of these wavelengths, and therefore an increased amount of corresponding compounds, is observed in samples of padouk (PA20, PA180), meranti (MT180, MT200, and MT220), and merbau (MB200). An increased amount of other suspended solids compared to the other samples is detected in the 810 nm region for the SC220 and PA20 samples.

3.5. Extractive Substances Determined in Tropical Wood Species by Other Authors

Tenorio et al. [76] published, in their work focused on Spanish cedar, a high value of soluble extractives in an ethanol–toluene mixture found in the 4- and 5-year-old trees, indicating a high presence of waxes, fats, resins, phytosterols, nonvolatile hydrocarbons, and polar extractives in the wood of this species. Among the water-extractable substances were tannins, gums, sugars, and coloring matters and starches. Likewise, high levels of soluble extractives in NaOH indicate a high presence of low-molecular-weight extractives, such as carbohydrates, consisting mainly of hemicellulose and degraded cellulose in wood.

Biwôlé et al. [77] identified chlorophorin, a highly hydrophobic substance, in iroko powder before and after 30 days of immersion in cold water. They also determined that cudraxanthone I and geranyl trihydroxystilbene exhibited low hydrophobicity in wet conditions. Biwôlé et al. [78] published that iroko wood is rich in polyphenols and used MALDI-TOF analysis instead of the abovementioned other compounds such as geranyl-trihydroxy-stilbene and neocyclomorusin.

The main extractive of padouk is homopterocarpin, directly recovered with a good purity via acetone extraction. It also contains pterocarpin and anethole, C5 and C6 carbohydrates, C14–C18 fatty acids, principally hexadecanoic acid, and triglycerides [53].

Yasuda et al. [79] studied the effect of meranti extractives on the manufacture of plywood. Using ^{1}H and ^{13}C NMR spectroscopy, they detected a high content of phenolic compounds, namely secondary oxidative polymers of resveratrol (e.g., hopeaphenol) and its related stilbenes. Other extractives of *Shorea* species, chrysophanol, scopoletin, and hexamethylcoruleoellagic acid, are also known [80].

Specifically, Kilic and Niemz [81] detected fatty acids, 3-hydroxybenzoic acid, vanillic acid, 3,4,5-trihydroxybenzoic acid, 4,4′-dihydroxy-3,3′-dimethoxystilbene, sitosterol, and sugars in merbau using GC/MS analysis. Malik et al. [82] state that merbau extractives are a promising material for the enhancement of wood properties because of their high content of phenolic compounds, particularly resorcinol.

Myo Aung [83] characterized the teak methanol extractives via HPLC and detected 2-methyl anthraquinone (tectoquinone), the compound responsible for the durability, i.e., mainly insecticidal properties, of teak wood along with a variety of extractives in lower quantities. The hydrophobicity, antioxidant properties, and oily nature of teak wood were mainly due to the caoutchouc compounds [84,85].

3.6. *Statistical Evaluation*

3.6.1. Principal Component Analysis (PCA)

The values of all variables for wood and extractives (except ISO whiteness and all calculated parameters, ΔL^*, Δa^*, Δb^*, and ΔE^*) were taken into consideration from Tables 2 and 3, from which the principal components were created. The original data were transformed into principal component space using PCA. Based on the results of the statistical processing, the number of components was chosen to be three, with the first component explaining 43.26%, the second component 20.94%, and the third component 8.15% of the variability. The dependence of the two most important components (1 and 2) can be seen in Figure 8.

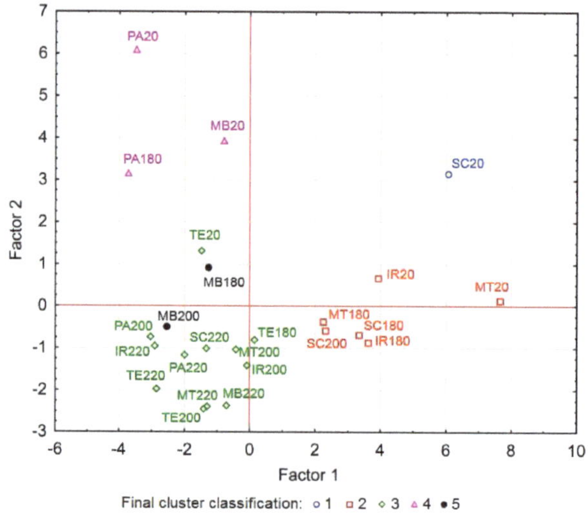

Figure 8. Projection of individual wood samples into the coordinate system of the 1st and 2nd principal components.

When evaluating the contributions of individual variables to the principal components, it can be observed that the L^* values and yellowness indices for wood (Y) and extractives (Yi), as well as the parameters a^*(Ac) and b^*(Ac) of extractives soluble in acetone have the greatest influence on component 1. The amount of all extractive substances w(Ac), w(Et) and w(Et-To) obtained from wood and the parameters a^* and b^* of the wood samples have the greatest influence on component 2. And the principal component 3 is influenced mainly by extractive substances soluble in ethanol, i.e., by the amount w(Et) of these extractive substances and the parameters L^*(Et), b^*(Et), and Yi(Et), and again by the parameters a^* and b^* of the wood samples.

A cluster analysis was subsequently performed on these three selected components. The method identified five clusters. The distribution of the clusters is shown in the graphs in Figure 9. Each graph shows a different combination of two out of the three principal components against each other, as some clusters are distinguished from each other only in their specific combination (e.g., cluster 5 is not distinguishable from cluster 3 in the space of the first and second principal components in Figure 8, but what makes it different are the coordinates in the direction of the third component, as can be seen in Figure 9).

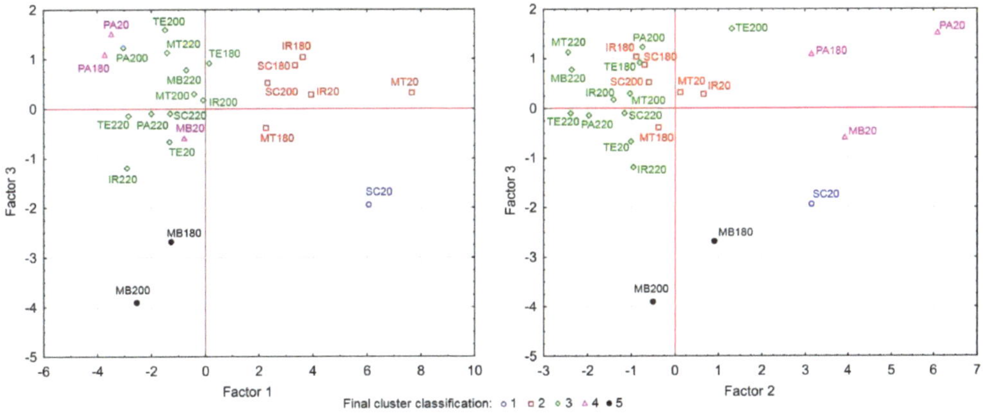

Figure 9. Cluster analysis shown in 1st and 3rd (**a**) and 2nd and 3rd (**b**) principal component spaces.

The first cluster consists of only one sample, which represents SC20. The second cluster includes six samples namely SC180, SC200, IR20, IR180, MT20, and MT180. The third cluster has the most samples and includes SC220, IR200, IR220, PA200, PA220, MT200, MT220, MB220, TE20, TE180, TE200, and TE220. The fourth cluster consists of three samples, namely PA20, PA180, and MB20. In the fifth cluster, there are the samples MB180 and MB200.

Based on this distribution, it can be seen that clusters 1, 2 (except SC200), 4, and 5 (except MB200) mainly consist of reference samples, where variability is obvious, and samples thermally modified at the lowest temperature, i.e., 180 °C. Conversely, cluster 5 includes all samples of thermally modified wood species at the highest temperature of 220 °C, most of the samples modified at 200 °C (except for the aforementioned SC200 in the second group and MB200 in the fifth group), and in addition, two teak samples (TE20 and TE180). This confirms the great similarity or homogeneity of color and related parameters of thermally modified tropical wood species starting at 200 °C, which were included in the cluster analysis.

If we approach the result of the cluster analysis from the point of view of individual wood species, the Spanish cedar samples are distributed in three clusters. SP20 is different to the others and no similarity is observed with any of the 24 samples (reference and thermally treated). Samples SC180 and SC200 show some similarity to iroko and meranti (reference and thermally treated samples at 180 °C). The SC200 sample belongs to the most numerous group 3, which also includes other thermally modified types of wood at the highest temperatures. A great similarity in optical properties can be seen in the iroko and meranti samples. As mentioned above, their reference and thermally modified samples at 180 °C are included in group 2, and the thermally modified samples at the two higher temperatures belong again identically in group 3. For PA20 and PA180, a certain agreement with MB20 (group 4) can be found, and the samples thermally modified at the two higher temperatures belong again to group 3. The merbau samples behave differently in terms of optical properties and, like the Sp. cedar, are distributed in three clusters, namely as follows: the reference sample MB20 together with PA20 and PA180 in a separate cluster 4, samples MB180 and MB200 in a separate cluster 5, and MB220 in the most numerous group 3. Teak is specific in that all samples from its series (TE20, TE180, TE200, and TE220) are assigned to cluster 3, which is typical for wood samples modified at the highest temperatures.

3.6.2. Functional Dependence of Yellowness

The measured yellowness of solid wood samples (Y) is an important evaluation parameter for thermally modified wood. It can be seen from the correlation analysis that the most correlated quantity is L^* with the value of the correlation coefficient r = 0.972 (Table 4). This value is statistically significantly non-zero and indicates a very high dependence. A specific comparison of the dependence of Y and L^* is shown in Figure 10a.

Table 4. The correlation coefficient of the wood yellowness variable (Y) concerning other parameters.

Variable	Y_i(Ac)	Y_i(Et)	Y_i(Et-To)
Y (Yellowness)	−0.7076	−0.3219	−0.6022
	L^*	a^*	b^*
	0.9719	0.2084	0.6141
	L^*(Ac)	a^*(Ac)	b^*(Ac)
	0.6502	−0.5561	−0.7579
	L^*(Et)	a^*(Et)	b^*(Et)
	0.2217	−0.2174	−0.2881
	L^*(Et-To)	a^*(Et-To)	b^*(Et-To)
	0.5746	−0.5467	−0.6007

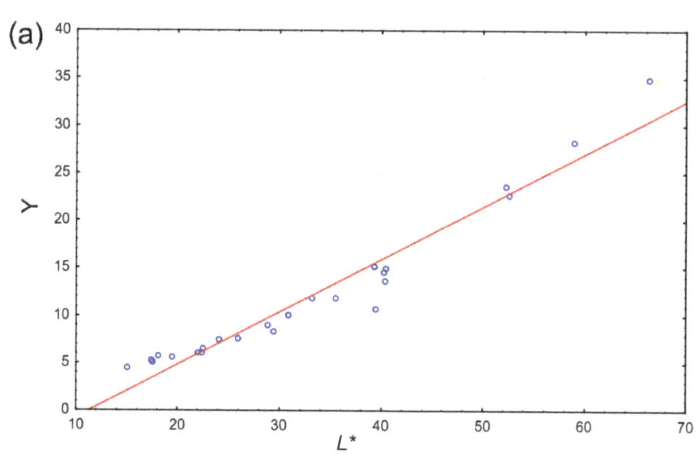

Figure 10. *Cont.*

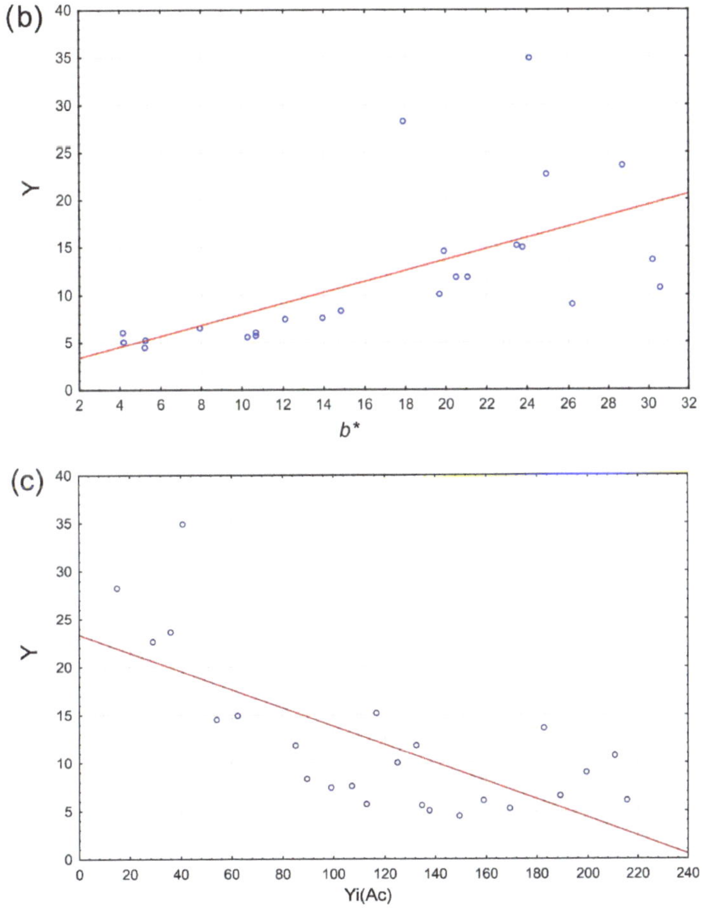

Figure 10. Relationship between wood yellowness (Y) and (**a**) L^* parameter, (**b**) b^* parameter, and (**c**) yellowness index of extractive substances soluble in acetone Yi(Ac).

When the functional dependence of wood yellowness (Y) on other variables is investigated using a regression analysis, then the only significant parameters are L^* and b^*. The regression model of this dependence is presented in Table 5. The coefficient of determination R2 is 0.973, which indicates a very high degree of explanation of the variable Y, with the help of parameters L^* and b^*. In other words, if we know the values of L^* and b^*, the value of Y can be estimated quite accurately using the regression equation. The dependence of Y and b^* is shown in Figure 10b.

Table 5. Regression analysis for the dependent variable wood yellowness (Y).

N = 24	**Regression Summary for Dependent Variable: Y R = 0.9866, R2 = 0.9733, Adjusted R2 = 0.9708 F (2.21) = 383.06, $p < 0.0000$, Std. The Error of Estimate: 1.3641**			
	b^*	**b**	**t(21)**	**p-Value**
Intercept		−5.7726 ± 0.7337	−7.8675	0.0000
L^*	1.1630 ± 0.0537	0.6633 ± 0.0306	21.6611	0.0000
b^*	−0.2555 ± 0.0537	−0.2388 ± 0.0502	−4.7595	0.0001

Table 5. *Cont.*

N = 24	Regression summary for dependent variable: Y R = 0.7141, R2 = 0.5099, adjusted R2 = 0.4364 F (3.20) = 6.9361, $p < 0.0022$, Std. The error of estimate: 5.9908			
Intercept		23.7871 ± 3.6857	6.4540	0.0000
Yi(Ac)	−0.6029 ± 0.2488	−0.0810 ± 0.03343	−2.4238	0.0250
Yi(Et)	0.0597 ± 0.1978	0.0080 ± 0.0264	0.3022	0.7656
Yi(Et-To)	−0.1702 ± 0.2786	−0.0214 ± 0.0351	−0.6109	0.5481

In search of connections between the yellowness of wood (Y) and the yellowness indices of the individual extractives Yi(Ac), Yi(Et), and Yi(Et-To), a regression analysis was performed, where Y was the dependent variable and individual yellowness indices were predictors. The results of this analysis are presented in Table 5. From the results, it can be seen that the only statistically significant factor is the yellowness index Yi(Ac). Even if we leave the insignificant factors Yi(Et) and Yi(Et-To) in the model, the coefficient of determination is only 0.510 compared to the previous analysis including the connection with L^* and b^*, i.e., 0.973. For completion, the relationship between Y and Yi(Ac) is also shown on the graph in Figure 10c, which is described by the equation Y = −0.0951 × Yi(Ac) + 23.3485.

4. Summary and Conclusions

The results have contributed to the understanding of the chemical degradation path of thermally modified tropical wood species by monitoring extractive content transformations in connection with their changes in color and the color of the wood itself. The importance of optimizing the treatment (temperature and duration of thermal modification or subsequent treatment related to increasing the stability of wood color in exterior conditions) for each wood species to make the best utilization of this material has been pointed out.

Solid Wood Samples

- The values of the L^*, a^*, and b^* parameters are in all cases positive, and with increasing temperature of thermal modification, they have a decreasing trend—the samples lose brightness and darken. The exception is merbau wood, where these values increased between 200 and 220 °C.
- Based on the cluster analysis in the CIELAB color space, the samples can be divided into six clusters according to similarity. A special group is made up of padauk samples PA20 and PA180, distinguished by their significant redness ($a^* \approx 26.5$–29.0).
- As the thermal modification temperature increases, ΔE^* increases for most samples except for merbau with the highest ΔE^* value for sample MB200 (for MB220 this value decreases by almost 20%). At a thermal modification temperature of 200 °C, all samples (except IR200) have similar ΔE^* values and at 220 °C, Spanish cedar, iroko, padouk, and meranti.

Extractive Samples

- In the case of the reference samples (except for SC and MB), the largest amount of extractives was obtained during extraction into acetone, into which mainly fatty and resinous acids and sterols are released.
- In the case of samples modified at higher temperatures, the highest yield of extractives (mainly phenolic compounds, glycosides, and dyes) was usually obtained using ethanol. Their representation in wood was mostly highest at a thermal modification temperature of 180 °C and decreased at 200 and 220 °C.

Assessment of Chromophoric Structures

- The increasing amount of chromophores with the thermal modification temperature is observed for Spanish cedar and meranti.
- The k/s intensity decreases along with increasing wavelength for all woods (except padouk).

- Padouk, merbau, and teak, and partially also iroko, modified at temperatures of 200 and 220 °C contain a comparable amount of chromophoric compounds detected at wavelengths of 360 nm and 457 nm.
- The discoloration of wood during thermal modification is due mainly to condensation and oxidation reactions. Both of these lead to the formation of extensive conjugation structures (alkenes and aromatics) while increasing the number of carbonyl or carboxyl groups.
- New types of extractive substances formed from hemicelluloses (2-furaldehyde, etc.), cellulose (lactones, etc.), and lignin (formic acid, methanol, monomer derivatives of phenols, etc.) are created already below a temperature of 180 °C when they are not subject to even more condensation reactions.

Statistical Evaluation

- Five clusters based on the similarity of the included parameters (L^*, a^*, and b^* of solid samples and extractives, yellowness of wood, yellowness indices of individual extractives, and intensity k/s at selected wavelengths and amount of extractives) were obtained by applying cluster analysis to the three main components.
- Clusters 1, 2, 4, and 5 comprise only the reference and thermally modified samples at 180 °C (except SC200 and MB200), whereas cluster 1 comprises only SC20. Cluster 3 consists of 12 samples, mostly thermally modified at 200 °C and 220 °C, plus TE20 and TE180 samples.
- The yellowness (Y) of the wood has a very high dependence (r = 0.972) on the brightness (L^*) of these samples.
- The coefficient of determination (r2 = 0.973) indicates a very high degree of explanation of the variable yellowness (Y) using the parameter L^*, and also b^*.
- A certain dependence (r = 0.714) is also noted between the yellowness (Y) and the yellowness index of extractives soluble in acetone Yi(Ac), whose relationship is described by the equation $Y = -0.0951 \times Y(Ac) + 23.3485$.

Due to thermal modification above 200 °C, tropical wood loses its specific color, which makes individual species unique and interesting in terms of design. Although the color becomes homogeneous, the wood significantly loses its brightness and yellowness, and from temperatures of 220 °C, there are almost no differences in their color, as confirmed by the cluster analysis within the PCA. This may not always be desirable, because in some cases, consumers select a wood based solely on its decorative properties. Of course, the use of this type of wood will always depend on the specific application. And whether the color of the wood itself will be important at all, or whether other requirements from the advantage of ThermoWood itself will be prioritized.

The next phase of this research aims to focus more closely on the characteristics of preserved extracted substances from selected tropical wood species thermally modified at different temperatures using gas chromatography (GC-MS) and infrared spectroscopy (FTIR).

Author Contributions: Conceptualization, T.J. (Tereza Jurczyková), O.Š. and F.K.; methodology, T.J. (Tereza Jurczyková), K.H., F.K., T.J. (Tomáš Jurczyk) and R.H.; software, T.J. (Tomáš Jurczyk); validation, T.J. (Tereza Jurczyková), O.Š. and T.J. (Tomáš Jurczyk); writing—original draft preparation, O.Š.; writing—review and editing, T.J. (Tereza Jurczyková), K.H., F.K., T.J. (Tomáš Jurczyk) and R.H.; supervision, T.J. (Tereza Jurczyková). All authors have read and agreed to the published version of the manuscript.

Funding: This research was funded by the Scientific Grant Agency of the Ministry of Education, Science, Research and Sport of the Slovak Republic and Slovak Academy of Sciences (Vedecká Grantová Agentúra MŠVVaŠ SR a SAV), grant number 1/0117/22.

Institutional Review Board Statement: Not applicable.

Data Availability Statement: Data are available from the authors on request.

Conflicts of Interest: The authors declare no conflict of interest.

References

1. Hon, D.N.S.; Minemura, N. Color and discoloration. In *Wood and Cellulosic Chemistry*; Hon, D.N.S., Shiraishi, N., Eds.; Marcel Dekker: New York, NY, USA, 2001; pp. 385–442.
2. Reinprecht, L.; Mamoňová, M.; Pánek, M.; Kačík, F. The impact of natural and artificial weathering on the visual, color, and structural changes of seven tropical woods. *Eur. J. Wood Wood Prod.* **2018**, *76*, 175–190. [CrossRef]
3. Panshin, A.J.; de Zeeuw, C. *Textbook of Wood Technology*; McGraw-Hill Book Company: New York, NY, USA, 1980; p. 722.
4. Rosenau, T.; Potthast, A.; Kosma, P.; Suess, H.U.; Nimmerfroh, N. Chromophores in aged hardwood pulp—Their structure and degradation potential. *Tappi J.* **2008**, *1*, 24–30.
5. Tribulová, T.; Kačík, F.; Evtuguin, D.V.; Čabalová, I. Assessment of Chromophores in Chemically Treated and Aged Wood by UV-Vis Diffuse Reflectance Spectroscopy. *Cellul. Chem. Technol.* **2016**, *50*, 659–667.
6. Bourgois, J.; Bartholin, M.C.; Guyonnet, R. Thermal treatment of wood: Analysis of the obtained product. *Wood Sci. Technol.* **1989**, *23*, 303–310. [CrossRef]
7. Hill, C.; Altgen, M.; Rautkari, L. Thermal modification of wood—A review: Chemical changes and hygroscopicity. *J. Mater. Sci.* **2021**, *56*, 6581–6614. [CrossRef]
8. Reinprecht, L.; Vidholdová, Z. *ThermoWood–Preparing, Properties and Applications*; Technical University in Zvolen: Zvolen, Slovakia, 2008; p. 89. (In Slovak)
9. Tjeerdsma, B.F.; Boonstra, M.; Pizzi, A.; Tekely, P.; Militz, H. Characterisation of thermally modified wood: Molecular reasons for wood performance improvement. *Holz Als Roh-Und Werkst.* **1998**, *56*, 149–153. [CrossRef]
10. Chen, Y.; Fan, Y.; Gao, J.; Stark, N.M. The effect of heat treatment on the chemical and color change of black locust (*Robinia pseudoacacia*) wood flour. *BioResources* **2012**, *7*, 1157–1170. [CrossRef]
11. Chen, Y.; Gao, J.; Fan, Y.; Tshabalala, M.A.; Stark, N.M. Heat-induced chemical and color changes of extractive-free black locust (*Robinia pseudoacacia*) wood. *BioResources* **2012**, *7*, 2236–2248. [CrossRef]
12. Huang, X.; Kocaefe, D.; Kocaefe, Y.; Boluk, Y.; Pichette, A. A spectrocolorimetric and chemical study on color modification of heat-treated wood during artificial weathering. *Appl. Surf. Sci.* **2012**, *258*, 5360–5369. [CrossRef]
13. Taraborelli, C.; Monteoliva, S.; Keil, G.; Spavento, E. Effect of heat treatment on hardness, density and color of *Populus* × *canadensis* 'I-214' wood. *For. Syst.* **2022**, *31*, e023. [CrossRef]
14. Bakar, B.F.A.; Hiziroglu, S.; Tahir, P.M. Properties of some thermally modified wood species. *Mater. Des.* **2013**, *43*, 348–355. [CrossRef]
15. Zaman, A.; Alén, R.; Kotilainen, R. Thermal behavior of Scots pine (*Pinus sylvestris*) and silver birch (*Betula pendula*) at 200–230. *Wood Fiber Sci.* **2000**, *32*, 138–143.
16. Yildiz, S.; Gezer, E.D.; Yildiz, U.C. Mechanical and chemical behavior of spruce wood modified by heat. *Build. Environ.* **2006**, *41*, 1762–1766. [CrossRef]
17. Majano-Majano, A.; Hughes, M.; Fernandez-Cabo, J.L. The fracture toughness and properties of thermally modified beech and ash at different moisture contents. *Wood Sci. Technol.* **2012**, *46*, 5–21. [CrossRef]
18. Brito, A.F.; Calonego, F.W.; Bond, B.H.; Severo, E.T.D. Color changes, EMC, and biological resistance of thermally modified yellow poplar. *Wood Fiber Sci.* **2018**, *50*, 439–446. [CrossRef]
19. Kačíková, D.; Kubovský, I.; Gaff, M.; Kačík, F. Changes of Meranti, Padauk, and Merbau Wood Lignin during the ThermoWood Process. *Polymers* **2021**, *13*, 993. [CrossRef]
20. Kačíková, D.; Kubovský, I.; Ulbriková, N.; Kačík, F. The impact of thermal treatment on structural changes of teak and iroko wood lignins. *Appl. Sci.* **2020**, *10*, 5021. [CrossRef]
21. Gašparík, M.; Gaff, M.; Kačík, F.; Sikora, A. Color and chemical changes in teak (*Tectona grandis* L. f.) and meranti (*Shorea* spp.) wood after thermal treatment. *BioResources* **2019**, *14*, 2667–2683. [CrossRef]
22. Gaff, M.; Kubovský, I.; Sikora, A.; Kačíková, D.; Li, H.; Kubovský, M.; Kačík, F. Impact of thermal modification on color and chemical changes of African padauk, merbau, mahogany, and iroko wood species. *Rev. Adv. Mater. Sci.* **2023**, *62*, 20220277. [CrossRef]
23. Korkut, S. Performance of three thermally treated tropical wood species commonly used in Turkey. *Ind. Crops Prod.* **2012**, *36*, 355–362. [CrossRef]
24. Mburu, F.; Dumarçay, S.; Huber, F.; Petrissans, M.; Gérardin, P. Evaluation of thermally modified *Grevillea robusta* heartwood as an alternative to shortage of wood resource in Kenya: Characterisation of physicochemical properties and improvement of bio-resistance. *Bioresour. Technol.* **2007**, *98*, 3478–3486. [CrossRef]
25. Lengowski, E.C.; Bonfatti Júnior, E.A.; Nisgoski, S.; Bolzon de Muñiz, G.I.; Klock, U. Properties of thermally modified teakwood. *Maderas Cienc. Y Tecnol.* **2021**, *23*. [CrossRef]
26. Esteves, B.; Videira, R.; Pereira, H. Chemistry and ecotoxicity of heat-treated pine wood extractives. *Wood Sci. Technol.* **2011**, *45*, 661–676.

27. Piernik, M.; Woźniak, M.; Pinkowski, G.; Szentner, K.; Ratajczak, I.; Krauss, A. Impact of the Heat Treatment Duration on Color and Selected Mechanical and Chemical Properties of Scots Pine Wood. *Materials* **2022**, *15*, 5425. [CrossRef] [PubMed]
28. Kačík, F.; Kubovský, I.; Bouček, J.; Hrčka, R.; Gaff, M.; Kačíková, D. Color and Chemical Changes of Black Locust Wood during Heat Treatment. *Forests* **2022**, *14*, 73.
29. Meier, E. *The Wood Dictionary: From Acacia to Ziricote, a Guide to the World's Wood;* Sierra's Ascent Publisher: San Francisco, CA, USA, 2021; p. 272.
30. Wagenfuhr, R. *Holzatlas*, 6th ed.; Fachbuchverlag Publisher: Leipzig, Germany, 2007; p. 816.
31. International Thermowood Association. *Thermowood Handbook;* International Thermowood Association: Helsinki, Finland, 2021; p. 55.
32. *TAPPI T 257 cm-02;* Sampling and Preparing Wood for Analysis. Technical Association of the Pulp and Paper Industry. Tappi Tappi Press Atlanta: Atlanta, GA, USA, 2011.
33. *ASTM D 1107–96;* Standard Test Method for Ethanol-Toluene Solubility of Wood. American Society for Testing and Materials: Philadelphia, PA, USA, 2021.
34. *ISO 187;* Paper, board, and pulps—Standard atmosphere for conditioning and testing and procedure for monitoring the atmosphere and conditioning of samples. International Organization for Standardization: London, UK, 1990.
35. *ISO 11664-4;* Colorimetry–Part 4: CIE 1976 L*a*b* colour space. International Organization for Standardization: London, UK, 2011.
36. *ISO/CIE 11664-6;* Colorimetry–Part 6: CIEDE2000 Color-difference formula. International Organization for Standardization: London, UK, 2014.
37. Kubelka, P. New contributions to the optics of intensely light-scattering materials. Part I. *J. Opt. Soc. Am.* **1948**, *38*, 448–457. [CrossRef]
38. *TAPPI T 280 pm-99;* Acetone extractives of wood and pulp. Technical Association of the Pulp and Paper Industry. Tappi Press Atlanta: Atlanta, GA, USA, 2000.
39. *TAPPI T 204 cm-97;* Solvent extractives of wood and pulp. Technical Association of the Pulp and Paper Industry. Tappi Press Atlanta: Atlanta, GA, USA, 1997.
40. *TAPPI T 264 cm-97;* Preparation of wood for chemical analysis. Technical Association of the Pulp and Paper Industry. Tappi Press Atlanta: Atlanta, GA, USA, 1997.
41. Lo Monaco, A.; Pelosi, C.; Agresti, G.; Picchio, R.; Rubino, G. Influence of thermal treatment on selected properties of chestnut wood and full range of its visual features. *Drewno* **2020**, *63*, 1–20.
42. Correal-Mòdol, E.; Wimmer, T.; Huber, H.; Schnabel, T. Approach for color homogenisation of chestnut (*Castanea sativa* [Mill.]) by thermal modification. *Int. Wood Prod. J.* **2014**, *5*, 69–73. [CrossRef]
43. Gonzáles-Peña, M.M.; Hale, M.D.C. Color in thermally modified wood of beech, Norway spruce and Scots pine. Part 1: Color evolution and color changes. *Holzforshung* **2009**, *63*, 385–393. [CrossRef]
44. Tudorović, N.; Popović, Z.; Milić, G.; Popadić, R. Estimation of heat treated wood properties by color change. *Bioresources* **2012**, *7*, 799–815. [CrossRef]
45. *ASTM E313;* Standard Practice for Calculating Yellowness and Whiteness Indices from Instrumentally Measured Color Coordinates. American Society for Testing and Materials: Philadelphia, PA, USA, 2015.
46. Yu, J.; Huang, X.; Wu, C.; Wu, X.; Wang, G.; Jiang, P. Interfacial modification of boron nitride nanoplatelets for epoxy composites with improved thermal properties. *Polymer* **2012**, *53*, 471–480. [CrossRef]
47. Baar, J.; Gryc, V. Color of tropical wood and discoloration due to simulated sunlight. *Acta Univ. Agric. Silvic. Mendel. Brun.* **2010**, *58*, 13–20. [CrossRef]
48. Cuccui, I.; Negro, F.; Zanuttini, R.; Espinoza, M.; Allegretti, O. Thermo-vacuum modification of teak wood from fast-growth plantation. *BioResources* **2017**, *12*, 1903–1915. [CrossRef]
49. Sehistedt-Persson, M. Color responses to heat-treatment of extractives and sap from pine and spruce. In Proceeding of the International IUFRO Wood Drying Conference, Brasov, Romania, 25 October 2003.
50. Sundqvist, B. Color Changes and Acid Formation in Wood During. Ph.D. Thesis, Luleå Tekniska Universitet, Luleå, Sweden, 2004.
51. Polcin, J.; Rapson, W.H. The interpretation of UV and visible spectrum of lignin. In Proceeding of the 55th Annual Meeting of the Technical Section, Montreal, QC, Canada; 1969; pp. 28–31.
52. Keating, J.; Johansson, C.I.; Saddler, J.N.; Beatson, R.P. The nature of chromophores in high-extractives mechanical pulps: Western red cedar (*Thuja plicata* Donn) chemithermomechanical pulp (CTMP). *Holzforschung* **2006**, *60*, 365–371. [CrossRef]
53. Chen, X.; Li, Y. Effect of heat treatment on microstructure and mechanical properties of high boron white cast iron. *Mater. Sci. Eng.* **2010**, *528*, 770–775. [CrossRef]
54. Diouf, P.N.; Stevanovic, T.; Boutin, Y. The effect of extraction process on polyphenol content, triterpene composition and bioactivity of yellow birch (*Betula alleghaniensis* Britton) extracts. *Ind. Crops Prod.* **2009**, *30*, 297–303. [CrossRef]
55. Bekhta, P.; Niemz, P. Effect of high temperature on the change in color, dimensional stability and mechanical properties of spruce wood. *Holzforschung* **2003**, *57*, 539–546. [CrossRef]

56. Čabalová, I.; Kačík, F.; Tribulová, T. The effects of heat treatment on the chemical alterations of oak wood. *Key Eng. Mater.* **2016**, *688*, 44–49. [CrossRef]
57. Lai, Y.Z. Chemical degradation. In *Wood and Cellulosic Chemistry*; Hon, D.N.S., Shiraishi, N., Eds.; Marcel Dekker: New York, NY, USA, 1991; pp. 455–524.
58. Saha, J.B.T.; Abia, D.; Dumarçay, S.; Ndikontar, M.K.; Gérardin, P.; Noah, J.N.; Perrin, D. Antioxidant activities, total phenolic contents and chemical compositions of extracts from four Cameroonian woods: Padouk (*Pterocarpus soyauxii* Taubb), tali (*Erythrophleum suaveolens*), moabi (*Baillonella toxisperma*), and movingui (*Distemonanthus benthamianus*). *Ind. Crops Prod.* **2013**, *41*, 71–77.
59. Diouf, P.N.; Merlin, A.; Perrin, D. Antioxidant properties of wood extracts and color stability of woods. *Ann. For. Sci.* **2006**, *63*, 525–534. [CrossRef]
60. Hafizoglu, H.; Holmbom, B. Chemical composition of extractives from Abies nordmanniana. *Holz Als Roh-Und Werkst.* **1995**, *53*, 273–275. [CrossRef]
61. Fengel, D.; Wegener, G. *Wood: Chemistry, Ultrastructure, Reactions*; Walter De Gruyter: Berlin, Germany, 1989; p. 626.
62. Stenius, P.; Gullichsen, J.; Paulapuro, H. Papermaking Science and Technology. In *Forest Products Chemistry*, 3rd ed.; Stenius, P., Ed.; Faper Oy: Helsinki, Finland, 2000; p. 350.
63. Nuopponen, M.; Vuorinen, T.; Jämsä, S.; Viitaniemi, P. Thermal modifications in softwood studied by FT-IR and UV resonance Raman spectroscopies. *J. Wood Chem. Technol.* **2005**, *24*, 13–26. [CrossRef]
64. Faix, O.; Meier, D.; Fortmann, I. Thermal degradation products of wood. Gas chromatographic separation and mass spectrometric characterization of monomeric lignin-derived products. *Holz Als Roh-Und Werkst.* **1990**, *48*, 281–285. [CrossRef]
65. Červenka, E.; Král, Z.; Tomis, B. *Chemistry of Wood and Cellulose I–III*; The University of Chemical Technology in Pardubice: Pardubice, Czech Republic, 1980; p. 228. (In Czech)
66. Chow, P.; Nakayama, F.S.; Blahnik, B.; Youngquist, J.A.; Coffelt, T.A. Chemical constituents and physical properties of guayule wood and bark. *Ind. Crops Prod.* **2008**, *28*, 303–308. [CrossRef]
67. Ayadi, N.; Lejeune, F.; Charrier, F.; Charrier, B.; Merlin, A. Color stability of heat-treated wood during artificial weathering. *Holz Als Roh-Und Werkst.* **2003**, *61*, 221–226. [CrossRef]
68. Van Nguyen, T.H.; Nguyen, T.T.; Ji, X.; Guo, M. Predicting Color Change in Wood During Heat Treatment Using an Artificial Neural Network Model. *BioResources* **2018**, *13*, 6250–6264. [CrossRef]
69. Cirule, D.; Sansonetti, E.; Andersone, I.; Kuka, E.; Andersons, B. Enhancing Thermally Modified Wood Stability against Discoloration. *Coatings* **2021**, *11*, 81. [CrossRef]
70. Palaiogiannis, D.; Chatzimitakos, T.; Athanasiadis, V.; Bozinou, E.; Makris, D.P.; Lalas, S.I. Successive Solvent Extraction of Polyphenols and Flavonoids from *Cistus creticus* L. Leaves. *Oxygen* **2023**, *3*, 274–286. [CrossRef]
71. Hon, D.N.; Shiraishi, N. *Wood, and Cellulosic Chemistry, Revised, and Expanded*, 2nd ed.; CRC Press: Boca Raton, FL, USA, 2000; p. 928.
72. Tjeerdsma, B.F.; Militz, H. Chemical changes in hydrothermal treated wood: FTIR analysis of combined hydrothermal and dry heat-treated wood. *Eur. J. Wood Wood Prod.* **2005**, *63*, 102–111. [CrossRef]
73. Lundquist, K. Acid degradation of lignin. *Acta Chem. Scand.* **1970**, *24*, 889–907. [CrossRef]
74. Dünisch, O.; Richter, H.G.; Koch, G. Wood properties of juvenile and mature heartwood in *Robinia pseudoacacia* L. *Wood Sci. Technol.* **2010**, *44*, 301–313. [CrossRef]
75. Charrier, B.; Haluk, J.P.; Metche, M. Characterization of European Oakwood Constituents Acting in the Brown Discolouration during Kiln Drying. *Holzforschung* **1995**, *49*, 168–172. [CrossRef]
76. Tenorio, C.; Moya, R. Evaluation of wood properties of four ages of *Cedrela odorata* trees growing in agroforestry systems with *Theobroma cacao* in Costa Rica. *Agrofor. Syst.* **2019**, *93*, 973–988. [CrossRef]
77. Biwôlé, J.J.E.; Biwolé, A.B.; Mfomo, J.Z.; Segovia, C.; Pizzi, A.; Chen, X.; Meausoone, P.J. Causes of differential behavior of extractives on the natural cold water durability of the welded joints of three tropical woods. *J. Adhes. Sci. Technol.* **2022**, *36*, 1314–1331. [CrossRef]
78. Biwôlé, J.J.E.; Biwolé, A.B.; Mfomo, J.Z.; Pizzi, A.; Segovia, C.; Abessolo, D.; Meausoone, P.J. Iroko wood (*Milicia excelsa* CC berg), a good candidate for high-speed rotation-induced wood dowel welding: An assessment of its welding potential and the water resistance of its welded joints. *Int. J. Adhes. Adhes.* **2023**, *123*, 103360. [CrossRef]
79. Yasuda, S.; Imai, T.; Fukushima, K.; Hamaguchi, E. Effect of the extractives of yellow meranti wood on the manufacture of plywood. *Eur. J. Wood Wood Prod.* **1998**, *56*, 87–89. [CrossRef]
80. Geevananda, Y.A.; Gunawardana, P.; Sultanbawa, M.U.S.; Balasubramaniam, S. Distribution of some triterpenes and phenolic compounds in the extractives of endemic *Dipterocarpaceae* species of Sri Lanka. *Phytochemistry* **1980**, *19*, 1099–1102. [CrossRef]
81. Kilic, A.; Niemz, P. Extractives in some tropical woods. *Eur. J. Wood Wood Prod.* **2012**, *70*, 79. [CrossRef]
82. Malik, J.; Santoso, A.; Ozarska, B. Polymerised Merbau Extractives as Impregnating Material for Wood Properties Enhancement. In Proceeding of the Materials Science and Engineering, Bogor, Indonesia, 1 September 2020.
83. Myo Aung, U. *A Preliminary Study of Anthraquinone Extractives in Teak. Leaflet No. 8/87-88*; Forest Research Institute: Yezin, Myanmar, 1988; p. 39.

84. Simatupang, M.H.; Yamamoto, K. Properties of teakwood (*Tectona grandis* L.f.) as influenced by wood extractives and its importance for tree breeding. In Proceedings of the Regional Seminar on Site, Technology and Productivity od Teak Plantations, Chiang Mai, Thailand, 26–29 January 1999.
85. Bhat, P.T.K. Chemical extractive compounds determining the brown-rot decay resistance of teak wood. *Holz Als Roh-Und Werkst.* **2007**, *65*, 121–124.

polymers

MDPI

Article

Effect of Silane Coupling Agent Modification on Properties of Brass Powder-Water-Based Acrylic Coating on *Tilia europaea*

Yan Han [1,2] and Xiaoxing Yan [1,2,*]

1 Co-Innovation Center of Efficient Processing and Utilization of Forest Resources, Nanjing Forestry University, Nanjing 210037, China
2 College of Furnishings and Industrial Design, Nanjing Forestry University, Nanjing 210037, China
* Correspondence: yanxiaoxing@njfu.edu.cn

Abstract: Fine art coating is usually created by the combination of metal filler and water-based coatings, decorated to the surface of wood structures, furniture, and crafts. However, the durability of the fine art coating is limited by its weak mechanical qualities. In contrast, the metal filler's dispersion and the coating's mechanical properties can be significantly improved by the coupling agent molecule's ability to bind the resin matrix with the metal filler. In this study, a brass powder-water-based acrylic coating was prepared, and three different silane coupling agents, 3-aminopropyltriethoxysilane (KH550), γ-(2,3-epoxypropoxy)propytrimethoxysilane (KH560), and γ-methacryloxypropyltrimethoxysilane (KH570), were used to modify the brass powder filler in orthogonal tests. The artistic effect and optical properties of the modified art coating induced by different proportions of brass powder, silane coupling agents, and pH were compared. The result demonstrated that the amount of brass powder and the kind of coupling agent used had a substantial impact on the coating's optical characteristics. Our results also determined how three different coupling agents affected the water-based coating with varying brass powder contents. The findings indicated that 6% KH570 concentration and pH 5.0 were the ideal conditions for brass powder modification. Better overall performance of the art coating applied to the surface of the Basswood substrates was provided by adding 10% of the modified brass powder into the finish. It had a gloss of 20.0 GU, a color difference of 3.12, a color main wavelength of 590 nm, a hardness of HB, an impact resistance of 4 kg·cm, an adhesion of grade 1, and better liquid resistance and aging resistance. This technical foundation for the creation of wood art coatings promotes the application of art coatings on wood.

Keywords: art coating; brass powder; optical properties; silane coupling agent; mechanical properties; liquid resistance; aging resistance

check for updates

Citation: Han, Y.; Yan, X. Effect of Silane Coupling Agent Modification on Properties of Brass Powder-Water-Based Acrylic Coating on *Tilia europaea*. *Polymers* **2023**, *15*, 1396. https://doi.org/10.3390/polym15061396

Academic Editors: Lubos Kristak, Roman Reh, Marius Cătălin Barbu and Eugenia Mariana Tudor

Received: 6 February 2023
Revised: 2 March 2023
Accepted: 6 March 2023
Published: 10 March 2023

1. Introduction

Art coating usually serves a decorative effect [1–3] but also has more robust protection than general coatings [4,5]. Studies [6–8] have shown that adding metal filler in the water-based coating and applying a modified art coating can lend the surface of buildings, furniture, and handicrafts a shiny metallic luster. Consequently, this modified coating provides a protective and decorative role to the applied products, causing an increase in added value [9]. Therefore, modified art coatings have a valued decorative material in the decorating industry, with a large annual consumption [10].

In coatings market segments like architecture, product finishes and specialty coatings consisting of acrylic resins are frequently used. The three primary categories of acrylic coatings are water-based acrylic coatings [11], solvent-based acrylic coatings [12], and acrylic powder coatings [13]. Water-based acrylic resin has a wide range of application possibilities for metal powder coating due to its superior hardenability, smooth finish, excellent stain resistance, fewer VOCs emission, low cost, and simple preparation [14–16].

The major issues are unavoidable shortcomings, such as being sticky when hot and brittle when cold and having poor anti-return adhesion, and inadequate thermal stability after drying into a film [17,18]. Kim et al. [19] successfully synthesized multifunctional acrylic polyurethane coating materials with graffiti, pollution, and adhesive resistance qualities by using low-viscosity single-ended silicone oil and acrylic polyols with high glass transition temperatures as raw materials. In order to increase the water resistance of the resin coating, Yu et al. [20] effectively synthesized water-based acrylic resin modified by itaconic acid and gamma-methacryloxypropyl triisopropoxidesilane by seeded emulsion polymerization.

Furthermore, a growing body of evidence [21–23] indicates that the requirement from consumers for a decorative appearance is increasing. Therefore, the demand for eco-friendly, individualized coatings, with low consumption, has turned into a development trend in fine art coating technology. Adding metal powder as filler into water-based coatings can achieve these goals. However, due to the poor compatibility with water-based coatings, pure metal powder filler is easily agglomerated and dispersed unevenly within the coating, which affects the optical properties and worsens the mechanical properties, making them brittle, prone to cracking, and less adhesive [24]. These deficiencies severely restrict their practical application and advancement.

The coupling agent, a kind of additive, has the ability to enhance the bonding performance between inorganic and organic substances [25]. Due to the fact that its structure contains two different properties of groups, one of which is pro-inorganic or easily reacts with the surface of inorganic materials, while the other is pro-organic or can cause reactions with organic polymers physico-chemically. Therefore, coupling agents, or "molecular bridges", could reinforce the interaction between organic and inorganic molecules, considerably enhancing the performance of composite materials. Currently, the coupling agents most frequently employed are silane-based. A silane-based coupling agent's general formula is Y-R-Si-X_3, where Y stands for the organic functional group and R for the alkylidene group, and X represents the group that can be hydrolyzed. Thus, the application of coupling agent molecules to metal powder coatings could open the door for their practical use. The adhesion behavior of polyurethane (PU), polyurethane acrylate (PUA), and two coats loaded with silicone rubber (SR) nanoparticles on tin pads was examined by Sivakumar et al. [26] in relation to the use of silane agents. On tin surfaces, it was discovered that silanes based on isocyanates offered outstanding adhesion for PU and PU + SR coatings, while silanes based on acrylic offered acceptable adhesion for PUA and PUA + SR coatings. For polyvinyl butyral coatings, Zhu et al. [27] used reduced graphene oxide and TiO_2-KH550 fillers to increase the corrosion resistance of 304 stainless steel in various conditions. Li et al. [28] produced antioxidant functionalized silica-coated TiO_2 nanorods utilizing KH550 as a bridge in a successful synthetic process. By means of melt mixing, the AO-KH550-SiO_2-TiO_2 was added to the polypropylene matrix. According to the findings, AO-KH550-SiO_2-TiO_2 might enhance polypropylene's photostability and thermal stability. Wang et al. [29] coated the surface of B4C particles with KH560 in order to accomplish the optimal dispersion of particles in the epoxy coating and increase the chemical interaction between particles and polymer coating. In addition to enhancing the interaction between B4C particles and epoxy resin, KH560 also demonstrates the capacity to greatly enhance epoxy resin's anti-corrosion performance. By using a unique silane coupling agent sensitization approach that is based on three distinct functional groups of silane coupling agents (KH550, KH570, and KH792), Zou et al. [30] examined the mechanism of chemical copper plating of intermediate-phase asphalt-based carbon fibers. In the copper-plated carbon fibers modified by KH550 grafting, larger grain size, reduced resistivity, and more uniform and continuous copper coating were seen.

Brass powders are inexpensive, easily accessible, and have vibrant, sparkling colors that are incredibly ornamental. There are, however, few studies on coating wood surfaces with water-based coatings that contain brass powder that has been treated with silane coupling agents. Coatings containing pristine brass powder have weak mechanical characteristics, poor age resistance, and poor liquid resistance, making it challenging for them to meet the standards of commercial coatings for wood surfaces [31,32]. The purpose of this paper is to modify copper powder by three different types of silane coupling agents, screen the suitable coupling agent type, optimize the copper powder modification process, and obtain the metal powder wood coatings with the best mechanical properties, liquid resistance, aging resistance, and, at the same time, good optical properties.

In the present study, brass powder was selected as a filler and three types of silane coupling agents (KH550, KH560, KH570) were used for modification to prepare brass powder-water-based acrylic coatings. An orthogonal test based on the optical properties was carried out to investigate the effects of four factors, coupling agent type, coupling agent content, pH value at modification, and brass powder content, on the coating's gloss, color difference, and chromaticity variation. After identifying the most significant influencing elements, single-factor experiments were designed to assess the optical, physical, dry-heat aging, ultraviolet (UV) photooxidation, and cold liquid resistance qualities of the films. These results establish the optimum method for preparing brass powder-water-based acrylic coating and provide technical evidence for the creation of fine art coatings.

2. Materials and Methods

2.1. Test Materials

Brass powder (size 1000 mesh, diameter about 13 μm) was provided by Nangong Xindun Alloy Welding Material Spraying Co., Ltd., Xingtai, China. 3-aminopropyltriethoxysilane (KH550, $C_9H_{23}NO_3Si$, M_w: 221.37 g/mol, CAS No.: 919-30-2), γ-(2,3-epoxypropoxy) propytrimethoxysilane (KH560, $C_9H_{20}O_5Si$, M_w: 236.36 g/mol, CAS No.: 2530-83-8), γ-methacryloxypropyltrimethoxysilane (KH570, $C_{10}H_{20}O_5Si$, M_w: 248.35 g/mol, CAS No.: 2530-85-0) were provided by Hangzhou Jessica Chemicals Co., Ltd., Hangzhou, China. Dulux water-based acrylic primer and topcoat were provided by Dulux Coatings Co., Ltd., Shanghai, China. Citric acid monohydrate ($C_6H_{10}O_8$, M_w: 210.14 g/mol, CAS No.: 5949-29-1) was acquired from Suzhou Changjiu Chemical Technology Co., Ltd., Suzhou, China. Anhydrous ethanol (C_2H_6O, M_w: 46.07 g/mol, CAS No.: 64-17-5) was acquired from Wuxi Jingke Chemical Co., Ltd., Wuxi, China. *Tilia europaea* (Basswood) boards ($100 \times 50 \times 5$ mm^3) were purchased from Beijing Yidimei Model Co., Ltd., Beijing, China. Detergent was obtained from Lishui Diaopai Chemical Co., Ltd., Lishui, China. Coffee was obtained from UCC Ueshima Coffee Co., Ltd., Kobe, Japan.

2.2. Silane Modification Method of Brass Powder

A water/alcohol mixture was prepared as a solvent with a mass ratio of ethanol-to-water of 4:1 to carry out the silane hydrolysis reaction [33], and the pH value was adjusted with citric acid monohydrate [34]. A certain quantity of silane coupling agent was weighed according to the different silane coupling agent concentrations ($m_{\text{silane coupling agent}}$: $m_{\text{brass powder}}$), put into the prepared water/alcohol mixture, and stirred at room temperature until it was fully hydrolyzed to obtain the silane modification solution. The brass powder was then added to the silane-modified solution and stirred for 3 h at a speed of 600 rpm in a DF-101s thermostatic heating magnetic stirrer (Gongyi Yingyu Instrument Co., Ltd., Gongyi, China) at 30 °C. After the reaction was terminated, the obtained material was rinsed and filtered repeatedly with anhydrous ethanol and deionized water, respectively, by an SHZ-D(III) circulating water multi-purpose vacuum pump (Zhengzhou Keda Machinery Instrument Equipment Co., Ltd., Zhengzhou, China) to remove the unreacted silane modifier, and then was put into a 60 °C DYJG-9023 precision blast drying oven (Hangzhou Yijie Technology Co., Ltd., Hangzhou, China) to obtain the modified brass powder. Table 1 shows the dosage of modified brass powder with different silane coupling agent contents.

Table 1. Dosage of brass powder modified with different silane coupling agent content.

Silane Coupling Agent Content (%)	Brass Powder (g)	Silane Coupling Agent (g)	Ethanol (g)	Deionized Water (g)
4	10	0.4	80	20
6	10	0.6	80	20
8	10	0.8	80	20

2.3. Preparation of Water-Based Acrylic Coating with Brass Powder

Firstly, the 2.0 g of water-based primer was weighed and evenly applied to the surface of the Basswood substrate three times. After each application, the sample was exposed to air for 20 min before being cured and dried in an oven at 60 °C. After each coat of primer is finished, it was sanded with 1000 mesh sandpaper. Next, the modified brass powder was added to the water-based acrylic topcoat at a certain mass concentration and mixed well with a glass rod to obtain a 2.0 g coating as the topcoat. The prepared topcoat coating was evenly applied on top of the primer film with the third layer two times, exposed to air for 20 min, and cured in a 60 °C oven until the quality no longer changed. Figure 1 depicts the structure of the prepared brass powder water-based acrylic coating.

Figure 1. The structure of the prepared brass powder water-based acrylic coating.

2.4. Orthogonal Experiment Design

A silane-based coupling agent can improve the metal filler's oxidation resistance and aging resistance in the coating system because of its wettability and dispersion with the metal filler. Common silane coupling agents used in metal-resin systems include KH550, KH560, KH570, etc. The concentration of the silane coupling agent, in addition to the type, has a significant impact on the coating effect with metal filler [35]. Meanwhile, the silane solution's pH influences the rate of the reaction between polycondensation and hydrolysis, which in turn impacts how much film forms on the surface of the metal filler [36]. While silane can hydrolyze more quickly in both acidic and alkaline settings, condensation of silane occurs more readily in an alkaline environment. In order to prevent significant amounts of silane condensation and flocculation during the reaction, the pH is often fixed between 3 and 5.

Consequently, the aforementioned variables were used to develop a four-factor, three-level orthogonal experiment of $L_9(3^4)$, with the following four factors: pH value, copper powder content, silane coupling agent type, and silane coupling agent content, according to Tables 2 and 3. The dosage details of the coating system of the orthogonal test samples are shown in Table 4.

Table 2. Four factors and three levels of the orthogonal test.

Level	Silane Coupling Agent Type	Silane Coupling Agent Content (%)	pH Value at Modification	Brass Powder Content (%)
1	KH550	4	3	20
2	KH560	6	4	40
3	KH570	8	5	60

Table 3. Arrangement table of the orthogonal test.

Sample (#)	Silane Coupling Agent Type	Silane Coupling Agent Content (%)	pH Value at Modification	Brass Powder Content (%)
1	KH550	4	3	20
2	KH550	6	4	40
3	KH550	8	5	60
4	KH560	4	4	60
5	KH560	6	5	20
6	KH560	8	3	40
7	KH570	4	5	40
8	KH570	6	3	60
9	KH570	8	4	20

Table 4. Different silane coupling agent content modified copper powder acrylic paint film dosage details.

Sample (#)	Brass Powder Content (%)	Brass Powder Mass (g)	Water-Based Topcoat Mass (g)	Water-Based Primer Mass (g)
1	20	0.4	1.6	2.0
2	40	0.8	1.2	2.0
3	60	1.2	0.8	2.0
4	60	1.2	0.8	2.0
5	20	0.4	1.6	2.0
6	40	0.8	1.2	2.0
7	40	0.8	1.2	2.0
8	60	1.2	0.8	2.0
9	20	0.4	1.6	2.0

2.5. Single-Factor Experiment Design

The prepared paint film samples were subjected to orthogonal analysis in terms of gloss, color difference, and chromaticity variation to determine the maximum influencing factors. The experimental process was optimized for its maximum influencing factor to investigate the effect of modifications by different silane coupling agents on the performance of brass powder water-based acrylic coatings. The materials and preparation process were kept unchanged, the concentration of the silane coupling agent was fixed at 6.0% and the pH value at modification was 5.0. The effect of brass powder modified by three different silane coupling agents (KH550, KH560, and KH570), and the content of unmodified brass powder on the overall qualities of the film was investigated (Table 5).

Table 5. Experimental arrangement of different amounts of silane coupling agent-modified copper powder on the comprehensive performance of the paint film.

Sample (#)	Silane Coupling Agent Type	Brass Powder Content (%)	Brass Powder Mass (g)	Water-Based Topcoat Mass (g)	Water-Based Primer Mass (g)
0	None	0	0	2.0	2.0
10		10	0.2	1.8	2.0
11		20	0.4	1.6	2.0
12		30	0.6	1.4	2.0
13		40	0.8	1.2	2.0
14		50	1.0	1.0	2.0
15		60	1.2	0.8	2.0
16		70	1.4	0.6	2.0
17	KH550	10	0.2	1.8	2.0
18		20	0.4	1.6	2.0
19		30	0.6	1.4	2.0
20		40	0.8	1.2	2.0
21		50	1.0	1.0	2.0
22		60	1.2	0.8	2.0
23		70	1.4	0.6	2.0
24	KH560	10	0.2	1.8	2.0
25		20	0.4	1.6	2.0
26		30	0.6	1.4	2.0
27		40	0.8	1.2	2.0
28		50	1.0	1.0	2.0
29		60	1.2	0.8	2.0
30		70	1.4	0.6	2.0
31	KH570	10	0.2	1.8	2.0
32		20	0.4	1.6	2.0
33		30	0.6	1.4	2.0
34		40	0.8	1.2	2.0
35		50	1.0	1.0	2.0
36		60	1.2	0.8	2.0
37		70	1.4	0.6	2.0

2.6. *Testing and Characterization*

2.6.1. Characterization of Micro-Morphology and Chemical Composition

The appearance of treated and untreated brass powder was observed using an EM Crafts VERITAS scanning electron microscope (SEM, Korea EM Crafts, Seoul, Korea). Chemical composition analysis of treated brass powder, untreated brass powder, and film samples was performed by a BOEN-85697 Fourier transform infrared spectrometer (FTIR, Fairborn Precision Instruments Co., Ltd., Shanghai, China).

2.6.2. Optical Performance Testing

According to the standard GB/T4893.6-2006 [37], the gloss of the coating was evaluated using a 3NH touchscreen triangle glossmeter NHG268 (Guangzhou Threenh Technology Co., Ltd., Guangzhou, China), which was utilized to measure the gloss at different incidence angles (20°, 60°, and 85°).

A Kutai MS3003 multi-angle spectrophotometer (Guangzhou Threenh Technology Co., Ltd., Guangzhou, China) was utilized to measure the color difference of the coating, and three groups of data L_1, a_1, and b_1 were outputted by the colorimeter at a random location on the film, and another group of data L_2, a_2, and b_2 were measured by the colorimeter at another location on the film. According to the CIE94 color difference Formula (1), ΔL is the difference between L_1 and L_2, Δa (red/green difference) is the difference between a_1 and a_2, and Δb is the difference between b_1 and b_2, to calculate the color difference value (ΔE) of the film [38,39].

$$\Delta E = \sqrt{(\Delta L)^2 + (\Delta a)^2 + (\Delta b)^2} \quad (1)$$

In addition, the colorimetric values of water-based paint films with 10%, 20%, 30%, 40%, 50%, and 60% brass powder content were measured using a SEGT-J portable colorimeter, and four points were taken for each film, and the means were calculated as the colorimetric values of the film, which were recorded as L^*, a^*, and b^*. Three colorimetric parameters were subtracted from the colorimetric values of pure water-based paint films without brass powder to obtain ΔL^*, Δa^*, Δb^*. The chromaticity variation of the film with different brass powder content was estimated according to Formula (1) and noted as ΔE^* [40–42].

The wavelength-reflectance curves, color dominant wavelengths, and color saturation of the Basswood board surface coatings in the visible wavelength range (380–780 nm) were obtained using a Hitachi Model U-3900/U-3900H UV spectrophotometer (Hitachi Limited, Tokyo, Japan) with a scanning speed of 600 nm/min. According to ASTM G173-03 [43], Formula (2) was used to calculate the solar reflectance (R) of the coating in the visible range, where $i(\lambda)$ is the standard solar radiation intensity and $r(\lambda)$ is the reflectance value obtained after testing.

$$R = \frac{\int_{380}^{780} r(\lambda)i(\lambda)d(\lambda)}{\int_{380}^{780} i(\lambda)d(\lambda)} \quad (2)$$

2.6.3. Mechanical Performance Testing

According to the standard GB/T6739-2006 [44], a QHQ-A portable pencil hardness tester (Liangchuang Instrument Co., Ltd., Suzhou, China) was utilized to fix a pencil (6B–6H) and pressed down on the surface of the film at 45° under a load of 750 g, gradually increasing the pencil hardness until the film had a lasting depression, at which time the pencil's hardness was recorded as the coating's hardness.

According to the standard GB/T4893.9-2013 [45], a BEVS 1601 impact tester (Beijing Shidai Shanfeng Technology Co., Ltd., Beijing, China) was utilized to assess the coating's impact resistance. The painted wooden board was put on the impact tester with the steel ball directly above the tested surface, then the steel ball was lifted to a certain height and allowed to free-fall impact on the board. The lowest height for the coating to be broken was used to calculate the impact resistance strength.

According to the standard GB/T4893.4-2013 [46], a QFH-HG600 hundred grid knife (Dongguan Zhongte Precision Instrument Technology Co., Ltd., Dongguan, China) was utilized to assess the coating's adhesion, using a cutting tool to apply evenly force perpendicular to the coating surface to form a set of parallel cutting lines on the coating, and then repeating the above operation to cut on the coating surface and intersecting with the original cutting line at 90° to form a grid pattern. Then the whole grid was covered with tape for 5 min. After smoothly tearing off the tape, the level of coating adhesion was determined with a total of 1–5 levels, with level 1 being the best.

Using the JB-6C roughness tester (Shanghai Gaozhi Precision Instrument Co., Ltd., Shanghai, China), the coated Basswood board was put on the testing table, and the probe was moved to make contact with it. The probe position was adjusted to make it stable at the 0 coordinate, and then the roughness R_a was recorded.

2.6.4. Test of Aging Resistance of the Film

The film samples were subjected to accelerated aging at 160 °C in an oven to carry out dry-heat aging resistance testing [47]. The UV photooxidation resistance testing was performed on the prepared paint samples by using a UVA-340 fluorescent lamp according to ASTM D4587-2011 [48] with a cycle of "4 h irradiation (0.89 $W \cdot m^{-2} \cdot nm^{-1}$) and 20 h condensation" three times. The coating's chromaticity variation ΔE^* was calculated by Formula (1) to evaluate its aging resistance.

2.6.5. Test of Cold Liquids Resistance of the Film

According to the standard GB/T4893.1-2021 [49], citric acid solution with a mass fraction of 10%, undenatured ethanol with a volume fraction of 96%, detergent, and coffee (freeze-dried 40 g instant coffee in 1 L boiling water) were selected as test solutions. Before starting the experiment, the coating's surface was gently wiped with a clean cloth, and a soft filter paper of 25 mm diameter without stain and adhesive was soaked in the test solution, removed, and placed on the surface of the coating, which was then placed in an airtight environment. After a certain period of time, the leftover liquid on the surface was discarded after the filter paper was removed, and the damage to the surface of the coating was observed after standing for 24 h, and the cold liquid resistance grade was evaluated with reference to Table 6. Before and after the liquid resistance test, the surface coating on the Basswood boards was measured for chromaticity using a color difference meter, and the chromaticity variation ΔE^* of cold liquid resistance of the film was computed using Formula (1).

Table 6. Grading table of cold liquid resistance of the film.

Grade	Description
1	No changes: the test area is indistinguishable from adjacent areas.
2	Slight changes: the test area is hardly distinguishable from adjacent areas unless the light source is projected onto the test site revealing indications such as fading, discoloration, and discoloration. No change in the structure of the test surface, such as swelling, fiber protrusion, cracking, or blistering.
3	Moderate changes: visible in several directions, the test area is distinguishable from adjacent areas through indications such as discoloration, discoloration, and discoloration. No change in test surface structure, such as swelling, fiber protrusion, cracking, or blistering.
4	Visible changes: visible in all visible directions, the test area is clearly distinguished from adjacent areas through indications such as discoloration, discoloration and discoloration, and/or slight changes in the structure of the test surface, such as swelling, fiber protrusion, cracking, or blistering.
5	Severe changes: significant changes in test surface structure, and/or fading, discoloration and discoloration, and/or the surface material is removed in whole or in part, and/or filter paper is stuck to the surface.

With an error of less than 5%, each of the aforementioned tests was performed four times.

3. Results and Discussion

3.1. Microscopic Morphology and Infrared Spectral Analysis of the Modified Brass Powder

To observe the encapsulating effect of the brass powder, SEM analysis was performed on the raw brass powder without silane coupling agent modification and the brass powder was modified under the optimal conditions of silane coupling agent content of 6.0% and at a pH of 5.0. From Figure 2A, the surface of the raw brass powder is flat and free of adhesion, while the brass powder's surface, which was modified by the silane coupling agent (Figure 2B–D) is covered with a dense film; the surface is not as smooth as that of the raw brass powder and a more obvious agglomeration phenomenon appears, demonstrating the effectiveness of the coating.

Figure 3 shows the infrared spectra of brass powder modified by different silane coupling agents. For the three spectra, the characteristic peaks generated by the stretching vibration of Si-OH can be observed at 1100 cm^{-1}, which is due to the conversion of the -$SiOC_2H_5$ group in KH550 and the $-SiOC_2H_3$ group in KH560 and KH570 to Si-OH during the hydrolysis process [50]. The stretching vibration peak of C-H in C-CH_3 has a peak at 2920 cm^{-1}. At 2853 cm^{-1}, there is the characteristic peak of =CH_2. The stretching vibration peak of C-C is at 900 cm^{-1}. The peak at 1476 cm^{-1} belongs to the deformation vibrations of

$=CH_2$ and $=CH_3$ [51]. The stretching vibration peak of Si-O occurs at 814 cm^{-1}. The Si-O-C vibration is responsible for the peak at 1180 cm^{-1} [52]. There is a distinctive peak of the stretching vibration of the amino group in KH550 at 3431 cm^{-1} [53]. The C=O in KH570 is represented by the peak at 1650 cm^{-1} [54]. The identified peaks indicate that the brass powder's surface has been successfully covered with the silane coupling agent.

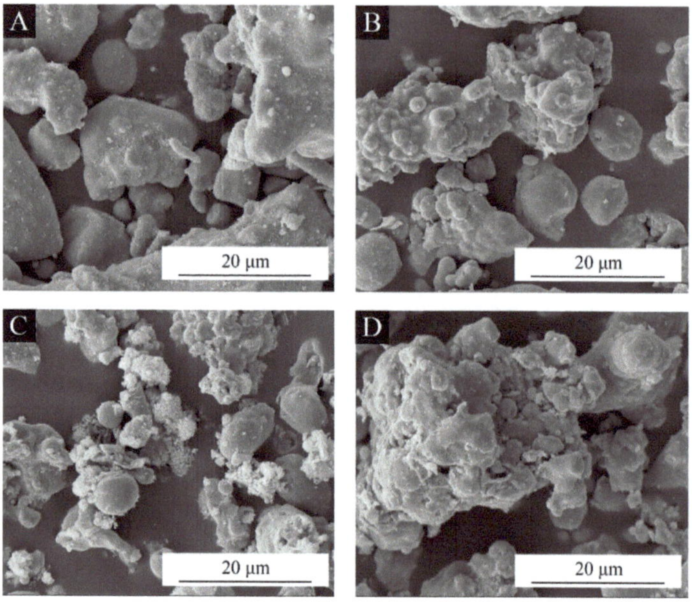

Figure 2. SEM images of brass powder modified by different silane coupling agents under the optimal conditions of silane coupling agent content of 6.0% and pH of 5.0: (**A**) unmodified, (**B**) KH550 modified, (**C**) KH560 modified, (**D**) KH570 modified.

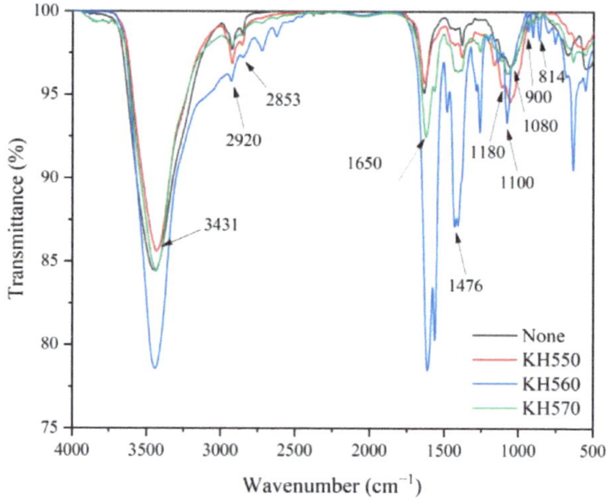

Figure 3. FTIR images of brass powder modified by different silane coupling agents.

3.2. Analysis of Orthogonal Test Results

3.2.1. Gloss Analysis

Gloss refers to the coating surface's capacity to specularly reflect light and serves as the foundation for determining how smooth the coating surface is. The findings are displayed in Table 7, where the gloss performance of the paint film surface is significantly indicated by the gloss corresponding to an incidence angle of 60° [55]. The visual analysis of the paint film's gloss at a 60° incidence angle is shown in Table 8. Sample 5# had the highest gloss level of 13.1 GU, followed by sample 9# with 10.6 GU and sample 1# with 10.5 GU. Figure 4 depicts a graph of the film's gloss effect at a 60° incidence angle. The range results show that the brass powder content has the greatest effect on the coating's gloss, while the coupling agent content, coupling agent type, and pH of the solution during the modification process have less effect on the coating's gloss. According to the results of the ANOVA of the gloss in Table 9, it was found that the ANOVA results of the above four factors were consistent with the results of the range, and the effect of brass powder content on the coating's gloss was significant. Combined with the results of the ANOVA, the better process parameters for the preparation of brass powder water-based coating were determined to be: silane coupling agent KH560, silane coupling agent content of 6%, pH value of the solution during the modification process of 4.0, and brass powder content of 20%.

Table 7. The gloss of brass powder coatings modified by different coupling agents.

Sample (#)	Gloss (GU)		
	20°	60°	85°
1	2.5	10.5	6.0
2	0.9	3.1	1.2
3	0.6	1.0	0.6
4	0.7	1.5	0.7
5	3.2	13.1	7.4
6	0.9	2.9	1.3
7	1.0	3.7	1.7
8	0.7	1.7	1.0
9	2.6	10.6	6.2

Table 8. Visual analysis table of gloss.

Sample (#)	Silane Coupling Agent Type	Silane Coupling Agent Content (%)	pH Value at Modification	Brass Powder Content (%)	Gloss (GU)
1	KH550	4	3	20	10.5
2	KH550	6	4	40	3.1
3	KH550	8	5	60	1.0
4	KH560	4	4	60	1.5
5	KH560	6	5	20	13.1
6	KH560	8	3	40	2.9
7	KH570	4	5	40	3.7
8	KH570	6	3	60	1.7
9	KH570	8	4	20	10.6
Mean 1	4.867	5.233	5.033	11.400	
Mean 2	5.833	5.967	5.067	3.233	
Mean 3	5.333	4.833	5.933	1.400	
Range	0.966	1.134	0.900	10.000	

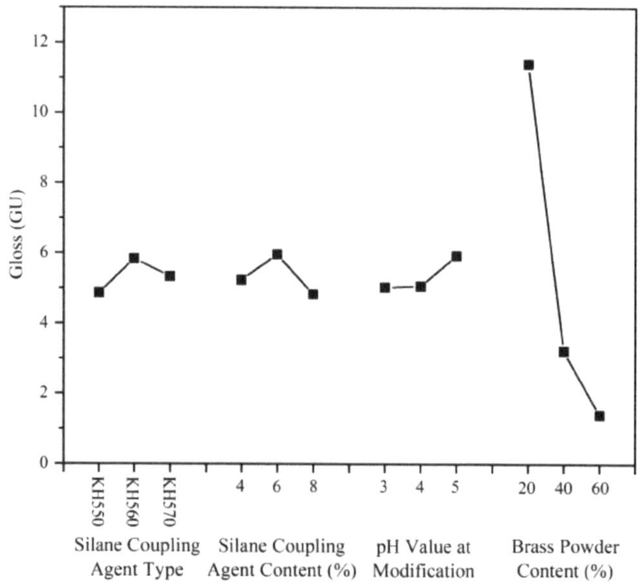

Figure 4. The graph of the film's gloss effect at a 60° incidence angle.

Table 9. Analysis of variance (ANOVA) of gloss.

Factor	Sum of Squares of Deviations	Freedom	F Ratio	F Critical Value	Significance
Silane Coupling Agent Type	1.402	2	1.000	19.000	
Silane Coupling Agent Content	1.982	2	1.414	19.000	
pH Value at Modification	1.562	2	1.114	19.000	
Brass Powder Content	170.056	2	121.295	19.000	*
Error	1.40	2			

Note: * in Table 9 means significant.

3.2.2. Color Difference Analysis

The primary factor used to determine whether the coating's color is uniform or not is its color difference value, and the results of the calculation of this parameter are displayed in Table 10. Table 11 shows the visual analysis of color difference. According to the color difference effect (Figure 5) and the color difference ANOVA (Table 12), it is the coupling agent type that has a significant effect on the coating color difference value compared to the brass powder content, coupling agent content, and pH of the solution during modification. Combined with the results of the range, the prepared brass powder water-based coating has a more uniform coating color under the conditions of silane coupling agent KH570, silane coupling agent content of 8%, pH value of the solution during the modification process of 4.0, and brass powder content of 40%.

Table 10. The color difference of brass powder coatings modified by different coupling agents.

Sample (#)	Chromaticity Parameter						ΔE
	L_1	a_1	b_1	L_2	a_2	b_2	
1	62.41	7.78	34.31	58.27	10.15	34.09	4.78
2	56.53	5.88	35.22	54.12	7.26	32.66	3.78
3	57.70	7.05	33.13	55.00	7.00	35.35	3.50
4	55.00	7.00	35.35	59.94	7.00	38.93	6.10
5	65.95	12.45	33.46	66.32	7.45	38.25	6.93
6	57.80	12.22	29.76	57.70	7.05	33.13	6.17
7	53.95	7.37	36.72	56.53	5.88	35.22	3.34
8	56.53	5.88	35.22	54.12	7.26	32.66	3.78
9	64.46	8.49	33.33	63.54	6.19	35.92	3.58

Table 11. Visual analysis table of the color difference.

Sample (#)	Silane Coupling Agent Type	Silane Coupling Agent Content (%)	pH Value at Modification	Brass Powder Content (%)	Gloss (GU)
1	KH550	4	3	20	4.78
2	KH550	6	4	40	3.78
3	KH550	8	5	60	3.50
4	KH560	4	4	60	6.10
5	KH560	6	5	20	6.93
6	KH560	8	3	40	6.17
7	KH570	4	5	40	3.34
8	KH570	6	3	60	3.78
9	KH570	8	4	20	3.58
Mean 1	4.020	4.740	4.910	5.097	
Mean 2	6.400	4.830	4.487	4.430	
Mean 3	3.567	4.417	4.590	4.460	
Range	2.833	0.413	0.423	0.667	

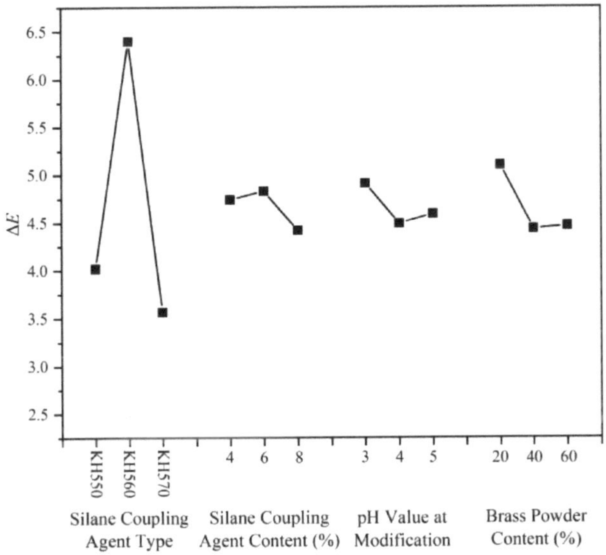

Figure 5. Effect of the color difference.

Table 12. ANOVA of the color difference.

Factor	Sum of Squares of Deviations	Freedom	F Ratio	F Critical Value	Significance
Silane Coupling Agent Type	13.898	2	49.110	19.000	*
Silane Coupling Agent Content	0.283	2	1.000	19.000	
pH Value at Modification	0.292	2	1.032	19.000	
Brass Powder Content	0.851	2	3.007	19.000	
Error	0.28	2			

Note: * in Table 12 means significant.

3.2.3. Chromaticity Variation Analysis

The chromaticity variation, i.e., the color change between the brass powder coatings and a pure water-based coating, can reflect the decorative color effect of the coating. The calculated results and visual analysis of the chromaticity variation of the coating are shown in Tables 13 and 14. Compared with other samples, the chromaticity parameter of sample 7# was the best, reaching 16.15, followed by samples 2 and 8, with 16.12 and 15.57, respectively. The chromaticity variation effect graph (Figure 6) and the chromaticity variation ANOVA table (Table 15) showed that among the coupling agent type, coupling agent concentration, pH value of the solution during coating modification, and brass powder content, brass powder content is the one that can significantly affect the chromaticity variation. Combined with the results of the range, the highest chromaticity variation of the prepared brass powder water-based coatings was obtained under the conditions of silane coupling agent KH550, silane coupling agent content of 4%, the pH value of the solution during the coating modification process of 4.0, and brass powder content of 40%.

Table 13. The chromaticity variation of brass powder coatings modified by different coupling agents.

Sample (#)	Chromaticity Parameter			ΔE^*
	L^*	a^*	b^*	
0	70.42	8.93	30.96	-
1	61.89	8.55	34.01	9.07
2	54.49	7.54	33.02	16.12
3	56.73	6.75	34.21	14.24
4	57.34	7.46	37.33	14.63
5	65.51	9.98	34.82	6.34
6	56.57	9.38	32.31	13.93
7	55.24	6.63	35.97	16.15
8	55.33	6.57	33.94	15.57
9	64.12	7.13	34.99	7.70

Table 14. Visual analysis table of the chromaticity variation.

Sample (#)	Silane Coupling Agent Type	Silane Coupling Agent Content (%)	pH Value at Modification	Brass Powder Content (%)	ΔE^*
1	KH550	4	3	20	9.07
2	KH550	6	4	40	16.12
3	KH550	8	5	60	14.24
4	KH560	4	4	60	14.63
5	KH560	6	5	20	6.34
6	KH560	8	3	40	13.93
7	KH570	4	5	40	16.15
8	KH570	6	3	60	15.57
9	KH570	8	4	20	7.70
Mean 1	4.020	4.740	4.910	5.097	
Mean 2	6.400	4.830	4.487	4.430	
Mean 3	3.567	4.417	4.590	4.460	
Range	2.833	0.413	0.423	0.667	

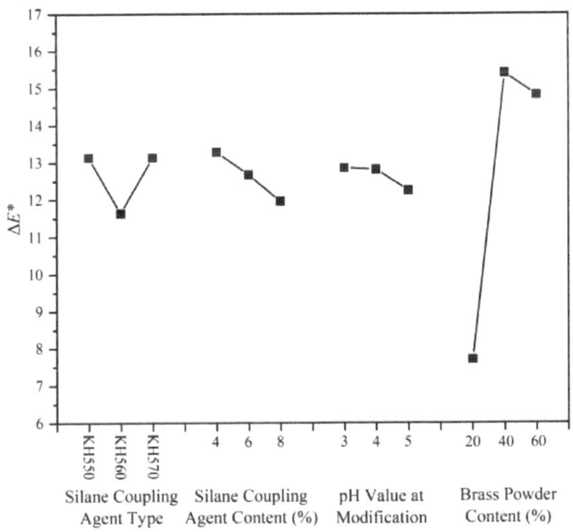

Figure 6. Effect of the chromaticity variation.

Table 15. Analysis of variance of the chromaticity variation.

Factor	Sum of Squares of Deviations	Freedom	F Ratio	F Critical Value	Significance
Silane Coupling Agent Type	4.550	2	6.445	19.000	
Silane Coupling Agent Content	2.646	2	3.748	19.000	
pH Value at Modification	0.706	2	1.000	19.000	
Brass Powder Content	110.135	2	155.999	19.000	*
Error	0.71	2			

Note: * in Table 15 means significant.

3.3. *Analysis of Single-Factor Experiment Results*

3.3.1. Optical Performance Analysis

Gloss

Table 16 and Figure 7 display the results of various coupling agent modifications on the gloss of the brass powder coatings. There is no discernible variation in gloss between unmodified and brass powder coatings modified by the three coupling agents, and the different modification treatments have no discernible impact on the gloss performance of the coatings. The gloss at incidence angles of 20°, 60°, and 85° revealed a negative association with the brass powder content, i.e., the higher the brass powder's concentration, the lower the coating's gloss, for both the unmodified and the brass powder coatings modified by the three coupling agents. This is so because brass powder is a solid powder, and adding more of it to the coating will make the coating surface rougher, which will reduce the coating's gloss due to diffuse light reflection [56]. The gloss of the coating will differ greatly depending on the brass powder percentage, which should be between 0 and 10%. The weakening effect of brass powder concentration on the gloss of the coating gradually reduces in the 20% to 50% range. The gloss curve trend tends to level out as the brass powder percentage reaches 50%, and at this point, the brass powder content begins to saturate. Therefore, the brass powder concentration in the modified coating should be kept to 0–10% in order to provide a nice gloss effect. And within this range, the gloss performance of the brass powder coating following the KH560 treatment (sample 24#) is best.

Table 16. Gloss of different coupling agent-modified brass powder coatings.

Sample (#)	Silane Coupling Agent Type	Brass Powder Content (%)	Gloss (GU)		
			20°	60°	85°
0	None	0	17.4	55.3	64.8
10		10	6.6	25.2	22.1
11		20	3.0	13.0	9.6
12		30	1.2	4.7	2.4
13		40	0.8	3.4	2.1
14		50	0.9	2.6	1.3
15		60	0.6	1.9	1.0
16		70	0.6	1.1	0.7
17	KH550	10	6.2	22.6	20.4
18		20	2.1	9.0	5.1
19		30	1.2	5.3	2.6
20		40	0.9	2.4	1.4
21		50	0.6	1.6	0.8
22		60	0.6	1.2	0.8
23		70	0.5	1.0	0.7
24	KH560	10	8.1	27.2	22.6
25		20	2.6	10.0	6.6
26		30	1.4	4.9	2.9
27		40	0.8	2.4	1.3
28		50	0.7	1.8	1.0
29		60	0.7	1.6	0.9
30		70	0.6	1.1	0.8
31	KH570	10	5.7	20.0	18.0
32		20	2.7	10.4	6.0
33		30	1.2	3.7	2.1
34		40	0.9	2.9	1.5
35		50	0.8	1.9	0.9
36		60	0.6	1.4	0.9
37		70	0.6	1.2	0.8

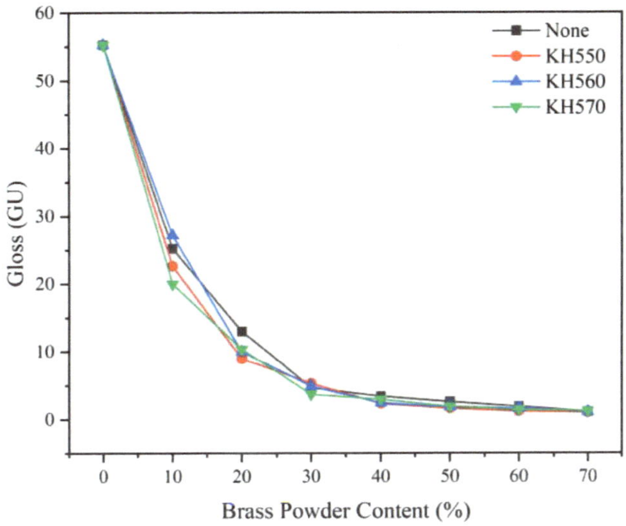

Figure 7. Gloss of different coupling agent-modified brass powder coatings with different brass powder content.

Color Difference

Figure 8 illustrates the impact of various coupling agent modifications on the brass powder coating's color difference. The color difference of the unmodified brass powder coating has the biggest change with an increase in the amount of brass powder. The coupling agent molecule has the structural formula Y-R-Si-X_3, where Y in KH550, KH560, and KH570 is an amino functional group, an epoxy hydrocarbon functional group, and a methacryloyloxyalkyl organic functional group, respectively. R is a carbon chain with saturated or unsaturated bonds, and X is easily combined with silicon atoms and can be hydrolyzed by chlorine, methoxy, ethoxy, acetoxy, and other groups. The brass powder is modified using a silane coupling agent to produce silicon hydroxyl (Si-OH), which is then further dehydrated and condensed between coupling agent molecules to produce an oligomeric siloxane that contains several Si-OH groups. The -OH on the surface of the brass powder combines with the Si-OH in the oligomeric siloxane to produce covalent or hydrogen bonds [57]. The organic functional groups of the polymer chemically react with the Y group. Brass powder is hydrophobic and has a low affinity for water-based coatings without surface treatment. It is essentially insoluble in water and forms huge aggregates in water-based coatings, making it challenging to distribute evenly. Although some agglomeration of the silane-modified brass powder is unavoidable, due to the surface's abundance of hydrophilic groups and its attraction for water-based coatings, the dispersion can be enhanced. Brass powder content does not significantly affect the coating's color fluctuation. The best dispersion at 10% brass powder concentration is found with KH560 (sample 24#), followed by KH570 (sample 31#) and KH550 (sample 17#), and the coatings produced have a more uniform color.

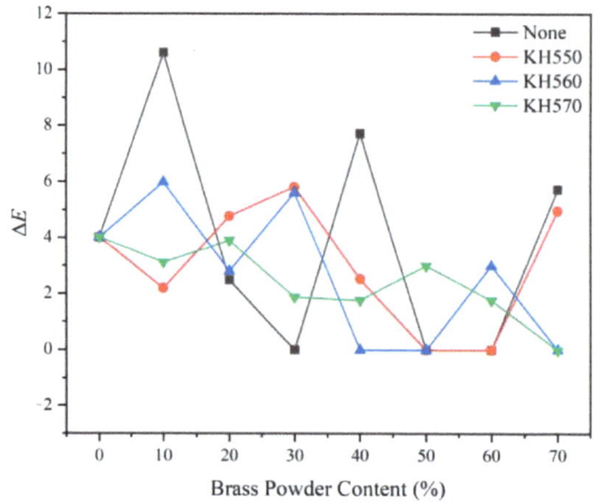

Figure 8. The color difference of different coupling agent-modified brass powder coatings with different addition contents.

Chromaticity Variation

Figure 9 depicts the impact of various coupling agent modifications on the chromaticity variation of brass powder coatings. The chromaticity variation between brass powder coatings modified by the three coupling agents is small when the brass powder concentration is the same, and the modification treatment of using different coupling agents has little effect on the coating's chromaticity variation. The chromaticity variation of unmodified and modified brass powder coatings with various coupling agents grew more pronounced with the increase in brass powder content. A brass powder content range of 10–40% was where the coatings' chromaticity variation was the most pronounced. Because the brass powder can essentially lie flat on the surface of the Basswood substrate once the brass powder content exceeds 40%, the chromaticity variation of the coating tends to level off after that point. Moreover, further increases in the brass powder content will have little impact on the coating's overall color. The most noticeable chromaticity difference with 10% brass powder content is found with KH560 (sample 24#), followed by KH570 (sample 31#) and KH550 (sample 17#).

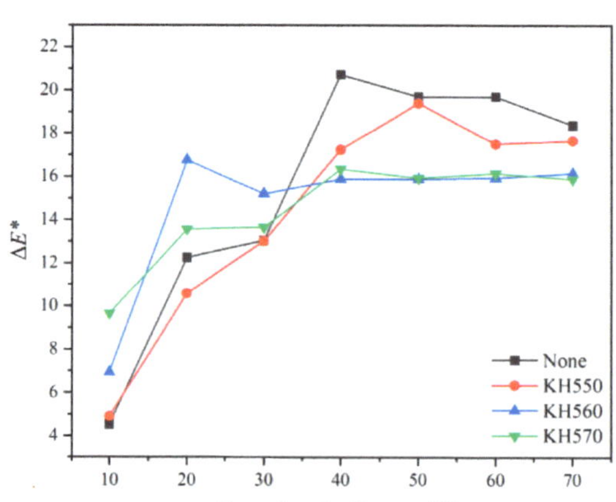

Figure 9. Variation curve of chromaticity variation of different coupling agent-modified brass powder coatings with different addition contents.

Visible Light Reflectance

When the outside temperature is high, a coating with high reflectivity applied to the surface of wood can prevent heat from penetrating to the interior of the wood, thereby preventing heat aging and other issues while extending the service life of wooden objects. The reflectance of the visible wavelength band for various coupling agent-modified brass powder coatings with various brass content is shown in Figures 10 and 11, and Table 17. The R-value of visible wavelength reflectance of the brass powder coating was not significantly affected by the silane coupling agent treatment. Brass powder concentration and visual reflectance R-value had a negative relationship. The R-value in the 0 to 40% range of brass powder composition significantly declines as the brass powder content rises. The R-value remained relatively unchanged once the brass powder percentage reached more than 40%. The coatings display the greatest visible light reflectance at a brass powder concentration of 10%.

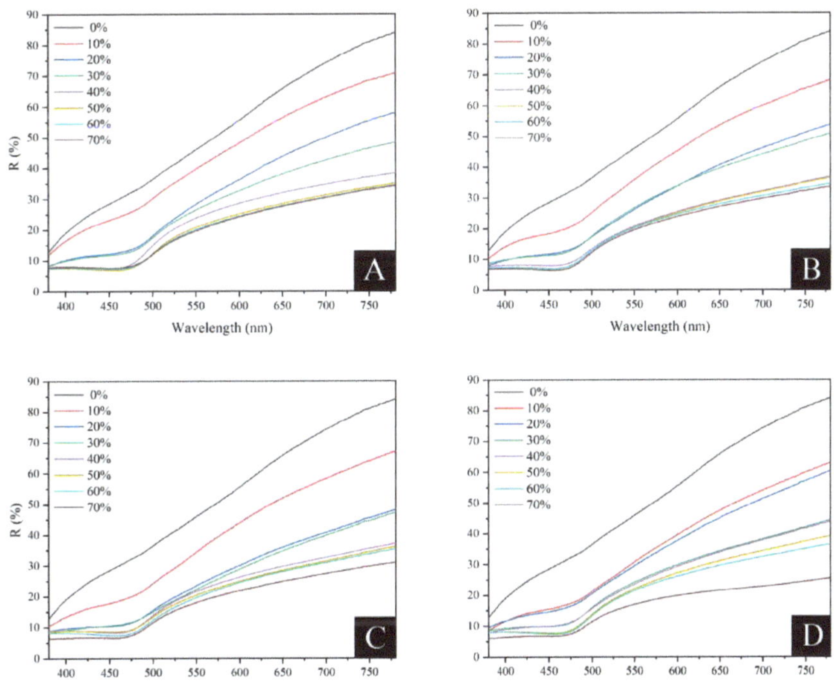

Figure 10. Visible wavelength reflectance of different coupling agent-modified brass powder coatings at different addition contents: (**A**) unmodified, (**B**) KH550 modified, (**C**) KH560 modified, (**D**) KH570 modified.

Figure 10 illustrates the coating samples' color saturation. The color saturation of the sample increases and decreases with the difference between the high and low points of the reflectance spectrum. More bright colors result from higher saturation levels, while darker colors result from lower saturation levels. The color saturation of the coated samples steadily decreased with the rise in brass powder content, and the color gradually became duller. This tendency was seen in both the unmodified and modified brass powder coatings. Among them, the coating's color brightness with a 10% brass powder concentration was the best.

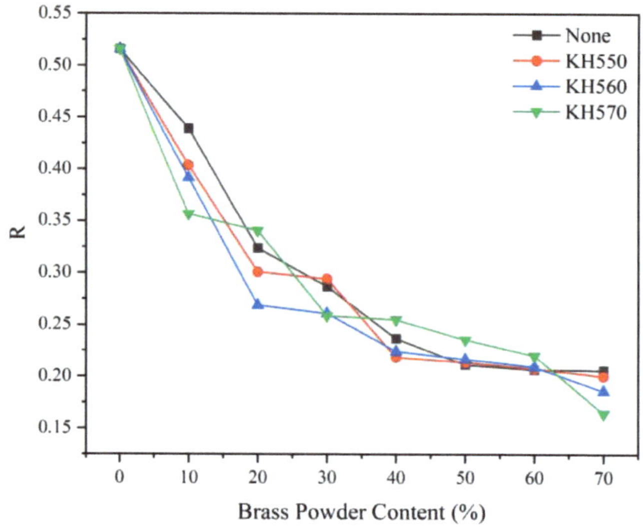

Figure 11. Visible wavelengths reflectance of different coupling agent-modified brass powder coatings with different brass content.

Table 17. Visible light reflectance of modified brass powder paint films with different coupling agent types with different brass content.

Brass Powder Content (%)	R			
	None	KH550	KH560	KH570
0	0.5158	0.5158	0.5158	0.5158
10	0.4389	0.4038	0.3919	0.3566
20	0.3238	0.3008	0.2688	0.3406
30	0.2867	0.2941	0.2610	0.2588
40	0.2367	0.2186	0.2245	0.2550
50	0.2115	0.2138	0.2169	0.2357
60	0.2068	0.2077	0.2094	0.2204
70	0.2062	0.2004	0.1860	0.1646

The primary wavelengths of modified brass powder paint films are shown in Table 18. The primary wavelength of the coating color can be utilized as a starting point for assessing the coating samples' overall visual color. All samples' primary color wavelengths fell within the range of 586–590 nm. The wavelength of the yellow component of visible light ranges from approximately 565 to 590 nm; the closer the wavelength is to 590 nm, the more orange the sample seems overall, and the closer it is to 565 nm, the greener it appears overall. The main wavelength of the coating is closer to 586 nm as the brass powder content rises, demonstrating that the coating's overall color is becoming more and more yellow, and that brass powder has a more pronounced effect on concealing the color and texture of the wood. These findings are in line with the findings of the chromaticity variation evaluation index.

Table 18. Main wavelengths of modified brass powder paint films with different coupling agent types.

Brass Powder Content (%)	Main Wavelength			
	None	KH550	KH560	KH570
0	589.89	589.89	589.89	589.89
10	589.34	589.56	589.88	590.00
20	589.29	589.47	589.78	589.68
30	588.30	588.57	589.21	588.15
40	587.64	587.56	587.90	587.80
50	587.25	587.40	587.68	587.40
60	586.88	587.06	587.26	586.09
70	586.41	587.00	587.13	587.05

3.3.2. Mechanical Performance Analysis

Hardness

According to Table 19, the amount of various silane coupling agent-modified brass powders in the coatings affected how hard the Basswood surface's coatings were. Due to the water-based coating's inherent softness, Basswood surface coatings often have a low hardness. The coating's hardness was 4B without the presence of brass powder. Because the thermoplastic acrylic self-drying technique primarily relied on self-crosslinking to generate a film, which is a physical film formation, its hardness performance was inferior. The film's hardness rose sharply to HB when 10% brass powder was added. The hardness of the film steadily rose as the amount of brass powder increased, but the difference was not substantial. The silane coupling agent greatly increased the coating's hardness compared to the unaltered brass powder coating. This is due to the fact that the Y group can interact with the resin matrix and serve as a coupling agent, improving the bonding efficiency of the resin and brass powder and significantly enhancing the mechanical qualities of the brass powder-resin system. The effect of the silane coupling agent's type on the coating's hardness is essentially the same when the coating contains 10–30% brass powder. When the coating contains 40% brass powder, the coating's hardness after being altered by various coupling agents appears to vary. Due to the cross-linking principle, the epoxy functional group of KH560 and the carboxyl group of water-based acrylic acid underwent an addition reaction to form a chemical bond, whereas the amino functional group of KH550 and the carboxyl group of water-based acrylic acid underwent a coordination reaction to form an ionic bond [58]. The hardness of the brass powder coating modified by KH560 and KH570 was 3H, which was higher than that of the brass powder coating modified by KH550. The lower hardness is caused by the readily broken ionic bonds.

Table 19. The hardness of modified brass powder film with different coupling agent types.

Brass Powder Content (%)	Hardness			
	None	KH550	KH560	KH570
0	4B	4B	4B	4B
10	HB	HB	HB	HB
20	H	H	H	H
30	H	2H	2H	2H
40	H	2H	3H	3H
50	H	3H	3H	3H
60	H	3H	3H	3H
70	H	5H	5H	5H

Impact Resistance

The amount of brass powder in the coating modified with various silane coupling agents affects the impact resistance of the Basswood surface coating according to Table 20. The amount of brass powder applied affects the coating's impact resistance, whereas the silane coupling agent's type and the amount of brass powder added both have an impact on the coating's impact strength. The modified brass powder coating has a much higher impact strength than the unmodified brass powder coating. The unmodified brass powder coating has a maximum impact strength of 5 kg·cm in the 0 to 70% addition range, while the modified brass powder coating has a maximum impact strength of 8 kg·cm. This is since, following the treatment of the brass powder, the coupling agent molecules can create a molecular layer that is covalently bound to the surface of the brass powder, and that layer can significantly hinder water infiltration. The resin matrix strength will diminish as a result of the water intrusion, and the bond between the resin matrix and the interface will eventually fail, drastically reducing the mechanical capabilities of the composite material. The impact strength of the unmodified and modified brass powder coatings tended to rise and then fall with the brass powder content. The coating's impact strength is optimum at 40% brass powder content for both unmodified and KH550-modified coatings. At a brass powder percentage of 50%, the KH560- and KH570-modified brass powder coatings showed the highest impact strength. Due to its double bonds, which boost strength, KH570 has the strongest impact resistance [59]. By reducing the freedom of the molecular chain and enhancing the impact resistance of the brass powder coating, the cross-linking reaction with the resin matrix increases the crosslinking density of the brass powder-acrylic resin.

Table 20. Impact resistance of modified brass powder films with different coupling agent types.

Brass Powder Content (%)	Impact Resistance (kg·cm)			
	None	KH550	KH560	KH570
0	1	1	1	1
10	1	2	3	4
20	3	3	4	4
30	3	3	5	5
40	5	8	6	7
50	4	7	8	8
60	3	6	7	7
70	4	8	5	6

Adhesion

According to Table 21, the amount of brass powder in coatings adjusted with various silane coupling agents affected how well they adhered to the surface of the Basswood. The adherence of the modified brass powder coating exhibits a more pronounced and significant improvement when compared to the unmodified brass powder coating. This is because of the coupling action of the silane coupling agent molecules with the covalently bonded molecular layer on the surface of the brass powder. The Y group resin molecules also effectively reduce the degradation of adhesion due to moisture intrusion and maintain or noticeably improve the mechanical properties of the composite [60]. Brass powder content and coating adhesion are inversely connected. The higher the brass powder concentration, the lower the coating adhesion. This is due to the fact that the coating's adhesion is also influenced by its own characteristics and how well it adheres to the Basswood substrate. The internal dense structure of the coating is damaged when the brass powder is mixed into the water-based acrylic coating. As a result, the coating's cohesion decreases, which is reflected in a decrease in the coating's adhesion to the wood substrate [61]. Therefore, the coating that contains an excessive amount of brass powder doesn't adhere as well. The coatings containing the modified brass powder showed the best adhesion (1 grade) when the content of metal fillers did not exceed 30%, which is consistent with the optimal samples (1 grade) in previous reports [62]. The coating modified by KH570 has the best adhesion of the three silane modifiers and can retain the adhesion of 2 grade even when the amount of brass powder is raised to 40%.

Table 21. Different coupling agent types modified brass powder paint film adhesion.

Brass Powder Content (%)	Adhesion (Grade)			
	None	KH550	KH560	KH570
0	0	0	0	0
10	1	1	1	1
20	2	1	1	1
30	2	1	1	1
40	4	3	3	2
50	4	3	4	3
60	4	3	4	3
70	5	4	4	4

Roughness

According to Table 22, the amount of brass powder in coatings modified with various silane coupling agents affected how rough the Basswood surface coatings were. The brass powder-acrylic coating's roughness was made worse by the coupling agent alteration. This is because the hydroxyl group reacts with the inorganic filler particles, causing the surface of the filler particles to progressively be replaced by the silane coupling agent condensate and making the particles' surface rougher (Figure 2). As the amount of brass powder in the coating increased, the coating's roughness frequently did as well. In the range of 0% to 10%, the growing trend was more pronounced. The lowest roughness of the coating was achieved when the content of metal filler was 10%, which was better than the roughness of 1.9 μm reported in reference [63].

Table 22. The roughness of modified brass powder coating film with different coupling agent types.

Brass Powder Content (%)	Roughness (μm)			
	None	KH550	KH560	KH570
0	0.480	0.480	0.480	0.480
10	1.635	1.209	1.022	1.322
20	2.318	1.742	1.612	1.965
30	2.679	2.481	1.413	3.554
40	2.983	3.654	3.554	3.210
50	2.766	2.988	2.824	2.504
60	3.908	2.782	3.037	3.453
70	4.138	3.860	3.050	4.094

3.3.3. Aging Resistance Analysis

High-Temperature Accelerated Aging

Figure 12 depicts the chromaticity variation of the coating on Basswood before and after the high-temperature accelerated aging test with the different brass powder content adjusted by various coupling agents. Both the coupling agent-modified brass powder coatings and the unmodified brass powder coating produced chromaticity changes during the aging process. Due to the prolonged treatment that accelerated the aging of the Basswood and the coating, the chromaticity variation of the coating became increasingly apparent as the aging period rose. Within 0–5 h of the high-temperature accelerated aging treatment, the coating's chromaticity variation was at its greatest. The brass powder concentration is roughly negatively connected with the chromaticity variation between the coating before and after aging throughout the same aging period, meaning that the higher the brass powder content, the less noticeable the coating's chromaticity variation after high-temperature accelerated aging. This is because the brass powder itself does not readily react chemically at high temperatures to alter the color. Additionally, a high concentration of brass powder coating on wood might cover the wood itself during a high-temperature carbonization reaction since the material's texture and color coverage are stronger. Because both the water-based coating and the wood are susceptible to denaturation at high temperatures, the weaker the color coverage of the wood substrate, and the higher the transparent water-based coating content in the coating, the greater the chromaticity variation on the coating with low brass powder content. After accelerated aging at high temperatures for 20 h, the coating with 10% unmodified brass powder had the biggest chromaticity variation of 38.22, which was greater than the coating with brass powder modified by coupling agents. This is because the silane coupling agent can coat the brass powder particles and raise the interfacial heat resistance of the brass powder particles, which to some extent retards the aging of the wood substrate and water-based coating substrate. Due to the addition of some KH550 internal amino groups to the reaction during polymerization, which reduces the coupling agent's strong alkalinity and increases its temperature resistance [64], the brass powder coating modified by KH550 shows the least amount of chromaticity variation after high-temperature aging.

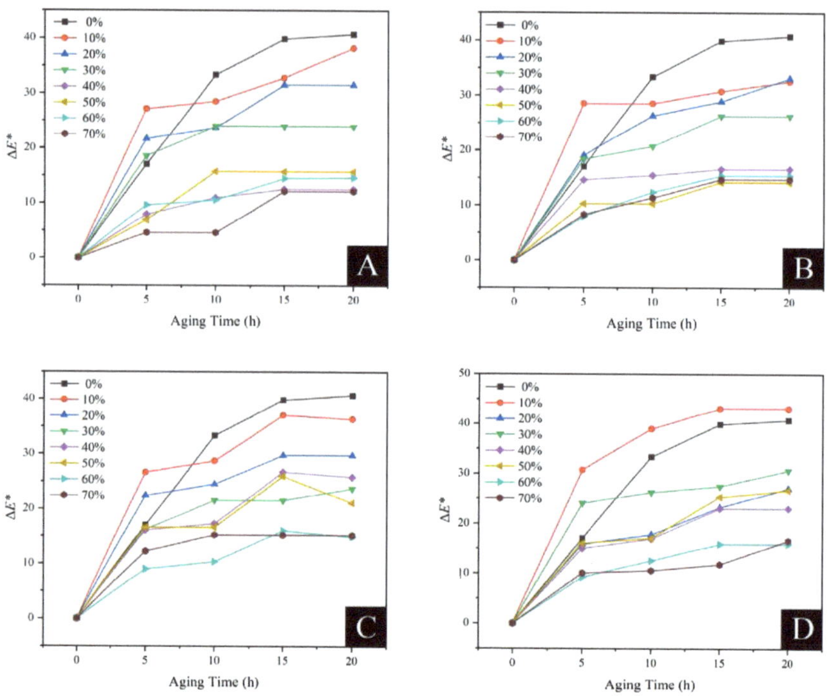

Figure 12. High-temperature accelerated aging resistance of different coupling agents-modified brass powder coatings at different addition contents: (**A**) unmodified, (**B**) KH550 modified, (**C**) KH560 modified, (**D**) KH570 modified.

UV Aging Resistance

The high UV energy of sunlight, in conjunction with oxygen, readily initiates the polymer materials' free radical chain photo-oxidative aging reaction. This results in the polymer macromolecular chain being broken or producing specific crosslinking, which directly affects the performance of polymer materials by causing color darkening, brittleness, hardening, surface cracking, and a decline in mechanical properties, among other aging phenomena, and eventually causes them to lose their value. Figure 13 depicts the chromaticity fluctuation of the Basswood surface coating with brass powder content in the coatings modified by various coupling agents. The amount of brass powder present and the coating's UV aging chromaticity variation are negatively correlated. The less noticeable the chromaticity variation of the coating after UV aging, the higher the concentration of brass powder filler. The coating color is primarily influenced by UV light when the brass powder content is between 10% and 20%. This is because the effect of UV aging on brass powder is less pronounced than it is on polymeric water-based coatings, and the coating's chromaticity fluctuation is more visible, the larger the proportion of water-based coatings in the coating. After the UV aging experiment, the unmodified brass powder coating had a more significant chromaticity variation. Brass powder coatings can be treated using silane coupling agents to increase their UV aging resistance. Among them, the coating modified with KH570 performed the best.

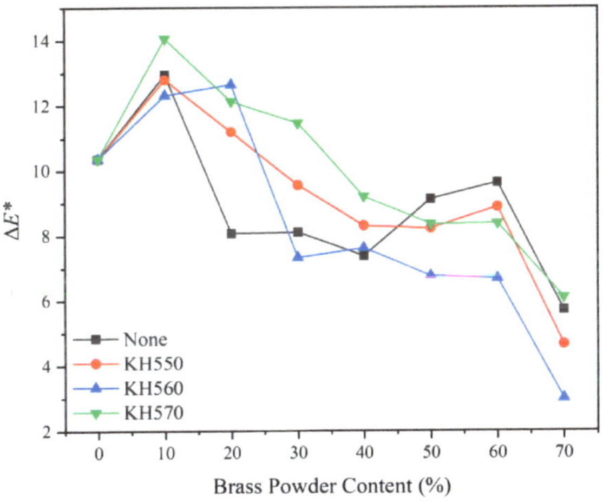

Figure 13. UV aging resistance of different coupling agent-modified brass powder coatings at different addition contents.

3.3.4. Cold Liquid Resistance Analysis

One of the crucial criteria for testing the physicochemical qualities of wood coatings is the measurement of the coating's liquid resistance. This parameter is based on a damage scenario induced by a liquid to which wooden objects may be exposed in daily life. The gloss of the Basswood surface coating after the cold liquid resistance test varied according to the amount of brass powder in the coating modified using various coupling agents, as indicated in Table 23. Following the evaluation of liquid resistance, all coatings displayed some degree of gloss degradation. At brass powder contents of 0 and 10%, the coatings' loss of gloss was most noticeable. Detergent and coffee, out of the four liquids examined, had the biggest impact on the gloss of the brass powder coatings. Detergent and coffee had the greatest impact on the gloss of the brass powder coating modified by KH560 at 10% concentration (sample 24#), decreasing it by 20.5 and 11.3, respectively, while detergent and coffee had the least impact on the gloss of the brass powder coating modified by KH550 at 10% concentration (sample 17#), decreasing it by 9.9 and 8.0, respectively. KH570-modified brass powder coatings demonstrated more consistent gloss in the liquid resistance.

Table 24 illustrates the chromaticity variation of the coated surface of Basswood before and after the cold liquid resistance tests with various coupling agent-modified brass powder content. After the cold liquid resistance tests, the brass powder acrylic coating that had been altered using coupling agents had a significant change in coating color. The acrylic coating with brass powder responded the least to the four test liquids in terms of color change, whereas the KH560-modified coating was more sensitive to the effects of the test liquids and created more pronounced color changes. The coating's chromaticity fluctuation was lowest at 10% brass powder content.

Table 23. Liquid resistance gloss of different coupling agent-modified brass powder coatings.

Sample (#)	Silane Coupling Agent Type	Brass Powder Content (%)	Gloss (GU)			
			Citric Acid	Ethanol	Detergent	Coffee
0	None	0	51.9	50.1	27.9	48.4
10		10	25.0	25.0	24.2	24.0
11		20	11.3	12.9	11.6	12.3
12		30	4.7	4.7	3.8	4.7
13		40	3.4	3.1	2.7	3.2
14		50	2.5	2.6	2.3	2.6
15		60	1.8	1.9	1.7	1.9
16		70	1.1	1.1	1.1	1.1
17	KH550	10	21.6	19.4	6.5	15.7
18		20	7.6	7.6	4.9	8.5
19		30	4.3	4.4	2.6	4.4
20		40	2.2	2.4	2.4	2.4
21		50	1.6	1.3	1.6	1.3
22		60	1.2	1.1	1.2	1.0
23		70	1.0	1.0	1.0	1.0
24	KH560	10	21.0	22.7	6.7	15.9
25		20	8.2	10.0	5.2	7.3
26		30	4.3	4.0	3.4	3.6
27		40	2.2	2.3	2.2	2.0
28		50	1.5	1.7	1.8	1.5
29		60	1.2	1.4	1.4	1.5
30		70	1.0	1.1	1.1	1.0
31	KH570	10	13.3	17.3	10.1	12.0
32		20	8.2	8.1	4.7	8.6
33		30	3.5	2.9	2.4	3.0
34		40	2.7	2.7	2.3	2.9
35		50	1.6	1.8	1.3	1.4
36		60	1.4	1.4	1.3	1.1
37		70	1.1	1.1	1.2	1.0

Table 24. Chromaticity variation of different coupling agent-modified brass powder coatings.

Sample (#)	Silane Coupling Agent Type	Brass Powder Content (%)	ΔE^*			
			Citric Acid	Ethanol	Detergent	Coffee
0	None	0	2.89	1.51	2.89	2.61
10		10	5.46	5.21	4.04	5.19
11		20	6.19	1.25	4.98	2.01
12		30	3.15	2.67	3.23	8.11
13		40	5.33	3.86	12.74	5.39
14		50	15.51	0.00	9.62	9.02
15		60	15.51	0.00	9.59	9.02
16		70	8.91	5.43	18.48	5.71
17	KH550	10	4.49	1.93	9.54	4.22
18		20	11.19	2.17	11.19	6.34
19		30	10.75	4.60	9.08	7.31
20		40	15.22	8.62	8.48	10.04
21		50	16.30	8.22	7.96	12.83
22		60	15.94	2.54	9.39	9.01
23		70	13.46	2.49	7.34	8.41

Table 24. *Cont.*

Sample (#)	Silane Coupling Agent Type	Brass Powder Content (%)	ΔE^*			
			Citric Acid	Ethanol	Detergent	Coffee
24		10	2.97	4.57	2.97	5.12
25		20	13.72	1.50	11.27	7.28
26		30	13.89	2.80	7.35	7.07
27	KH560	40	14.77	9.19	11.82	10.56
28		50	14.94	9.19	14.77	10.08
29		60	20.39	1.50	11.93	10.35
30		70	15.23	3.00	15.23	3.69
31		10	6.64	4.99	5.75	8.80
32		20	1.27	4.51	4.61	3.07
33		30	5.79	1.17	3.38	6.75
34	KH570	40	8.33	2.54	2.98	2.98
35		50	6.05	7.32	9.38	6.05
36		60	21.34	2.95	6.84	9.23
37		70	18.12	0.00	11.82	5.87

Table 25 displays the findings of the grading of the Basswood surface coatings following the measurement of liquid resistance. All of the coatings were essentially grade 1 for ethanol, leaving no mark on the coating film surface; grade 2 for coffee, leaving a minor discoloration mark on the coating film surface. The primary cause is that coffee is a dark brown color, which makes it easier for the coating to stain. The liquid resistance level for detergent is primarily grade 2, whereas the grade 3 performance of unmodified brass powder coating can be effectively improved by silane coupling agent modification. After the citric acid test, the brass powder coating modified by KH570 has improved liquid resistance and can still retain grade 2 liquid resistance even when the amount of brass powder increases to 50%. Based on the information in Table 25, silane coupling agent KH570 and a brass powder content of 10% resulted in the best liquid resistance performance of brass powder acrylic coating.

Table 25. Liquid resistance grade of different coupling agent-modified brass powder coatings.

Sample (#)	Silane Coupling Agent Type	Brass Powder Content (%)	Liquid Resistance Grade (Grade)			
			Citric Acid	Ethanol	Detergent	Coffee
0		0	1	1	2	1
10		10	1	1	2	1
11		20	2	1	2	1
12		30	2	1	3	2
13	None	40	2	1	3	2
14		50	3	1	3	2
15		60	3	1	3	2
16		70	3	1	3	2
17		10	2	1	2	2
18		20	2	1	2	2
19		30	2	1	2	2
20	KH550	40	3	1	2	2
21		50	3	1	2	2
22		60	3	1	2	2
23		70	3	1	2	2

Table 25. *Cont.*

Sample (#)	Silane Coupling Agent Type	Brass Powder Content (%)	Liquid Resistance Grade (Grade)			
			Citric Acid	Ethanol	Detergent	Coffee
24	KH560	10	1	1	1	2
25		20	2	1	2	2
26		30	3	1	2	2
27		40	3	1	2	2
28		50	3	1	2	2
29		60	4	1	2	2
30		70	4	1	2	2
31	KH570	10	1	1	1	2
32		20	1	1	2	2
33		30	2	1	2	2
34		40	2	1	2	2
35		50	2	1	2	2
36		60	3	1	2	2
37		70	4	1	2	2

3.3.5. Microscopic Morphology and Infrared Spectral Analysis of the Coating

The brass powder-acrylic acid coating modified with 10% KH570 has the best overall performance, as can be seen from the findings above. Although the color difference is smaller, the chromaticity fluctuation is more noticeable, and the primary wavelength of color is closer to orange, its gloss is not significantly different from the pure brass powder water-based coating. It also has higher levels of hardness, impact resistance, and adherence. Its resistance to aging and fluids is also stronger. The microscopical characterizations of the brass powder-acrylic acid coatings with and without 10% KH570 modification are illustrated in Figure 14. The surface of the modified coating is essentially flat when compared to the pure brass powder water-based coating, and the minor protrusion is compatible with the findings of the roughness tests (Table 22). The material shows excellent surface condition.

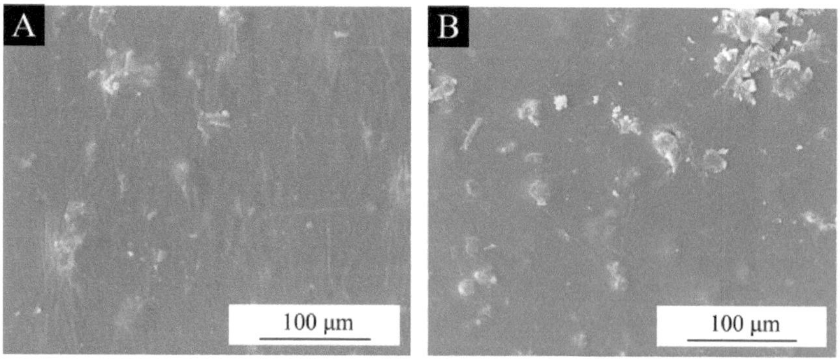

Figure 14. SEM images of Basswood surface films with 10% brass powder: (**A**) unmodified (Sample 10#), (**B**) KH570 modified (Sample 31#).

Figure 15 shows the FTIR spectra of the brass powder-acrylic acid coating modified by 10% KH570 and the pristine brass powder-acrylic acid coating. The spectrum of the modified brass powder-acrylic acid coating clearly contains the characteristic peaks of KH570. The peaks for methylene are at 2927 cm^{-1} and 2857 cm^{-1}. The Si-O-M (M = Cu, Zn, Si) asymmetric stretching mode is thought to be responsible for the strong adsorption band near 1100 cm^{-1} [65], demonstrating that a covalent link between KH570 and the brass powder was successfully formed. Si-C bonds have significant distinctive peaks at 842 cm^{-1}

and 754 cm^{-1}, and C-H in C=CH_2 has an in-plane bending vibration peak at 987 cm^{-1}. The asymmetric stretching vibration peak of silyl ether is 1037 cm^{-1}. The deformation vibration peak of the C=O is at 1727 cm^{-1}, and the deformation vibration peak of the C=C is at 1645 cm^{-1}. The double bond is a significant element that can enhance the coating's mechanical capabilities. The intensity of the C=C peak of the unmodified coating was greatly reduced in comparison to the FTIR spectra of the KH570-modified brass powder, and KH570 was successful in polymerizing acrylic acid.

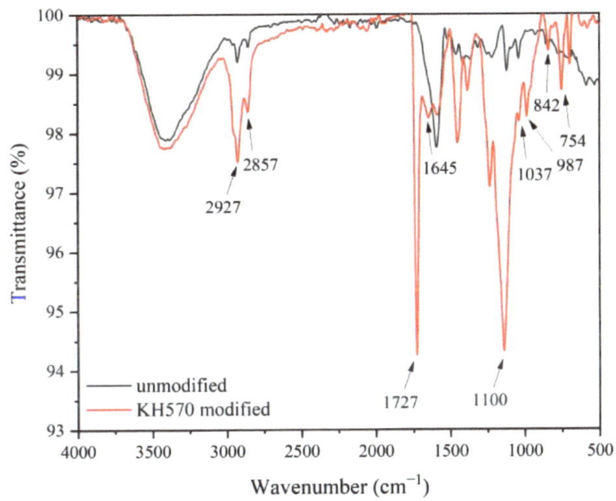

Figure 15. FTIR spectra of Basswood surface films with 10% brass powder.

4. Conclusions

Orthogonal tests were used to examine the effects of various factors (brass powder content, silane coupling agent type, silane coupling agent content, pH value) on the optical properties of modified brass powder-water-based acrylic coatings on a Basswood surface. The results showed that the coupling agent type and brass powder content had the greatest effects on the optical properties, with coating gloss, color difference, and chromaticity variation serving as the reference basis. Brass powder content had a negative correlation with coating gloss and a positive correlation with coating chromaticity variation under various silane coupling agent modifications. The silane coupling agent-modified coatings had a less pronounced color difference. The silane coupling agent can greatly increase the coating's hardness, impact resistance, and adhesion in terms of mechanical properties. Aging resistance and liquid resistance tests reveal that the addition of a silane coupling agent can increase these properties. Combining all data, it is clear that the water-based acrylic acid coating containing 10% brass powder modified by KH570 (pH 5 of the solution at modification, coupling agent content 6%) has the best overall performance, improving the coating's mechanical characteristics, aging resistance, and liquid resistance in addition to ensuring good optical properties, with a gloss of 20.0 GU, a color difference of 3.12, a hardness of HB, an impact resistance of 4 kg·cm, an adhesion of grade 1, and better liquid resistance and age resistance. These results provide the technological foundation for the development of artistic coatings for wood.

Author Contributions: Conceptualization, methodology, validation, resources, data management, supervision, formal analysis, and investigation, Y.H.; writing-review and editing, X.Y. All authors have read and agreed to the published version of the manuscript.

Funding: This project was partly supported by the Postgraduate Research & Practice Innovation Program of Jiangsu Province (KYCX22_1098) and the Natural Science Foundation of Jiangsu Province (BK20201386).

Institutional Review Board Statement: Not applicable.

Informed Consent Statement: Not applicable.

Data Availability Statement: The data presented in this study are available upon request from the corresponding author.

Conflicts of Interest: The authors declare that there is no conflict of interest.

References

1. Ellahi, R.; Zeeshan, A.; Hussain, F.; Abbas, T. Study of shiny film coating on multi-fluid flows of a rotating disk suspended with nano-sized silver and gold particles: A comparative analysis. *Coatings* **2018**, *8*, 422. [CrossRef]
2. Liu, C.; Xu, W. Effect of Coating Process on Properties of Two-Component Waterborne Polyurethane Coatings for Wood. *Coatings* **2022**, *12*, 1857. [CrossRef]
3. Liu, Q.Q.; Gao, D.; Xu, W. Influence of the bottom color modification and material color modification process on the performance of modified poplar. *Coatings* **2021**, *11*, 660. [CrossRef]
4. Wang, C.; Lin, X.; Schafer, C.C.; Hirseman, S.; Ge, J.P. Spray synthesis of photonic crystal based automotive coatings with bright and angular-dependent structural colors. *Adv. Funct. Mater.* **2021**, *31*, 2008601. [CrossRef]
5. Volokitin, G.G.; Skripnikova, N.K.; Sinitsyn, V.A.; Volokitin, O.G.; Shekhovtsov, V.V.; Vaschenko, S.P.; Kuz'min, V.I. Plasma treatment of wood. *Thermoghys. Aeromech.* **2016**, *23*, 119–124. [CrossRef]
6. Tsai, S.C.; Chen, L.H.; Chu, C.P.; Chao, W.C.; Liao, Y.C. Photo curable resin for 3D printed conductive structures. *Addit. Manuf.* **2022**, *51*, 102590. [CrossRef]
7. Bauer, F.; Mehnert, R. UV curable acrylate nanocomposites: Properties and applications. *J. Polym. Res.* **2005**, *12*, 483–491. [CrossRef]
8. Abd El-Wahab, H.; Hassan, A.M.; Naser, A.M.; Fouad, O.A.; El-Din, A.M.; Wahba, O.A.G. Preparation and evaluation of nanosized mixed calcium iron oxide ($CaFe_2O_4$) as high heat resistant pigment in paints. *Pigment. Resin Technol.* **2015**, *44*, 172–178. [CrossRef]
9. Meshalkin, V.P.; Abrashov, A.A.; Vagramyan, T.A.; Grigoryan, N.S.; Utochkina, D.S. Development of composition and investigation of properties of a new, environmentally friendly molybdenum-containing decorative protective conversion coating on ZinN-plated surfaces. *Dokl. Chem.* **2018**, *480*, 125–132. [CrossRef]
10. Liang, S.T.; Wang, H.Z.; Liu, J. Spray printing and encapsulated liquid metal as a highly reflective metallic paint for packing products. *Sci. China Technol. Sci.* **2019**, *62*, 1577–1584. [CrossRef]
11. Yan, X.X.; Tao, Y.; Qian, X.Y. Preparation of microcapsules for core materials and their effects on properties of waterborne coatings on basswood. *J. For. Eng.* **2022**, *7*, 186–192.
12. Licbarski, A.; Bartkowiak, M.; Czech, Z. Influence of selected crosslinking agents and selected unsaturated copolymerizable photoinitiators referring to the shrinkage resistance of solvent-based acrylic pressure-sensitive adhesives. *Polymers* **2022**, *14*, 5190. [CrossRef] [PubMed]
13. Lee, D.E.; Lee, J.J.; Cho, H.J.; Lee, J.S. Investigating the correlation between the curing behavior and properties of acrylic powder coatings. *J. Coat. Technol. Res.* 2023; *online*. [CrossRef]
14. Ling, K.L.; Feng, Q.M.; Huang, Y.H.; Li, F.; Huang, Q.F.; Zhang, W.; Wang, X.C. Effect of modified acrylic water-based paint on the properties of paint film. *Spectrosc. Spect. Anal.* **2020**, *40*, 2133–2137.
15. Huang, Y.H.; Feng, Q.M.; Ye, C.Y.; Nair, S.S.; Yan, N. Incorporation of ligno-cellulose nanofibrils and bark extractives in water-based coatings for improved wood protection. *Prog. Org. Coat.* **2020**, *138*, 105210. [CrossRef]
16. Li, Z.X.; Lv, X.; Che, Z.X.; Liao, H.L.; Zheng, J. Performance test and pyrolysis kinetics of water-based acrylic acid-amino resin. *Prog. Org. Coat.* **2022**, *172*, 107101. [CrossRef]
17. Oncel, M.; Vurdu, H.; Kaymakci, A.; Ozkan, O.E.; Aydogan, H. Coating performances of crimean juniper (Juniperus excelsa M. BIEB.) wood. *Cerne* **2019**, *25*, 36–43. [CrossRef]
18. Akkus, M.; Akbulut, T.; Candan, Z. Application of electrostatic powder coating on wood composite panels using a cooling method. Part 1: Investigation of water intake, abrasion, scratch resistance, and adhesion strength. *Bioresources* **2019**, *14*, 9557–9574.
19. Kim, Y.S.; Cho, H.J.; Lee, H.; Kim, W.Y.; Jung, Y.C.; Lee, S.Y. Development of a multi-functional acrylic urethane coating with high hardness and low surface energy. *Prog. Org. Coat.* **2020**, *147*, 105748. [CrossRef]
20. Yu, Z.H.; Yan, Z.Y.; Zhang, F.H.; Wang, J.X.; Shao, Q.; Murugadoss, V.; Alhadhrami, A.; Mersal, G.A.M.; Ibrahim, M.M.; El-Bahy, Z.M.; et al. Waterborne acrylic resin co-modified by itaconic acid and gamma-methacryloxypropyl triisopropoxidesilane for improved mechanical properties, thermal stability, and corrosion resistance. *Prog. Org. Coat.* **2022**, *168*, 106875. [CrossRef]
21. Wang, C.; Zhou, Z. Optical properties and lampshade design applications of PLA 3D printing materials. *BioResources* **2023**, *18*, 1545–1553. [CrossRef]
22. Hegemann, D.; Amberg, M.; Ritter, A.; Heuberger, M. Recent developments in Ag metallised textiles using plasma sputtering. *Mater. Technol.* **2009**, *24*, 41–45. [CrossRef]

23. Liu, L.; Shan, H.R.; Jia, X.K.; Duan, S.J.; Gao, Y.J.; Xu, C.Y.; Wei, S.Y. Study on UV curable tung oil based waterborne polyurethane wood coatings. *J. For. Eng.* **2022**, *7*, 115–121.
24. Akpinar, S.; Yazici, Z.O.; Abbak, S.; Can, M.F. Preparation of surface-modified borax powders for their application in low-temperature vitreous coatings. *J. Aust. Ceram. Soc.* **2021**, *57*, 1137–1144. [CrossRef]
25. Liu, X.T.; Ma, X.J.; Zhu, L.Z.; Zhu, L.Z. Fabrication of poly(3-hydroxybutyrate-co-4-hydroxybutyrate) biocomposite reinforced wood fiber modified with mixed coupling agents. *Ind. Crops Prod.* **2021**, *164*, 113352. [CrossRef]
26. Sivakumar, P.; Du, S.M.; Selter, M.; Daye, J.; Cho, J. Improved adhesion of polyurethane-based nanocomposite coatings to tin surface through silane coupling agents. *Int. J. Adhes. Adhes.* **2021**, *110*, 102948. [CrossRef]
27. Zhu, Z.W.; Chen, S.F.; Zhang, Y.; Wang, W. Corrosion resistance of polyvinyl butyral/reduced graphene oxide/titanium dioxide composite coatings for stainless steel in different environments. *Prog. Org. Coat.* **2022**, *173*, 107226. [CrossRef]
28. Li, G.; Wang, F.; Liu, P.; Gao, C.; Ding, Y.F.; Zhang, S.M.; Yang, M.S. Antioxidant functionalized silica-coated TiO_2 nanorods to enhance the thermal and photo stability of polypropylene. *Appl. Surf. Sci.* **2019**, *476*, 682–690. [CrossRef]
29. Wang, J.Z.; He, Y.; Xie, Z.F.; Chen, C.L.; Yang, Q.B.; Zhang, C.L.; Wang, B.Y.; Zhan, Y.Q.; Zhao, T.H. Functionalized boron carbide for enhancement of anticorrosion performance of epoxy resin. *Polym. Adv. Technol.* **2018**, *29*, 758–766. [CrossRef]
30. Zou, J.L.; Shi, J.F.; Yang, J.X.; Li, J.; Liu, Y. Electroless copper plating mechanism of mesophase pitch-based carbon fibers by the grafting modification of silane couple agents. *Mater. Today Commun.* **2022**, *32*, 104053. [CrossRef]
31. Hu, W.; Zhang, J. Effect of growth rings on acoustic emission characteristic signals of southern yellow pine wood cracked in mode I. *Constr. Build. Mater.* **2022**, *329*, 12709. [CrossRef]
32. Li, S.; Hu, W. Study on mechanical strength of cantilever handrail joints for chair. *Bioresources* **2023**, *18*, 209–219. [CrossRef]
33. Salon, M.C.B.; Belgacem, M.N. Hydrolysis-condensation kinetics of different silane coupling agents. *Phosphorus Sulfur Silicon Relat. Elem.* **2011**, *186*, 240–254. [CrossRef]
34. Pantoja, M.; Martinez, M.A.; Abenojar, J.; Encinas, N.; Ballesteros, Y. Effect of EtOH/H_2O ratio and pH on Bis-Sulfur silane solutions for electrogalvanized steel joints based on anaerobic adhesives. *J. Adhes.* **2011**, *87*, 688–708. [CrossRef]
35. Zhang, C.S.; Wang, X.; Hu, Z.C.; Wu, Q.S.; Zhu, H.J.; Lu, J.F. Long-term performance of silane coupling agent/metakaolin based geopolymer. *J. Build. Eng.* **2021**, *36*, 102091. [CrossRef]
36. Soontornvatin, V.; Prasansuttiporn, T.; Thanatvarakorn, O.; Jittidecharaks, S.; Hosaka, K.; Foxton, R.M.; Nakajima, M. Bond strengths of three-step etch-and-rinse adhesives to silane contaminated dentin. *Dent. Mater. J.* **2021**, *40*, 385–392. [CrossRef]
37. *GB/T 4893.6-2013*; Test of Surface Coatings of Furniture. Part VI: Determination of Gloss Value. Standardization Administration of the People's Republic of China: Beijing, China, 2013.
38. *GB/T 11186.3-1989*; Methods for Measuring the Color of Paint Films. Part III: Calculation of Color Differences. Standardization Administration of the People's Republic of China: Beijing, China, 1989.
39. Li, R.R.; He, C.J.; Wang, X.D. Evaluation and modeling of processability of laser removal technique for bamboo outer layer. *JOM-US* **2021**, *73*, 2423–2430. [CrossRef]
40. Liu, Q.Q.; Gao, D.; Xu, W. Effect of Polyurethane Non-Transparent Coating Process on Coating Performance Applied on Modified Poplar. *Coatings* **2022**, *12*, 39. [CrossRef]
41. Zhou, J.C.; Xu, W. Toward interface optimization of transparent wood with wood color and texture by silane coupling agent. *J. Mater. Sci.* **2022**, *57*, 5825–5838. [CrossRef]
42. Han, Y.; Yan, X.X.; Yin, T.Y. Effects of different proportions of mixed microcapsules on aging resistance and self-healing properties of waterborne coatings. *J. For. Eng.* **2022**, *7*, 183–189.
43. Marzo, A.; Ferrada, P.; Beiza, F.; Besson, P.; Alonso-Montesinos, J.; Ballestrin, J.; Roman, R.; Portillo, C.; Escobar, R.; Fuentealba, E. Standard or local solar spectrum? Implications for solar technologies studies in the Atacama Desert. *Renew. Energ.* **2018**, *127*, 871–882. [CrossRef]
44. *GB/T 6739-2006*; Paint and Varnishes-Determination of Film Hardness by Pencil Test. Standardization Administration of the People's Republic of China: Beijing, China, 2006.
45. *GB/T 4893.9-2013*; Determination of Impact Resistance of Film. Standardization Administration of the People's Republic of China: Beijing, China, 2013.
46. *GB/T 4893.4-2013*; Test of Surface Coatings of Furniture. Standardization Administration of the People's Republic of China: Beijing, China, 2013.
47. Yan, X.X.; Han, Y.; Yin, T.Y. Synthesis of urea-formaldehyde microcapsule containing fluororesin and Its effect on performances of waterborne coatings on wood surface. *Polymers* **2021**, *13*, 1674. [CrossRef]
48. *ASTM D4587-2011*; Standard Practice for Fluorescent UV-Condensation Exposures of Paint and Related Coatings. American Society for Testing and Materials: West Conchhocken, PA, USA, 2011.
49. *GB/T 4893.1-2021*; Test of surface coatings of furniture—Part 1: Determination of surface resistance to cold liquids. Standardization Administration of the People's Republic of China: Beijing, China, 2021.
50. Yan, X.X.; Xu, G.Y. Synergy effect of silane and CTAB on corrosion-resistant property of low infrared emissivity Cu/polyurethane coating formed on tinplate. *Surf. Coat. Technol.* **2010**, *204*, 1514–1520. [CrossRef]
51. Li, B.; Li, S.M.; Liu, J.H.; Yu, M. The heat resistance of a polyurethane coating filled with modified nano-$CaCO_3$. *Appl. Surf. Sci.* **2014**, *315*, 241–246. [CrossRef]

52. Gui, C.M.; Ma, H.D.; Zhang, R.X.; Liu, Y.L.; Li, H.L.; Huang, J.J.; Li, P. Stability analysis: Rational design of a Ag nanoparticle/polymer brush for fabricating Cu coating on a PET surface. *Langmuir* **2021**, *37*, 5673–5681. [CrossRef]
53. Panicker, C.Y.; Varghese, H.T.; John, A.; Philip, D.; Nogueira, H.I.S. Vibrational spectra of melamine diborate, C(3)N(6)H(6)2H(3)BO(3). *Spectrochim. Acta A Mol. Biomol. Spectrosc.* **2002**, *58*, 1545–1551. [CrossRef]
54. Liu, Z.G.; Xie, Y.T.; Yang, L.R.; Wang, C.M.; Feng, X.X.; Tang, Z.Z.; Yu, Y.X. The mechanical strength of sulphoaluminate cement-gypsum composite material reinforced with polypropylene fibers synergistically modified by KH570 and nano-SiO_2. *Mater. Compos. Interfaces* **2022**, *29*, 1249–1265. [CrossRef]
55. Meng, Y.Y.; Yong, Q.W.; Liao, B.; Zeng, W.; Pang, H. Synthesis, characterization and formation mechanism of acrylate emulsion-based self-matting coatings. *New J. Chem.* **2020**, *44*, 13971–13978. [CrossRef]
56. Yong, Q.W.; Xu, D.; Liu, Q.; Xiao, Y.; Wei, D.D. Advances in polymer-based matte coatings: A review. *Polym. Adv. Technol.* **2022**, *33*, 5–19. [CrossRef]
57. Uetsuji, Y.; Fukui, N.; Yagi, T.; Nakamura, Y. The effect of number of chemical bonds on intrinsic adhesive strength of a silane coupling agent with metals: A first-principles study. *J. Mater. Res.* **2022**, *37*, 923–932. [CrossRef]
58. Wang, Q.T.; Zhang, Y.; Liang, W.K.; Wang, J.J.; Chen, Y.X. Effect of silane treatment on mechanical properties and thermal behavior of bamboo fibers reinforced polypropylene composites. *J. Eng. Fibers Fabr.* **2020**, *15*, 1558925020958195. [CrossRef]
59. Wang, L.L.; Li, R.M.; Wang, C.L.; Hao, B.R.; Sha, J.Z. Surface grafting modification of titanium dioxide by silane coupler KH570 and its influences on the application of blue light curing ink. *Dyes Pigm.* **2019**, *163*, 232–237. [CrossRef]
60. Min, Y.H.; Fang, Y.; Huang, X.J.; Zhu, Y.H.; Li, W.S.; Yuan, J.M.; Tan, L.G.; Wang, S.Y.; Wu, Z.J. Surface modification of basalt with silane coupling agent on asphalt mixture moisture damage. *Appl. Surf. Sci.* **2015**, *346*, 497–502. [CrossRef]
61. Yan, X.X.; Li, W.B.; Han, Y.; Yin, T.Y. Preparation of melamine/rice husk powder coated shellac microcapsules and effect of different rice husk powder content in wall material on properties of wood waterborne primer. *Polymers* **2022**, *14*, 72. [CrossRef] [PubMed]
62. Yan, X.X.; Xu, G.Y. Corrosion and mechanical properties of epoxy-polyurethane/bronze composite coatings with low infrared emissivity. *Surf. Coat. Technol.* **2010**, *205*, 2307–2312. [CrossRef]
63. Yu, H.J.; Xu, G.Y.; Shen, X.M.; Yan, X.X.; Cheng, C.W. Low infrared emissivity of polyurethane/Cu composite coatings. *Appl. Surf. Sci.* **2009**, *255*, 6077–6081. [CrossRef]
64. Wang, D.Y.; Huang, Z.X.; Shi, S.J.; Ren, J.Y.; Dong, X.B. Environmentally friendly plant-based waterborne polyurethane for hydrophobic and heat-resistant films. *J. Appl. Polym. Sci.* **2022**, *139*, e52437. [CrossRef]
65. Dang, Z.M.; Zhang, B.; Li, J.G.; Zha, J.W.; Hu, G.H. Copper particles/epoxy resin thermosetting conductive adhesive using polyamide resin as curing agent. *J. Appl. Polym. Sci.* **2012**, *126*, 815–821. [CrossRef]

polymers

MDPI

Article

Analysis of the Anisotropy of Sound Propagation Velocity in Thin Wooden Plates Using Lamb Waves

Dagmar Faktorová [1], Mariana Domnica Stanciu [2,*], Michal Krbata [1], Adriana Savin [2,3,*], Marcel Kohutiar [1], Milan Chlada [4] and Silviu Marian Năstac [2,5]

1 Faculty of Special Technology, Alexander Dubcek University of Trenčín, 911 06 Trenčín, Slovakia; dagmar.faktorova@tnuni.sk (D.F.); michal.krbata@tnuni.sk (M.K.); marcel.kohutiar@tnuni.sk (M.K.)
2 Faculty of Mechanical Engineering, Transilvania University of Brașov, 29 Eroilor Blvd., 500036 Brașov, Romania; silviu.nastac@ugal.ro
3 National Institute of Research and Development for Technical Physics, 47 D. Mangeron Blvd., 700050 Iasi, Romania
4 Institute of Thermomechanics of the CAS, v. v. i. Dolejškova 1402/5, 18200 Prague, Czech Republic; chlada@it.cas.cz
5 Faculty of Engineering and Agronomy, Braila, 810017, "Dunarea de Jos" University of Galati, 810017 Galati, Romania
* Correspondence: mariana.stanciu@unitbv.ro (M.D.S.); asavin@phys-iasi.ro (A.S.)

Abstract: The objective of the study was to analyze the influence of coating treatments on sound propagation speeds in thin boards, along the longitudinal and radial directions of resonance wood. The samples studied were thin boards made of spruce and maple wood with dimensions of 240 mm × 80 mm × 4 mm (length × width × thickness) subjected to different coating treatments (oil-based varnish and alcohol varnish) as well as unvarnished samples, exposed to radiation UV, and specimens treated in the saline fog. The test method consisted of evaluating the propagation speeds of Lamb waves applied to thin plates, according to a semicircular test model, so that the results highlighted both the acoustic response in the longitudinal and radial directions as well as the variation in the anisotropy of the samples with the change in the sound propagation direction relative to wood fibers. Based on the statistical analysis, sound propagation speed profiles were obtained in each of the 38 directions analyzed for all wood samples. The results highlighted that the oil-based varnish led to a decrease in the speed of propagation in the radial direction, compared to the alcoholic varnish, whose major effect was in the longitudinal direction, on the spruce wood. On maple wood, increasing the number of varnish layers, regardless of the type of varnish, led to a decrease in the anisotropy ratio between the longitudinal and radial directions.

Keywords: varnish; anisotropy; Lamb waves; coating treatments; spruce; maple

check for updates

Citation: Faktorová, D.; Stanciu, M.D.; Krbata, M.; Savin, A.; Kohutiar, M.; Chlada, M.; Năstac, S.M. Analysis of the Anisotropy of Sound Propagation Velocity in Thin Wooden Plates Using Lamb Waves. *Polymers* **2024**, *16*, 753. https://doi.org/10.3390/polym16060753

Academic Editors: Ľuboš Krišťák, Roman Reh, Eugenia Mariana Tudor and Marius Cătălin Barbu

Received: 17 January 2024
Revised: 6 March 2024
Accepted: 7 March 2024
Published: 9 March 2024

1. Introduction

Wooden material, due to its remarkable properties, traditionally presents wide applicability. Wood, through its cellular structure, shape and way of organizing the wood cells in the three main directions, and content of lignin, cellulose, and hemicelluloses, confers optimal physical, mechanical, and acoustic properties for musical instruments with strings and a resonator body. The physical properties (density) and mechanical properties of wood (elastic constants and damping) influence the acoustic properties of wood (speed of sound, characteristic impedance, and sound radiation) because sound is produced by the vibrations of the material [1–3]. According to [2], the parameters that reflect the vibrational and acoustic quality of wood are the relative acoustic conversion efficiency as the ratio of acoustic energy radiated to the vibration energy, the anisotropy of wood calculated as the ratio between dynamic Young's modulus and dynamic shear modulus. Thus, a high specific dynamic modulus in the longitudinal direction, a high anisotropy factor, and a low damping are indicators of the tone quality of the soundboard. From the point of view of the fiber orientation compared to the cutting directions of the wooden boards, [3] considers that the grain angle influences the anisotropy of viscoelastic vibrational properties, just as [2,4] correlates the acoustic qualities of wood with the effect of the angle of the microfibril on the damping and specific dynamic Young's modulus.

On the other hand, the accessibility of this resource since ancient times and its easy workability have consecrated wood as a lignocellulose composite material for numerous musical instruments [1].

The reaction of wood in the interaction with sound waves depends on the transmitted sound energy, as well as the nature and condition of the wood: the macro- and microscopic structure of the wood–cell, membrane construction, fiber size, cohesion, humidity, elastic properties, etc. Spruce and maple wood (resonance wood) are wood with a very fine structure and physical properties suitable for the construction of stringed musical instruments [5,6].

According to [7,8], who identified the origin of the resonance wood of the musical instruments belonging to the Cherubini Conservatory Collection in Florence, the most important areas for the production of soundboards were the Eastern Alps (with regions from Italy and Slovenia), Bavarian forests, and the Carpathian Mountains (Poland, Romania, and southern Russia). The resonance wood from the central–northern area of the Eastern Carpathians presents special physical and acoustic properties that have been studied, with them being correlated with the anatomical quality classes used by luthiers [8–11]. The choice of raw material by luthiers is based on rigorous physical and acoustic criteria, with the resonance wood thus being classified in correlation with the quality and price of the musical instrument. The properties of the materials used in the construction of stringed musical instruments show great variability, and that is why the first stage in the manufacture of musical instruments is the choice of wood [12–14]. The basic characteristic that distinguishes resonance wood from the ordinary type is the very homogeneous anatomical structure, obtained from the small and uniform width of the annual rings as well as the small proportion of late wood. The microscopic structure of spruce wood consists of vessels, resin canals, and cellular composition. The length of the tracheids shows a significant increase when moving away from the pith and parcel. The vessels in the ring assembly are 1.1–6.0 mm long. Tracheid length decreases with latitude. The resin canals are visible only with a magnifying glass and are rare (25–35 pcs/cm^2). The cellular composition consists of 95% tracheids, 3.6% resiniferous formations, 1.4% radial parenchyma, and a proportion of wood voids of 71% [15–17].

From an anatomical point of view, maple wood is made up of vessels (trachea), wood parenchyma, radial parenchyma, and fibers. The vessels are small and rare, and the fibers occupy about 75% of the volume of the wood substance. The porosity (proportion of voids) is 60% of the apparent volume. Maple wood is part of the category of hardwoods with scattered pores, with these being small, visible with a magnifying glass, and quite rare. They are, therefore, unevenly scattered among the early and late wood and their size and thickness decrease towards the outer limit of the annual ring.

Producers choose wood according to its cost and mechanical, artistic, and mechanical processing criteria without, however, having the guarantee of the performance of the final product, even in the conditions of the best acoustic performance of the wood [3,8]. The wood used in the construction of musical instruments is covered with lacquer films that have the role of protecting the musical instrument from wear and fluctuations in ambient humidity. Depending on the type of varnish (the solvent), the thickness of the film, the degree of penetration into the wood, the physical and mechanical properties of the newly formed system (wood–wood/varnish interface and the varnish film) show elastic, acoustic, and dynamic properties different from those of untreated wood [2–4].

Lately, there has been increased interest in scientific research on these materials by applying direct and indirect methods and the appropriate use of measuring instruments [16]. Non-destructive techniques (NDT/E) involve the identification of properties without measuring them and preserving the structural integrity of the wood [11–13]. Simulation methods using the finite element method (FEM) can be used for the evaluation techniques of material properties even when they have complex behavior [11–13]. Ultrasonic (US) and ultrasonic resonance (RUS) methods as part of NDT/E methods successfully achieve the results obtained by standard static methods [14–17].

The results obtained in determining the propagation speeds of US in wood depend on the orientation of the structure in relation to the sound source (longitudinal—L, radial—R, and tangential—T) [18–21], according to the axes that identify its spatial directions. The propagation speed of US through wood is a fundamental parameter in the characterization of resonant wood, with it being dependent on the density of the wood as well as the modulus of elasticity [22,23]. From a mechanical point of view, to understand the elastic behavior of resonant wood, it is necessary to determine the propagation speeds in the three directions. The method based on US propagation as a non-destructive method involves nine propagation speeds in accordance with the directions of mechanical oscillation [22–24]. According to [20,21], the structural anisotropy of wood can be determined using different methods: either by calculating the ratios between the longitudinal and transverse wave velocities in the three main directions of symmetry or based on the ratios between shear velocities, or through the estimation of wood anisotropy deals with velocity invariants. The analysis of the anisotropy of wooden plates based on the propagation of Lamb waves provides useful information not only about the propagation velocity of horizontal shear waves but also the dispersion phenomena that depend on the geometry of the structure, the dependence between the wave number, and the frequency of the particular mode that can be obtained by a numerical solution of the Rayleigh–Lamb frequency equation [25–27]. According to [18,28,29], for bulk waves, it is already well known that the anisotropy implies a difference between the group velocity and the phase velocity. It is shown here that this phenomenon also occurs for Lamb waves, with it affecting both the speed and direction of propagation.

This paper aims to emphasize the effect of the acoustic anisotropy of wood as a result of the preferential orientation of the anatomical elements and the different surface treatments applied to spruce and maple wood. The novelty of the study consists of the analysis of the acoustic anisotropy of wood subjected to different surface treatments (artificial aging, salt spray, and different types of finishes).

2. Materials and Methods

2.1. Materials

Spruce (*Picea abies Karst* L.) and maple (*Acer pseudoplatanus* L.) wood specimens were included in the tests, with dimensions 80 × 4 × 240 mm^3 corresponding to the radial × tangential × longitudinal directions, which were tested (according to ISO13061-1:2014/Amd1:2017 [30]) under laboratory conditions (20 °C and relative humidity (RH) of 65%). In Figure 1, the geometry and physical aspects of the samples are presented. From the point of view of the anatomical structure of the wood, two quality classes were analyzed: class A representing wood with a fine anatomical structure, with narrow and regular annual rings (for spruce wood) and wavy fibers (for maple wood), and class D representing wood with a coarser anatomical structure, with annual rings wider and irregular in width (for spruce wood) without wavy fibers (for maple wood) [10,23].

The samples were subjected to different surface treatments to check to what extent these treatments influence the degree of anisotropy of the wood. Table 1 shows the characteristics of the investigated samples. The moisture content was monitored using a Merlin HM8-WS1 (Tumeltsham, Austria) moisture meter and the mass was measured with a Kern & Sohn Weighing Balance, EWJ 600-2SM, Balingen, Germany.

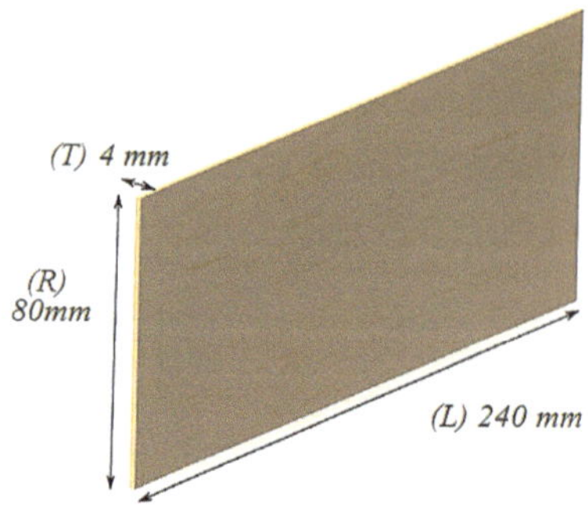

Figure 1. The geometry of the samples.

Table 1. Physical properties: wood species (WS), density (ρ), equilibrium moisture content (EMC), and type of treatment (T).

Samples Code	WS	ρ (kgm^{-3})	EMC (%)	AML (g/m^2)	T
MDHD15.1	spruce	391.01 (33.42)	6.4 (0.4)	0.000	Untreated
MAHD10.3	spruce	470.83 (18.68)	7.9 (0.5)	0.000	
PDNC15.1	maple	605.72 (17.54)	6.1 (0.3)	0.000	
PANC10.1	maple	656.90 (15.82)	7.3 (0.5)	0.000	
MDUV2	spruce	421.50 (27.92)	6.8 (0.5)	−10.318	UV exposure 300 h
MAUV2	spruce	473.45 (18.32)	6.2 (0.4)	−13.979	
PDUV3	maple	565.40 (17.62)	6.2 (0.2)	−12.432	
PAUV3	maple	640.36 (11.75)	6.4 (0.7)	−17.813	
MDSAL2	spruce	429.33 (10.19)	8.3 (1.3)	16.047	Salt fog exposure NaCl 3.5% 72 h
MASAL2	spruce	491.16 (13.81)	8.8 (1.4)	21.563	
PDSAL1	maple	679.17 (35.24)	9.1 (1.2)	27.135	
PASAL1	maple	669.77 (24.71)	8.5 (1.8)	30.146	
MALU5.3	spruce	399.31 (45.60)	7.4 (0.9)	11.755	5 layers of oil-based varnish
MDLU5.2	spruce	407.85 (37.50)	7.8 (0.7)	9.807	
PALU5.1	maple	662.57 (33.14)	8.4 (0.6)	8.375	
PDLU5.3	maple	652.23 (19.20)	7.4 (0.5)	6.938	
MALU10.2	spruce	556.98 (29.40)	7.8 (0.3)	15.969	10 layers of oil-based varnish
MDLU10.2	spruce	496.42 (32.12)	7.3 (0.7)	17.526	
PALU10.1	maple	705.36 (41.24)	8.7 (0.3)	14.708	
PDLU10.1	maple	611.77 (32.15)	8.2 (0.4)	14.380	
MALU15.2	spruce	522.55 (21.35)	8.0 (0.5)	22.979	15 layers of oil-based varnish
MDLU15.2	spruce	511.99 (22.31)	7.9 (0.4)	23.599	
PALU15.1	maple	744.69 (34.32)	8.3 (0.3)	23.443	
PDLU15.1	maple	694.91 (21.12)	7.8 (0.6)	24.229	
MALS5.2	spruce	430.08 (36.23)	7.2 (0.4)	2.865	5 layers of spirit varnish
PALS5.2	maple	657.79 (15.43)	7.3 (0.2)	11.453	
MALS10.2	spruce	417.60 (18.45)	7.5 (0.2)	7.042	10 layers of spirit varnish
PALS10.2	maple	672.47 (21.34)	7.1 (0.3)	12.896	
MALS15.2	spruce	491.42 (35.82)	7.6 (0.5)	13.234	15 layers of spirit varnish
PALS15.2	maple	621.58 (28.12)	7.7 (0.2)	16.651	

The samples were obtained by cutting radially longitudinally from the boards, which, after natural drying, were conditioned in a drying chamber up to a moisture content of 6–8%. Both nowadays and in the past, stringed musical instruments were covered with varnish to protect them from variations in atmospheric humidity and dirt [31–33]. Current studies with modern methods of determining the chemical fingerprint of the varnish layers on old violins have highlighted the fact that luthiers used recipes based on alcohol-solvent resins, with oils, or the wood was treated with salts like borax and the sulfates of Zn, Cu, Cr, Na, and Fe [32,34]. Based on these considerations, currently, the most well-known types of varnishes used for maestro- and professional-level musical instruments are those with an oil-based solvent and those with an alcoholic solvent (spirit) [32–35]. Not only is the chemical composition of the surface treatment important but also the thickness of the varnish film and the varnish–wood interface, aspects highlighted in studies [36–38]. The type of varnish and its thickness correlated with the number of layers applied in the finishing process influence the tonality of the musical instrument as a result of not only the difference in stiffness between the wood and the varnish film but also as an effect of the increase in the mass of the layered material (wood–varnish) [31,32,38]. Thus, in the presented study, spruce and maple wood samples covered with a number of layers of lacquer were analyzed according to the procedures applied at a violin factory (5, 10, and 15 layers). Therefore, some of the samples were kept as control samples, others were exposed to artificial aging with UV radiation, another set of samples were varnished with oil-based varnish and spirit varnish with a different number of layers (5, 10, 15), and another set of samples were exposed to the salt fog, as can be seen in Table 1.

According to [36,38], to compare the mass changes of the spruce and maple boards before and after surface treatments (VS varnish system, UV exposure, and salt fog exposure), the changes were evaluated as area mass loading (AML) calculated based on mathematical relation (1):

$$AML = \frac{m_t - m_w}{b * l}, \quad (1)$$

where AML is area mass loading induced by treatment (g·m^{-2}); m_t—the final mass of the treated wood; m_w—the initial mass of the wooden sample; b—the width of the plate; and l—length of the plate [36,38].

2.2. Methods

2.2.1. Experimental Set-Up

Basically, ultrasound (US) non-destructive evaluation consists of applying physical elastic waves to the sample tested and analyzing the interaction between the material samples and the field [19,39–41].

A problem associated with the propagation of Lamb waves is the coexistence of several modes of oscillation in the plate, which leads to difficulties in interpreting the results. The higher-order modes of oscillation start to propagate when the thickness of the plate becomes comparable to the wavelength of the longitudinal and transverse waves. Lamb waves are guided elastic plate waves that can be described as a combination of compression waves (P-waves) and shear waves (S-waves) [22,42]. For the excitation of Lamb waves in the lowest mode, transducers with Hertzian contact are used. A diagram of the system that generates and detects the Lamb waves for measurement is presented in Figure 2.

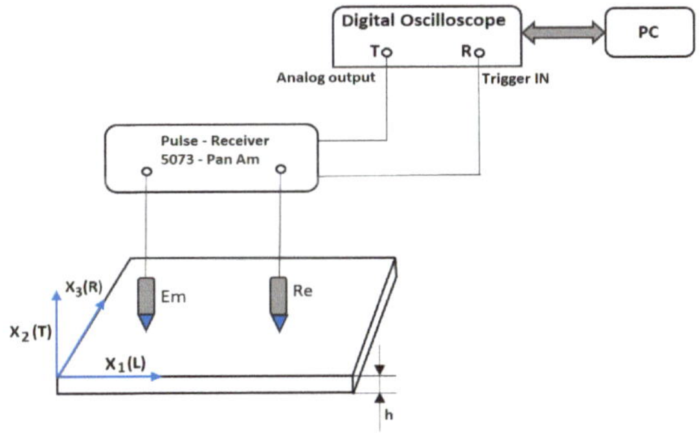

Figure 2. Experimental set-up.

A piezo-electric transducer, of very low eigenfrequency (tens of kilohertz), is coupled to a buffer rod made of a material of elasticity module E_1 and Poisson coefficient ν_1 which has a spherical bumped head of radius R at one end. The buffer rod is compressed onto a plate (with elasticity module E_2 and the Poisson coefficient ν_2). To increase the reliability and assurance of the quality of measurement, the ultrafine force presses the US sensors at constant value. The speed of the Lamb waves generated by Hertzian contact was determined, placing the emission and the reception transducer on the sample taken into study, measuring with ± 10 μm precision the distance between the centers of the transducer and measuring with 0.1μs precision the time between the generation of the Lamb wave and its reception. The two buffer rods of the US sensors used at the emission and reception are both identical, with them being made of the 7075-T6 aluminum–magnesium alloy, with a density of 2.7×103 kg/m^3, Young's modulus of 7×1010 N/m^2, Poisson coefficient of 0.34, and point curvature radius of 2 mm. The US sensors are connected to a 5073PR Pulse Receiver—equipment from Panametrics NDT USA (Olympus Corporation, Waltham, MA, USA). The visualization of the signal and the measurement of the time of propagation were made using a Le Croy Wave Runner 64Xi digital oscilloscope (LeCroy Corporation, Chestnut Ridge, NY, USA), with it having a sampling frequency of 10G S/s [23,24].

The signals emitted by the transmitter positioned at point E (Figure 3) were successively measured at points B1–B19 and A1–A19 as points on a semicircle, located at an interval of 10 degrees and a radius of 70 mm. The triangle symbol in points denoted E represents the position of the transmitter, and the circle symbol arranged in the form of a semicircle or quarter circle represents the successive position of the receiver. In order to avoid the dissipative behavior of the Lamb waves in the wood specimens, an equidistant distribution of the position of the transducers was chosen so that the effect of the anisotropy of the wood structure could be highlighted [18,27–29].

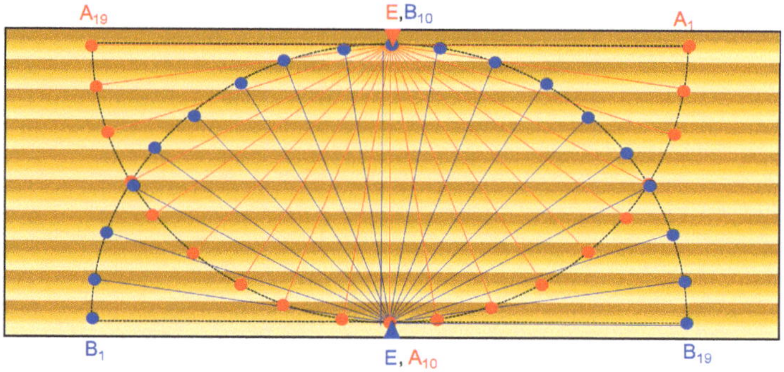

Figure 3. The measured points on the samples.

2.2.2. Statistical Analysis

The data were analyzed to evaluate the anisotropy of the material samples in terms of the change in the speed of sound with respect to the direction of propagation. The distribution of resonance points on a board made of a given type of material seems to be largely influenced by the changes in the speed of sound propagation in different directions and can affect the acoustic impression of the musical instrument produced. The question remains as to the overall subjective assessment of the quality of the instrument, which is influenced by other factors but cannot be captured by the distribution of sound velocities alone. In this limited sense, the data were analyzed with a focus on selecting the most significant wood samples. Factor analysis (FA) was used as a basic tool to identify possible dependencies in the data. Based on the Q-factor analysis (QFA), several clusters of samples with similar anisotropy were extracted.

3. Results and Discussion

3.1. Grain Angle Dependence of Lamb Wave Velocity in Wooden Plates

Since wood is an anisotropic material both in terms of morphology and properties, the changes regarding the wave propagation capacity vary gradually with the direction of the fibers. In Figure 4, it can be seen that for all samples, the speed of Lamb waves in the radial direction ranges between 1.4 to 3 times lower than in the longitudinal direction, depending on the species, the structure of the wood, and the type of treatment applied.

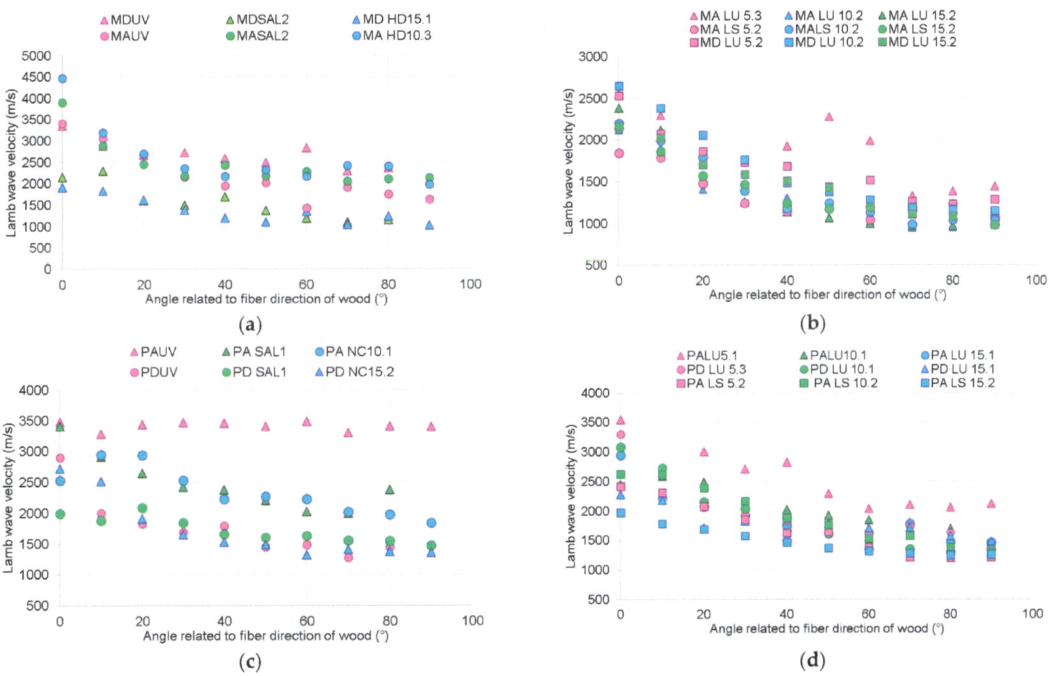

Figure 4. The variation in Lamb wave velocity in wooden plates with fiber angle variations: (**a**) spruce wood sample untreated, UV aged, and salt fog exposed; (**b**) spruce wood sample finished with oil-based varnish and spirit varnish; (**c**) maple wood sample untreated, UV aged, and salt fog exposed; (**d**) maple wood sample finished with oil-based varnish and spirit varnish.

The treatments applied to the boards influence their response to the propagation of waves in the board. Thus, artificial aging via exposure to UV radiation and slight thermal degradation for 300 h led to a decrease in both longitudinal (24%) and radial (1.7%) propagation speeds compared to the untreated wood. On the other hand, the coating with varnish films produced a wave dispersion phenomenon, with their speed decreasing in the longitudinal direction by 42–58% and the alcoholic varnish affecting the propagation of sounds the most in both spruce and maple wood (Figure 5). In the radial direction, the oil-based varnish applied on spruce wood is the one that presents the lowest values of the wave propagation speed, with it being 60% lower than the speed of the control samples. In the case of maple wood, the film of alcoholic varnish reduces the sound propagation speed the most, by 32% compared to the control sample (untreated).

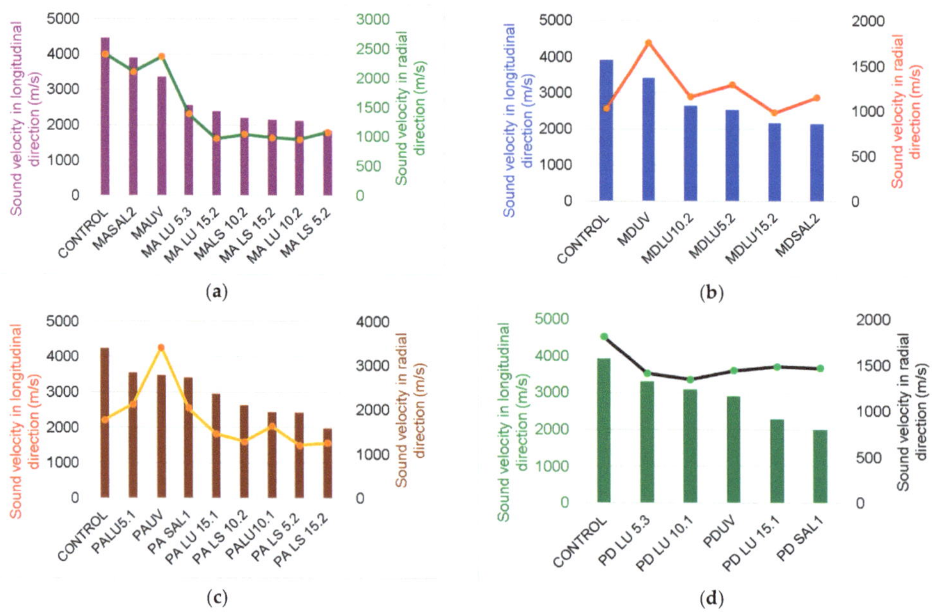

Figure 5. The influence of surface treatment on longitudinal and radial wave velocity in wooden plates: (**a**) spruce wood sample for maestro musical instruments; (**b**) spruce wood sample for school musical instruments; (**c**) maple wood sample for maestro musical instruments; (**d**) maple wood sample for school musical instruments.

Sound propagation in spruce and maple wood depends on the type, organization, and three-dimensional shape of the anatomical elements of each species. In spruce wood (*Picea abies Karst* L.) in the longitudinal direction, tracheids constitute 90% of the wood volume and are approximately 100 times longer than they are wide, with them overlapping with adjacent cells both in the upper and lower parts by 20–30% of their length. This construction ensures the rapid propagation of waves in softwood [42–45]. In Figure 6a,c, the microscopic view in the tangential direction of spruce wood in relation to the measurement principle can be noted. In the radial direction, the transmitted waves meet the tracheids whose walls show thicknesses that increase from earlywood to latewood, so it can be considered a layered material (with each annual ring being a layer) (Figure 6b).

In the radial direction, spruce wood contains radial parenchyma cells that have a geometry like rectangular prisms with dimensions of 15 μm high by 10 μm wide and 150–250 μm long in the radial or horizontal direction [8,21]. In the case of maple wood (*Acer pseudoplatanus* L.), the anatomical architecture is more complex than that of spruce wood, with it being made up of fibers, vessels, and axial parenchyma cells, with different shapes, organization, and quantity [42,43]. The vessels are small and rare, and the fibers occupy about 75% of the volume of the woody substance, the proportion of voids is 60% of the apparent volume, and from the point of view of organization, maple wood is part of the hardwoods with diffused, small pores, visible with a magnifying glass and quite rare, with their size and thickness decreasing towards the outer limit of the annual ring (Figure 6d–f) [44,45]. The vessels have average lengths between 350–800 μm, with small diameters, <50 μm. As a result, the propagation of sounds in maple wood differs from that of spruce wood; the waves propagate through a structure with a different organization and construction.

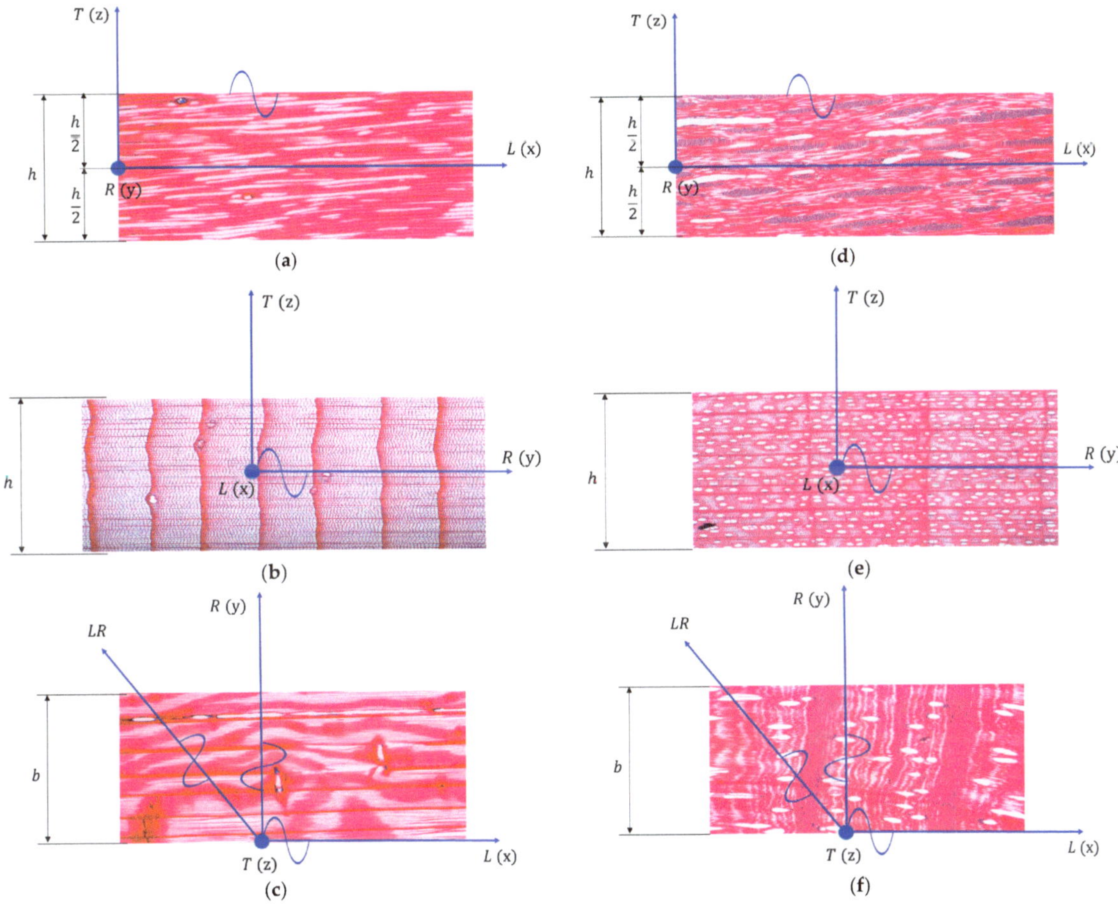

Figure 6. The microscopic views of the main section of wood related to the Lamb wave direction in the studied plates: (**a**) tangential section of spruce wood; (**b**) cross-section of spruce wood; (**c**) radial section of spruce wood; (**d**) tangential section of maple wood; (**e**) cross-section of maple wood; (**f**) radial section of maple wood.

Figure 7 highlights the variation in Lamb wave speeds measured on a circular trajectory applied to the wooden board. The angles of 0 and 180 degrees correspond to the longitudinal axis of the wood, and the angle of 90 degrees corresponds to the radial axis. Thus, it can be observed that between the two directions, L and R, the structural anisotropy of the wood is at its maximum, results highlighted in other studies [18–21].

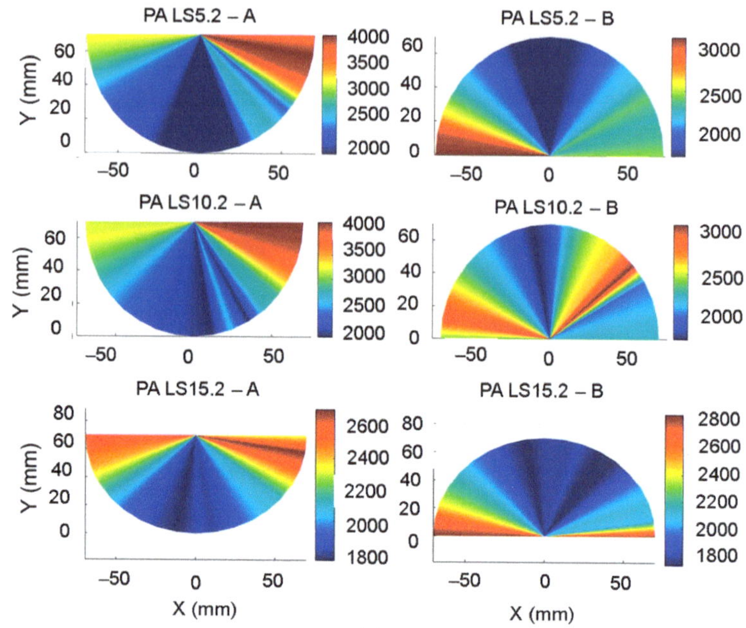

(a)

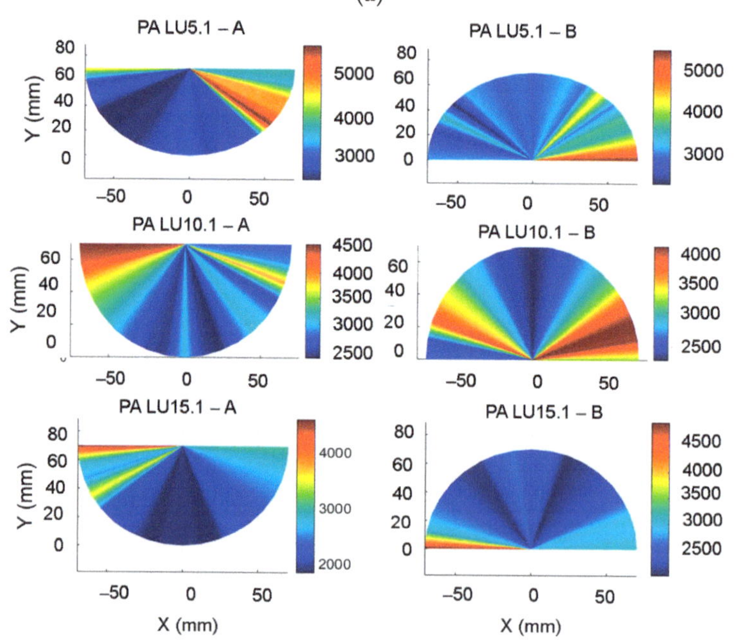

(b)

Figure 7. *Cont.*

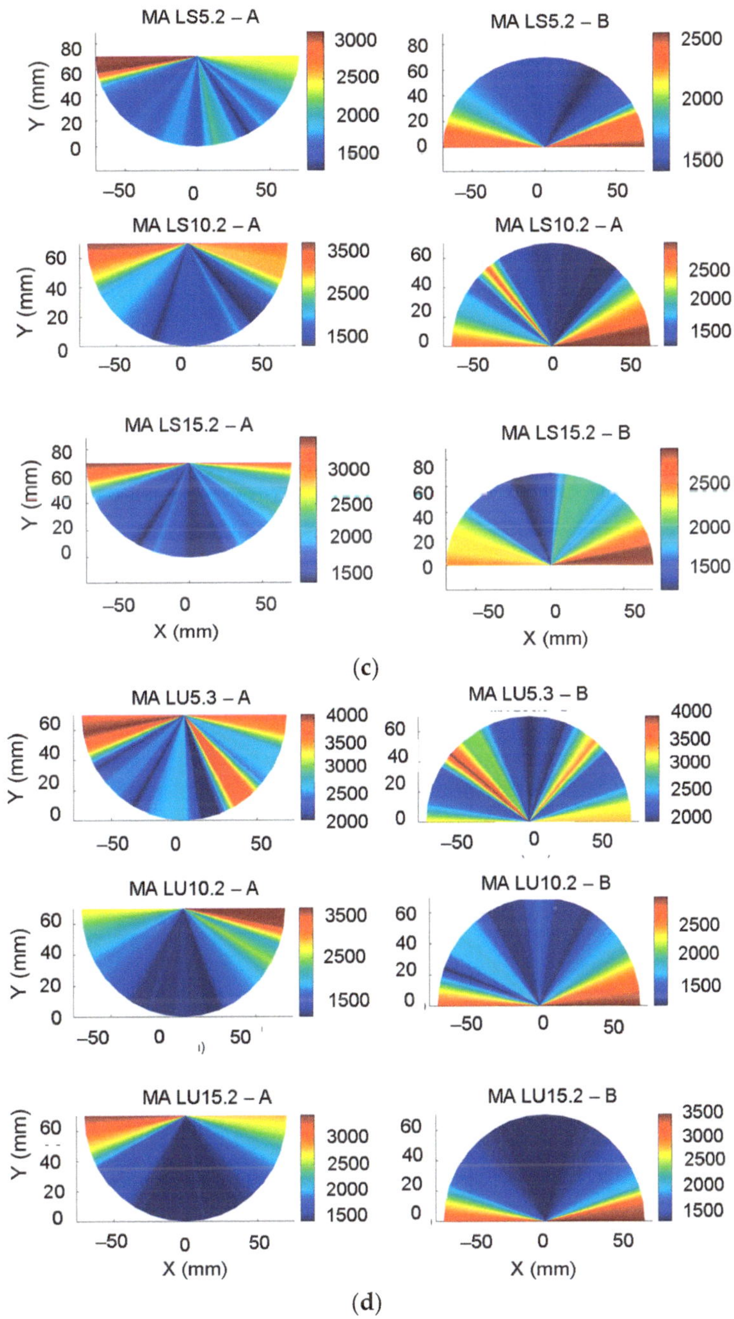

Figure 7. The variation in Lamb wave speeds with the rotation of the receiver position: (**a**) maple samples covered with spirit varnish; (**b**) maple samples covered with oil-based varnish and spirit varnish; (**c**) spruce samples covered with spirit varnish; (**d**) spruce wood sample finished with oil-based varnish.

Similar results were reported by [46], who analyzed spruce and sycamore wood samples from three types of quality classes: the highest, the moderate quality class, and the control samples, obtaining the following values for the investigated elastic and acoustic parameters: for high-quality spruce wood samples versus a control wood sample, the sound velocity in the longitudinal c_L and radial c_R directions of $c_L = 5103 \pm 280$ m/s and $c_R = 1365 \pm 201$ m/s versus the control wood sample $c_L = 5388 \pm 134$ m/s and $c_R = 1124 \pm 93$ m/s. For sycamore, ref. [46] obtained, $c_L = 3894 \pm 310$ m/s and $c_R = 1662 \pm 97$ m/s. Estimations of the elastic and acoustic properties were made using methods based on ultrasound [21], reported for resonance spruce with the following longitudinal velocities $c_L = 5600$ m/s; tangential velocity $c_T = 1600$ m/s; and radial velocity $c_R = 2000$ m/s, and for shear velocity, $c_{LT} = 1425$ m/s; $c_{RT} = 298$ m/s; and $c_{LR} = 1374$ m/s and for common spruce, $c_L = 5353$ m/s; $c_T = 1146$ m/s; and $c_R = 1580$ m/s; and for shear velocity, $c_{LT} = 1230$ m/s; $c_{RT} = 477$ m/s; and $c_{LR} = 1322$ m/s. For curly maple, the values determined using ultrasonic method are: $c_L = 4350$ m/s; $c_T = 1914$ m/s; and $c_R = 2590$ m/s, and for shear velocity $c_{LT} = 1468$ m/s; $c_{RT} = 812$ m/s; and $c_{LR} = 1744$ m/s, since for common Sycamore, the values are $c_L = 4695$ m/s; $c_T = 1878$ m/s; and $c_R = 2148$ m/s, and for shear velocity, $c_{LT} = 1148$ m/s; $c_{RT} = 630$ m/s; and $c_{LR} = 1354$ m/s. Ref. [47] noticed that for the ratio of the sounds velocities in wood in the two longitudinal and radial directions, within the same wood species, for the spruce samples, the ratio CL:CR was 3.257, and for the samples of maple wood, the ratio was 2.17.

3.2. Statistical Results

In order to verify the continuity of the measurement points, a correlation analysis of the values collected for all material samples was performed, considering the values measured on the two semicircles A and B as forming a fictitious circle. In the correlation matrix in Figure 8, a "butterfly" distribution of correlation values can be observed, with a diagonal symmetry that corresponds to the orientation symmetry of the wood fibers.

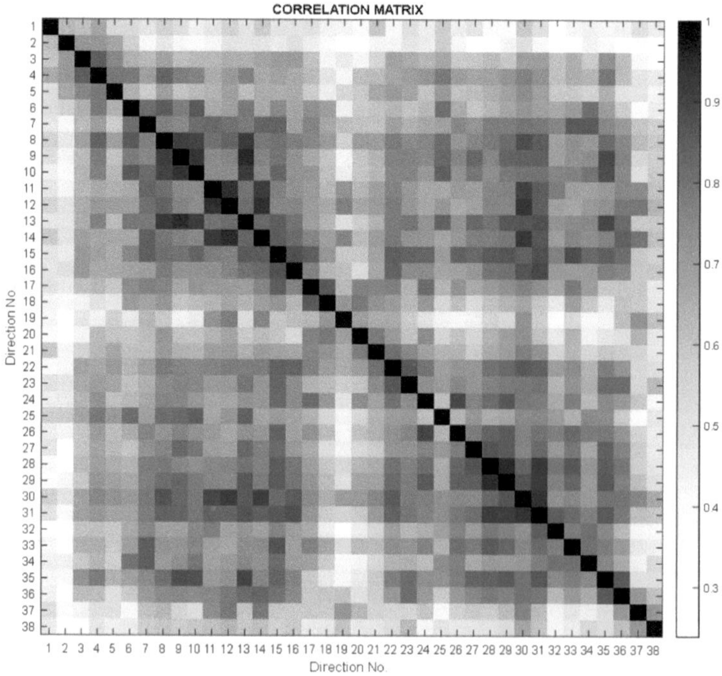

Figure 8. "Butterfly" distribution of correlation values.

The anisotropy of the sound propagation speeds in the wood samples is highlighted in the polar graph of the speeds in each of the 38 directions for all of the wood samples (Figure 9).

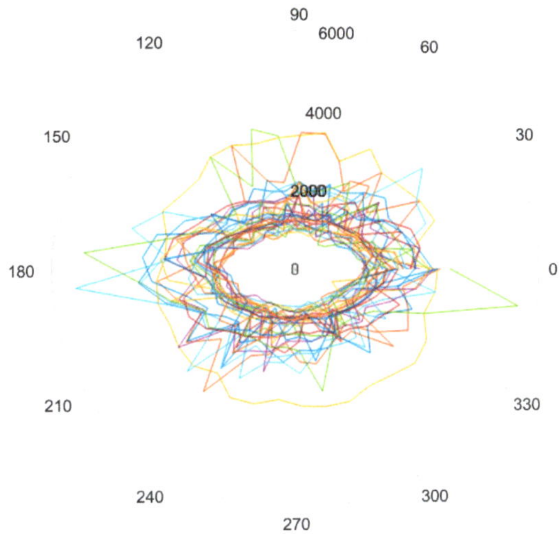

Figure 9. Polar graphs of sound velocities for each sample.

This summary shows the variability of the measured values with respect to the different directional dependencies, as well as the relatively large variation in the maximum speed measured in individual samples. Using the Q-variant of FA (based on a transposed data matrix and resulting in a clustering scheme, see the left part of Figure 8), it was possible to point to certain samples that had a particular pattern of anisotropy. Based on the Q-factor analysis (QFA), several groups of samples with similar anisotropy were extracted, depending on the wood species, the applied treatment, or the anatomical characteristics. In some of these groups, only a specific sample belongs to a cluster and therefore can be declared exceptional in its own way. These selected samples are due to the type of wood and different treatment methods. Using QFA, the measured parameters are ordered into a so-called individual profile, which helps to identify similar characteristics. By applying the RMS (the square root of the mean square) and STD (the standard deviation) complementary diagrams, it was found that the PA UV3 sample appears as a very unique sample with only one large affiliation (the dark highlighted value in Figure 10a) in the third group. This is a sample with a very high mean velocity (Figure 10b) and, in contrast, a small anisotropy (low STD value—Figure 10c). PD sample SAL1 has a similar small anisotropy. However, it does not have such high speeds and therefore represents a specific type in itself, defining another cluster, number 10. The sample MA HD10.3, which has an exceptional anisotropy (see the maximum measured standard deviation), defines cluster 1, to which samples such as PD LU5.3, PA SAL1, and several others with less similarity are assigned. Based on the above, several samples can be identified that are representative of "their" characteristic group. These are as follows: MA HD 10.3, PA LS 10.2, PA UV 3, MA UV 2, PA LU 5.1, PA LU 10.1, MA SAL 2, and MA LU 5.3. Cluster No. 7 is interesting in this respect, where two very similar velocity profiles are obtained from different wood species (Maple PA NC 10.1 and Spruce MD LU 15.2).

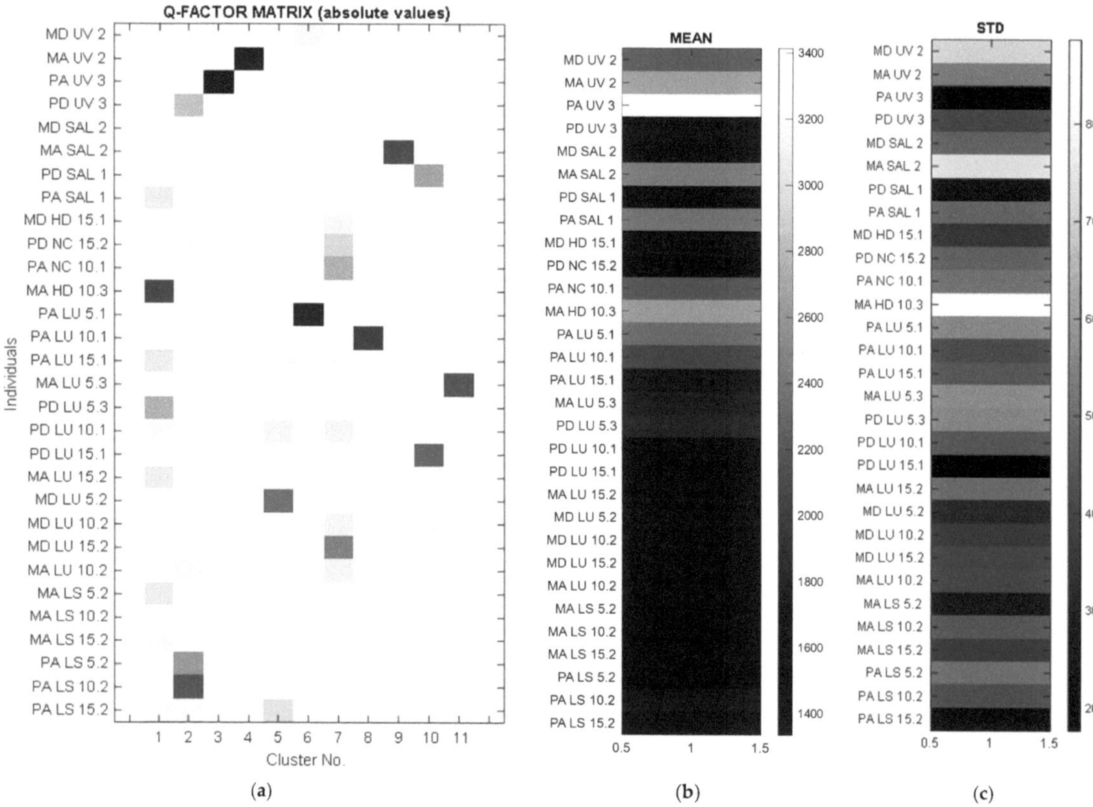

Figure 10. Comparative analysis of statistical data: (**a**) Q-FA clustering scheme; (**b**) the square root of the mean square values; (**c**) STD values for each sample.

4. Conclusions

The use of non-destructive control to determine the sound propagation speeds in different stages of the finishing of musical instruments can be a modern and precise method of optimizing the finishing technology to obtain the desired acoustic features, depending on the quality of the wood, the type of varnish and the treatments applied. The results of this study help instrument manufacturers to integrate new technologies with the art of luthiers and also apply innovative solutions for treating resonance wood, which they can later verify using the ultrasound method. Based on the obtained experimental data, the following conclusions were drawn:

1. The treatment applied to the surface of the wooden resonance plates influences the sound propagation speed both in the longitudinal and radial directions. The acoustic behavior of the wood depends both on the type of treatment and the anatomical characteristics of the wood, with differences in propagation speeds being recorded between samples of the same woody species but with different anatomical structures.
2. In spruce wood samples, the film of alcohol varnish reduces the speed of propagation of sounds by approximately 42–58% in the longitudinal direction, and in the radial direction, the oil-based varnish reduces the speed of propagation by 60%.
3. In the maple wood samples, the anisotropy ratio between the longitudinal and radial directions decreases with the increase in the thickness of the varnish film.

Author Contributions: Conceptualization, D.F., A.S., M.D.S. and M.C.; methodology, A.S., M.K. (Michal Krbata), M.K. (Marcel Kohutiar) and M.C.; software, D.F., M.D.S., S.M.N. and M.C.; validation, M.K. (Michal Krbata) and M.K. (Marcel Kohutiar); formal analysis, A.S., M.C., S.M.N. and M.D.S.; investigation, D.F., A.S. and M.C.; resources, M.D.S., M.K. (Michal Krbata) and M.K. (Marcel Kohutiar); data curation, D.F., A.S. and M.C.; writing—original draft preparation, A.S., M.D.S. and M.C.; writing—review and editing, D.F., M.K. (Michal Krbata) and M.K. (Marcel Kohutiar); visualization, M.C. and S.M.N.; supervision, M.D.S. and A.S.; project administration, M.D.S., D.F. and M.C.; funding acquisition, M.K. (Michal Krbata), M.K. (Marcel Kohutiar) and M.D.S. All authors have read and agreed to the published version of the manuscript.

Funding: This research was supported by a grant from the UEFISCDI (Unitatea Executiva pentru Finantarea Cercetării, Dezvoltării si Învățământului Superior din Romania), project number 61PCE/2022, PN-III-P4-PCE2021-0885, ACADIA—Qualitative, dynamic and acoustic analysis of anisotropic systems with modified interfaces.

Institutional Review Board Statement: Not applicable.

Data Availability Statement: Data are contained within the article.

Acknowledgments: We are also grateful to the technical staff of Gliga Musical Instruments, Reghin, Romania, for supplying the specimens. We are grateful to Cătălin Roibu and his colleagues for the microscopic specimens prepared and analyzed in the Forest Biometrics Laboratory, "Stefan cel Mare" University of Suceava, Romania. Milan Chlada would like to acknowledge the support of the Institute of Thermomechanics of the Czech Academy of Sciences, grant No. RVO:61388998. Adriana Savin would like to acknowledge the support of the Ministry of Research, Innovation and Digitalization, Romania, through Project PN 23-11-01-02.

Conflicts of Interest: The authors declare no conflicts of interest.

References

1. Wegst, U.G.K. Wood for sound. *Am. J. Bot.* **2006**, *93*, 1439–1448. [CrossRef] [PubMed]
2. Obataya, E.; Ono, T.; Norimoto, M. Vibrational properties of wood along the grain. *J. Mater. Sci.* **2000**, *35*, 2993–3001. [CrossRef]
3. Brémaud, I.; Gril, J.; Thibaut, B. Anisotropy of wood vibrational properties: Dependence on grain angle and review of literature data. *Wood Sci. Technol.* **2011**, *45*, 735–754. [CrossRef]
4. Ono, T. Effect of grain angle on dynamic mechanical properties of wood. *J. Soc. Mater. Sci. Jpn.* **1983**, *32*, 108–113. [CrossRef]
5. Sedighi Gilani, M.; Pflaum, J.; Hartmann, S.; Kaufmann, R.; Baumgartner, M.; Schwarze, F.W.M.R. Relationship of vibro-mechanical properties and microstructure of wood and varnish interface in string instruments. *Appl. Phys. A* **2016**, *122*, 260. [CrossRef]
6. Bucur, V. *Handbook of Materials for String Musical Instruments*; Springer: Berlin/Heidelberg, Germany, 2016. [CrossRef]
7. Bernabei, M.; Bontadi, J. Determining the resonance wood provenance of stringed instruments from the Cherubini Conservatory Collection in Florence, Italy. *J. Cult. Herit.* **2011**, *12*, 196–204. [CrossRef]
8. Schmidt-Vogt, H. *Die Fichte Band I. Taxonomie-Verbreitung-Morphologie-Ökologie-Waldgesellschaften*; Paul Parey: Hamburg/Berlin, Germany, 1977.
9. Dinulică, F.; Albu, C.T.; Borz, S.A.; Vasilescu, M.M.; Petritan, I.C. Specific structural indexes for resonance Norway spruce wood used for violin manufacturing. *Bioresources* **2015**, *10*, 7525–7543. [CrossRef]
10. Dinulică, F.; Stanciu, M.D.; Savin, A. Correlation between Anatomical Grading and Acoustic–Elastic Properties of Resonant Spruce Wood Used for Musical Instruments. *Forests* **2021**, *12*, 1122. [CrossRef]
11. Viala, R.; Placet, V.; Cogan, S. Simultaneous non-destructive identification of multiple elastic and damping properties of spruce tonewood to improve grading. *J. Cult. Herit.* **2020**, *42*, 108–116. [CrossRef]
12. Carlier, C.; Alkadri, A.; Gril, J.; Bremaud, I. Revisiting the notion of "resonance wood" choice: A decompartementalised approach from violin makers' opinion and perception to characterization of material properties variability. wooden musical instruments. Wooden musical instruments—Different forms of knowledge: Book of end of WoodMusICK COST Action FP1302. 2018, pp. 119–142. Available online: https://hal.science/hal-02051004/document (accessed on 16 January 2024).
13. Caldersmith, G. Vibrations of orthotropic rectangular plates. *Acta Acust. United Acust.* **1984**, *56*, 144–152.
14. Mclntyre, M.; Woodhouse, J. On measuring wood properties, part 1–3. *J. Catgut Acoust. Soc.* **1984**, *42*, 1984–1986.
15. Warguła, Ł.; Wojtkowiak, D.; Kukla, M.; Talaśka, K. Symmetric nature of stress distribution in the elastic-plastic range of Pinus L. pine wood samples determined experimentally and using the finite element method (FEM). *Symmetry* **2021**, *13*, 39. [CrossRef]
16. Wang, R.; Liang, F.; Wang, B.; Mou, X. ODCA-YOLO: An Omni-Dynamic Convolution Coordinate Attention-Based YOLO for Wood Defect Detection. *Forests* **2023**, *14*, 1885. [CrossRef]

17. Gonçalves, R.; Trinca, A.J.; Cerri, D.G.P. Comparison of elastic constants of wood determined by ultrasonic wave propagation and static compression testing. *Wood Fiber Sci.* **2011**, *43*, 64–75.
18. Duquenme, L.; Moulin, E.; Assood, J.; Grandel, S. Transient modeling of Lamb wave generated in viscoelastic materials by surface bonded piezoelectric transducers. *J. Acoust. Soc. Am.* **2004**, *116*, 133–141. [CrossRef]
19. Fang, Y.; Lin, L.; Feng, H.; Lu, Z.; Emms, G.W. Review of the use of air-coupled ultrasonic technologies for nondestructive testing of wood and wood products. *Comput. Electron. Agric.* **2017**, *137*, 79–87. [CrossRef]
20. Bucur, V.; Archer, R.R. Elastic constants for wood by an ultrasonic method. *Wood Sci. Technol.* **1984**, *18*, 255–265. [CrossRef]
21. Bucur, V. Varieties of resonance wood and their elastic constants. *J. Catgut Acoust. Soc.* **1987**, *47*, 42–48.
22. Bucur, V. Wood structural anisotropy estimated by acoustic invariants. *IAWA Bull.* **1988**, *9*, 67–74. [CrossRef]
23. Dinulică, F.; Savin, A.; Stanciu, M.D. Physical and Acoustical Properties of Wavy Grain Sycamore Maple (*Acer pseudoplatanus* L.) Used for Musical Instruments. *Forests* **2023**, *14*, 197. [CrossRef]
24. Schmerr, L.W., Jr. *Fundamentals of Ultrasonic Nondestructive Evaluation—A Modeling Approach*; Plenum Press: New York, NY, USA, 1998.
25. Grimberg, R.; Savin, A.; Steigmann, R.; Stanciu, M.D.; Grum, J. Determination of Elastic Properties of CFRP Using Lamb Waves Resonant Spectroscopy. In Proceedings of the 2nd International Symposium on NDT in Aerospace, Hamburg, Germany, 22–24 November 2010.
26. Alleyne, D.N.; Cawley, P. A two dimensional Fourier transform method for the measurement of propagating multimode signals. *J. Acoust. Soc. Am.* **1991**, *89*, 1159–1168. [CrossRef]
27. Neau, G.; Deschamps, M.; Lowe, M.J.S. Group velocity of Lamb waves in anisotropic plates: Comparison between theory and experiments. *AIP Conf. Proc.* **2001**, *557*, 81–88. [CrossRef]
28. Kuznetsov, S.V. Resonance phenomena in layered media: Merging proper and quasi-resonances. *Z. Angew. Math. Phys.* **2023**, *74*, 245. [CrossRef]
29. Nowak, T.; Hamrol-Bielecka, K.; Jasienko, J. Non-destructive testing of wood–correlation of ultrasonic and stress wave test results in glued laminated timber members. Annals of Warsaw University of Life Sciences-SGGW. *For. Wood Technol.* **2015**, *92*, 317–324.
30. *ISO 13061-1:2014/Amd 1:2017*; Physical and Mechanical Properties of Wood. Test Methods for Small Clear Wood Specimens. Part 1: Determination of Moisture Content for Physical and Mechanical Tests. ISO: Geneva, Switzerland, 2017.
31. Schelleng, J.C. Acoustical Effects of Violin Varnish. *J. Acoust. Soc. Am.* **1968**, *4*, 1175–1183. [CrossRef]
32. Schleske, M. Contemporary violin making. Analysis of design, materials, varnish and normal modes. *Catgut Acoust. Soc. J.* **2002**, *4*, 50–65.
33. Nagyvary, J.; Guillemette, R.N.; Spiegelman, C.H. Mineral Preservatives in the Wood of Stradivari and Guarneri. *PLoS ONE* **2009**, *4*, e4245. [CrossRef] [PubMed]
34. Cai, W.; Tai, H.-C. String Theories: Chemical Secrets of Italian Violins and Chinese Guqins. *AsiaChem* **2020**, *11*, 10–17. [CrossRef]
35. Skrodzka, E.B.; Linde, B.J.B.; Krupa, A. Modal Parameters of Two Violins with Different Varnish Layers and Subjective Evaluation of Their Sound Quality. *Arch. Acoust.* **2013**, *38*, 75–81. [CrossRef]
36. Lämmlein, S.L.; Van Damme, B.; Mannes, D.; Schwarze, F.W.M.; Burgert, I. Violin varnish induced changes in the vibro-mechanical properties of spruce and maple wood. *Holzforschung* **2020**, *4*, 765–776. [CrossRef]
37. Odlyha, M.; Lucejko, J.J.; Lluveras-Tenorio, A.; di Girolamo, F.; Hudziak, S.; Strange, A.; Bridarolli, A.; Bozec, L.; Colombini, M.P. Violin Varnishes: Microstructure and Nanomechanical Analysis. *Molecules* **2022**, *27*, 6378. [CrossRef]
38. Stanciu, M.D.; Cosnita, M.; Gliga, G.V.; Gurau, L.; Timar, C.M.; Guiman, M.V.; Năstac, S.M.; Roșca, I.C.; Bucur, V.; Dinulică, F. Tunable Acoustic Properties Using Different Coating Systems on Resonance Spruce Wood. *Adv. Mat. Interfaces* **2024**, *1*, 2300781. [CrossRef]
39. Stobbe, D.M.; Grünsteidl, C.M.; Murray, T.W. Propagation and Scattering of Lamb Waves at Conical Points in Plates. *Sci. Rep.* **2019**, *9*, 15216. [CrossRef]
40. Bucur, V. A Review on Acoustics of Wood as a Tool for Quality Assessment. *Forests* **2023**, *14*, 1545. [CrossRef]
41. Fathi, H.; Kazemirad, S.; Nasir, V. A nondestructive guided wave propagation method for the characterization of moisture-dependent viscoelastic properties of wood materials. *Mater. Struct.* **2020**, *53*, 147. [CrossRef]
42. Wiedenhoeft, A.C.; Miller, R.B. Structure and Function of Wood. In *Handbook of Wood Chemistry and Wood Composites*; Rowell, R.M., Ed.; CRC Press: Boca Raton, FL, USA, 2005.
43. Dinulică, F. *Forest Products: Structure and Properties of Wood (in Romanian Language)*; Publishing House Transilvania University of Brasov: Brasov, Romania, 2021.
44. Gurău, L.; Timar, M.C.; Coșereanu, C.; Cosnita, M.; Stanciu, M.D. Aging of Wood for Musical Instruments: Analysis of Changes in Color, Surface Morphology, Chemical, and Physical-Acoustical Properties during UV and Thermal Exposure. *Polymers* **2023**, *15*, 1794. [CrossRef] [PubMed]
45. Guiman, M.V.; Stanciu, M.D.; Roșca, I.C.; Georgescu, S.V.; Năstac, S.M.; Câmpean, M. Influence of the Grain Orientation of Wood upon Its Sound Absorption Properties. *Materials* **2023**, *16*, 5998. [CrossRef]

46. Spycher, M.; Schwarze, F.W.M.R.; Steiger, R. Assesment of resonance wood quality by comparing its physical and histological properties. *Wood Sci. Technol.* **2008**, *42*, 325–342. [CrossRef]
47. Crețu, N.; Roșca, I.C.; Stanciu, M.D.; Gliga, V.G.; Cerbu, C. Evaluation of Wave Velocity in Orthotropic Media Based on Intrinsic Transfer Matrix. *Exp. Mech.* **2022**, *62*, 1595–1602. [CrossRef]

polymers

MDPI

Article

Energy Consumption for Furniture Joints during Drilling in Birch Plywood

Weronika Pakuła, Barbara Prałat, Zbigniew Potok, Krzysztof Wiaderek and Tomasz Rogoziński *

Department of Furniture Design, Faculty of Forestry and Wood Technology, Poznań University of Life Sciences, ul. Wojska Polskiego 38/42, 60-627 Poznan, Poland; 70655@student.up.poznan.pl (W.P.); barbara.pralat@up.poznan.pl (B.P.); zbigniew.potok@up.poznan.pl (Z.P.); krzysztof.wiaderek@up.poznan.pl (K.W.)

* Correspondence: tomasz.rogozinski@up.poznan.pl

Abstract: The purpose of this study is to support eco-design ideas and sustainable manufacturing techniques by examining the energy consumption related to drilling holes for different furniture connections. The experimental model is a simple piece of furniture made from birch plywood with three different types of joints. Eccentric joints, confirmat screws, and dowel measurements of energy consumption with a CNC drilling and milling machine show different values for every kind of connector. The energy consumption was measured using a portable power quality analyzer, specifically the PQ-box 150 manufactured by A:Eberle GmbH & Co. KG Nürnberg, Germany. This device likely adheres to industry standards for energy measurement, ensuring accurate and reliable results. The measurement process involved recording energy consumption at different stages of the machining process, allowing for the analysis of specific cutting work and total energy consumption for various joint types. Dowels exhibit the lowest energy consumption at 0.105 Wh for one furniture joint, confirmat screws at 0.127 Wh, while eccentric joints, despite their higher energy consumption (0.173 Wh), offer enhanced transportability and assembly flexibility of a piece of furniture. Specific cutting power for one selected piece of furniture was 227.89 J/mm^3 for dowels, 190.63 J/mm^3 for eccentric joints and 261.68 J/mm^3 for confirmat screws.

Keywords: furniture joints; drilling; energy consumption

check for updates

Citation: Pakuła, W.; Prałat, B.; Potok, Z.; Wiaderek, K.; Rogoziński, T. Energy Consumption for Furniture Joints during Drilling in Birch Plywood. *Polymers* **2024**, *16*, 1045. https://doi.org/10.3390/polym16081045

Academic Editor: Hiroshi Yoshihara

Received: 26 February 2024
Revised: 4 April 2024
Accepted: 8 April 2024
Published: 10 April 2024

1. Introduction

Circular economy principles and a product's environmental impact should be considered from the very beginning of its design. The term "eco-design" refers to the integration of various disciplines, including ecological engineering, ecological restoration, green architecture, and others [1]. Eco-design must consider all aspects of the product's life cycle, including shipping, machining, waste generation overall, and the product's end, in addition to materials and aesthetics [2].

The optimization of the production process brings benefits to the environment with an almost immediately achievable goal: reducing the environmental impact of individual stages of material processing by changing various process factors, combined with the development of more advanced technologies and efficient processing methods [3]. The manufacturing sector is leading the way in lowering energy consumption because of the industry's constant rise in energy, the usage of energy receivers, and the increasing amount of energy used in the production process. This includes the fact that variables influencing energy consumption have an impact on cutting parameters [4]. Over the past fifteen years, many energy consumption models have been published, which is particularly significant when it comes to energy-minimization machining. The relationship between energy consumption and the material removal rate (MRR) in a typical machining operation was initially theorized by Gutowski [5].

Specific cutting energies can be employed as machining-process efficiency indicators in most material removal operations. The energy required to create new surfaces, the energy expended in the primary and secondary deformation zones, and the interfacial friction activities at the tool–workpiece interfaces are all examples of specific cutting energies. During the cutting process, this energy is transferred from the cutting tool to the chip, workpiece, and heat via the cutter rake and flank surfaces. The ratio of the volume of workpieces to the energy consumed during machining operations is known as the specific cutting energy. The following factors affect the specific cutting energy: growl production speed, cutting force, and cutting speed [6]. Furthermore, the energy efficiency of the machining process is often determined using that specific cutting energy [7]. To achieve process-level energy efficiency in processing, it is essential to comprehend the impact of material properties and process variables on the required cutting energy factor [8]. In grinding processes, the selection of cutting conditions can result in notable alterations and an increased specific cutting energy [9].

When drilling holes, energy consumption is the most important parameter, but the quality of holes made for selected furniture joints is equally essential. The degree of dimensional accuracy in machining results in a higher vulnerability to errors during the assembly of furniture components. Because of the characteristics of the workpiece, the drill tip may slip during the initial drilling phase, potentially leading to inaccurate machining [10]. When compared to the holes created in two adjacent layers of veneer, the holes created in the adhesive layer displayed deviations that were roughly twice as large. The errors in the centers' positions did not significantly correlate with the deviations of the holes' axis angles. Wood shrinkage and swelling have an impact on how effectively furniture pieces work together. Thus, they should be considered while designing. This study investigated how moisture variations affected the drilled holes' effective diameter [11]. The intricate variations in hole shape caused by moisture are as follows: depending on the depth and wood grain pattern, the holes ovalized with varying ranges, and at the bottom, their diameters were slightly enlarged [12].

In other industries, there has also been an attempt to improve the efficiency of cutting-energy utilization. The aircraft industry makes considerable use of metal materials, particularly multilayer stacks made of CFRP (carbon fiber-reinforced polymer) and Ti6Al4V (Titanium-64, or Grade 5 Titanium Alloy), because of their high specific stiffness, outstanding corrosion resistance, and exceptional structural efficiency. In CFRP/Ti6Al4V stack drilling, a comparative analysis between various cooling conditions was conducted. To ascertain the impact of various cooling techniques on the torque, specific cutting energy, and surface morphology of the processed composite material during CFRP/Ti6Al4V stacks, a series of drilling tests were conducted under dry circumstances and minimum-quantity lubrication. Technical advice for choosing a cooling technique and enhancing energy efficiency in the titanium-composite process machining operations can be obtained from this work [13]. Other research on energy modeling and processing visualization have been acknowledged as useful and efficient methods for identifying areas for energy savings and enhancing energy efficiency. The examination focused on the study's drilling visualization goals and energy modeling. Drilling procedures revealed that the suggested drilling energy model had an average forecast accuracy of 96.2%. Additionally, the results demonstrated a 12.6% increase in energy efficiency and 7417.8 J of energy savings. The suggested approach helped the drilling process save energy by guaranteeing the energy model's high accuracy, identifying possible energy savings, and enhancing energy efficiency [14].

In the wood industry, cutting forces are also examined. The impact of introducing SBR gum granulate (Styrene-Butadiene-Rubber gum granulate) on particleboard machinability was examined in a study, with a particular emphasis on cutting forces during drilling. Various formulations, including 0% to 30% SBR, were investigated. The tests, which were carried out on a CNC machine using a 10 mm diameter polycrystalline diamond drill, showed that adding more SBR significantly improved machinability. Relative machinability indicators based on axial force (MIF) and torque (MIM), which show improved machinability, particularly at increasing SBR content, are included in the study. The results demonstrate how SBR gum granulation reduces cutting forces during drilling, which raises the possibility that it could enhance wood-based composite materials used in machining [15]. The analysis of cutting forces during the thermally modified ash-wood drilling process was the main goal of another study. Two sets of workpieces were used in the experiment; one set underwent heat alteration, while the other did not. Both sets of through-holes were drilled, and piezoelectric sensors were used to measure the thrust force and torque data. When thermally changed ash wood was drilled, the results showed a statistically significant increase in thrust force and a statistically significant decrease in torque when compared to the unmodified equivalent. Regarding yearly rings, the tool feed orientation did not have a statistically significant effect on cutting forces. In conclusion, it was discovered that thermally altering ash wood changed the cutting forces, particularly by decreasing torque and raising thrust force, throughout the drilling operation [16].

According to a not fully specified definition, cutting can be considered as changing the density of the workpiece by the mechanical removal of material. Parameters such as cutting speed, feed per revolution, cutting depth, and cross-sectional area of the cut layer are needed to determine energy consumption [17]. The cutting force can be calculated using the above parameters, which allows for energy estimation. As of 2016, Poland was 30.3% dependent on the import of energy raw materials [18]. A piece of furniture with the same structure and function in large-scale production may have different costs if you take a closer look at it, considering the type of connection. Holes for appropriate connections may differ in drilling depth, execution time, and, therefore, the tool's operating time, diameter, and quantity [19]. The study presented herein addresses a critical gap in the existing body of scientific literature concerning the energy consumption associated with drilling holes for furniture joints. Surprisingly, there remains a paucity of comprehensive research in this domain, highlighting the significance of this paper for both academic researchers and the furniture industry. By delving into the energy usage during the machining process for various furniture connections, this study not only fills an important knowledge void but also offers valuable insights into eco-design practices and sustainable manufacturing techniques. The findings have the potential to inform future research endeavors and guide decision-making processes within the furniture sector, ultimately advancing the pursuit of environmentally friendly production methods. For a furniture designer, how much energy is consumed when drilling holes should be known. If the designer wants their product to be as environmentally friendly as possible, energy consumption when drilling holes for furniture joints should be as low as possible with the assumed functionality. Therefore, the purpose of this work is to verify how selected furniture connections influence energy consumption for drilling holes.

2. Materials and Methods

2.1. Furniture Construction

For the study, a piece of furniture with a simple structure was designed. The shelf was designed to use three different furniture joints while maintaining the same type of material, shape, and usability. The furniture design consists of two wreaths, two sides, a vertical partition, and two horizontal partitions. The furniture dimensions are: 800 mm length, 1000 mm height, and 400 mm depth. Moreover, 15 mm thick birch *Betula pendula* plywood (Paged Pisz Sp. z o.o., Pisz, Poland) was used to make the furniture.

2.2. Stiffness Test

The stiffness assessment of birch plywood was performed following the PN–EN 310 [20].

The load was applied to the canter of the sample supported by two supports. The sample dimensions were 350 mm × 50 mm × 15 mm. The marking was performed based on 10 plywood samples with external layers arranged longitudinally, and 10 plywood samples with external layers arranged transversely. The elastic modulus was calculated from the sample deflection in the linear range of deflection and force dependence.

The furniture body stiffness was calculated from the formula: G—Kirchhoff modulus, d—plate thickness, $l_{1,2}$—dimensions of the i-th plate, and α—α coefficient, calculated differently depending on the element.

$$k = \sum\nolimits_{i=1}^{n} \frac{G_{i\,i}^{3}}{3(l_1 l_2)_i}\alpha_i^2 \tag{1}$$

The global stiffness of the tested structure exceeded the values recommended by standards by more than twice, which allowed for the adoption of a solution in engineering practice for furniture subject to higher loads, such as library shelves. Regardless of the selected connection, the global stiffness remained unchanged.

2.3. Numerical Study

A numerical study was performed in the Inventor Nastran 2024 program, as shown in Figure 1. The numerical analysis was carried out to determine the correctness of the structure in terms of stiffness and to determine the maximum stresses regardless of the type of furniture joints used. A simplified structure was made reflecting the designed product by entering the corresponding parameters for the selected material. In the case of Poisson's ratio, it was taken from the article "Theory of elasticity of an anisotropic body". The calculations confirmed the observations for constant material values. The stiffness K [N/mm] depends on the dimensions of the individual structural elements of the furniture. In the case of the analyzed structure, the furniture elements press on each other, exerting contact stresses of no more than 0.6 MPa.

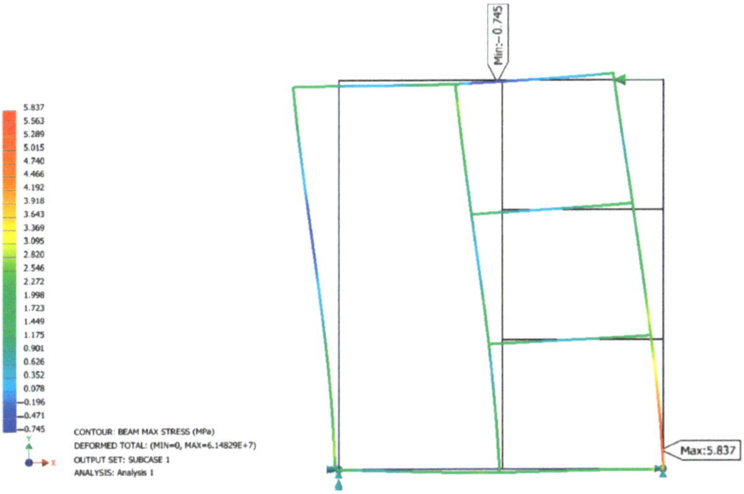

Figure 1. Inventor Nastran tension of the structure when a force was applied.

2.4. Types of Joints

Three types of joints (Häfele, Nagold, Germany) were selected and tested. Dowel joints use glue as a permanent connection. The second type was the confirmat screw, which is used as a conditionally detachable connection, enabling the furniture to be unscrewed and reassembled. After reassembling the furniture, it does not have the same strength due to losses in the material caused by the opening of the furniture connection. The last joint considered for testing is the eccentric joint, as a detachable joint. Drilling was performed for 30 furniture joints of each type, as shown in Figure 2. In Figure 3, drilled holes for each joint are shown.

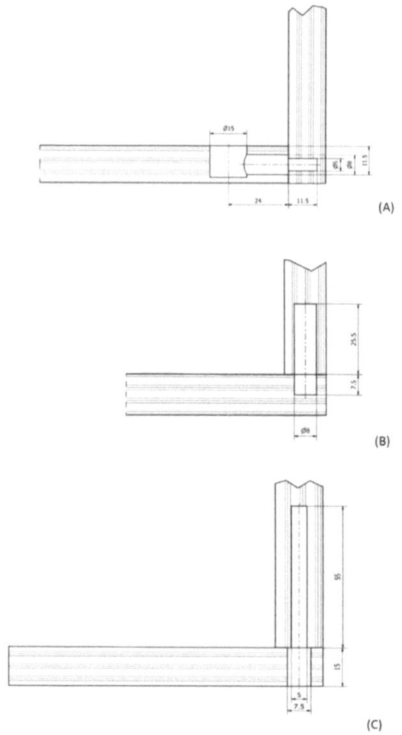

Figure 2. (**A**): drilling diagram for an eccentric joint (three holes); (**B**): drilling diagram for a confirmat screw (two holes); (**C**): drilling diagram for a dowel connector (two holes). Measurements are mm.

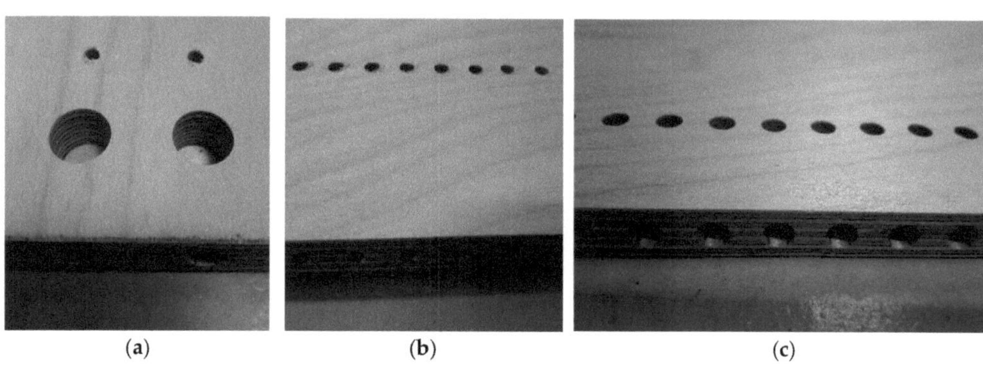

Figure 3. (**a**): drilled holes for an eccentric joint; (**b**): drilled holes for a confirmat screw (two holes); (**c**): drilled holes for a dowel connector (two holes).

2.5. Tools and Machine Used

To make holes for the pins, drills with a diameter of 8 mm were used: a screw drill with a centering spike and cutters that was non-uniform, single, two-edged, and single-stage, with a cylindrical shank; an ordinary drill with a diameter of 8 mm; and a screw drill with a centering spike and cutters, which was non-uniform, single, two-edged, and single-stage, with a cylindrical shank. To make the holes for the eccentrics, drills with a diameter of 5 mm were used: screw drills with a centering spike and cutters that were non-uniform, single, two-point, and single-stage, with a cylindrical shank, and ordinary drills with a diameter of 15 mm that were straight with a centering spike and cutters, non-uniform, and had a cylindrical shank. Drills with a diameter of 8 mm were used, including screw drills with a centering spike and cutters that were ordinary, non-uniform, single, two-edged, and single-stage, with a cylindrical shank, For drilling confirmats screws, drills with a diameter of 5 mm were used, including screw drills with a centering spike and cutters that were ordinary, uniform, single, two-point, single-stage, and had a cylindrical shank, as well as drills with 7 mm diameter screws with a centering spike that were uniform, single, two-point, and single-stage, with a cylindrical, ordinary shank. The drills were provided by ITA Tools Sp. z o.o. (Cracow, Poland). A CNC drilling and milling machine was used for testing Pass-through CNC creator 950 by Felder Group (Hall in Tirol, Austria) with a feed of 2 m/min, and revolutions of 6000 rpm. The energy consumption was measured using a portable power quality analyzer PQ-box 150 (manufacturer: A:Eberle GmbH & Co. KG, Nürnberg, Germany). This type of quality meter was used for the first time with good results for this purpose. So, this is a documentation of the new and effective method for measuring energy consumption during individual operations in the processing of wood-based materials.

2.6. Tests Performed

For each of the tested joints, after preheating the machine, the first stage involved performing so-called zero tests. Their aim was to determine the amount of energy used to operate the machine during the machining program (e.g., drive and movement of the drill, material movement, or computer operation) without the energy used during actual cutting. This was achieved by "tricking" the machine by using a smaller workpiece so that the machine executed the correct machining program, but the tool had no contact with the material. Three zero tests were performed for each type of connection as Test 0, and the average of these measurements, denoted *E0*, was used for calculations. The next stage involved conducting actual tests, where the same program ran, but this time, the tool drilled holes in the material. Before running the CNC program each time, the energy consumption was measured. The recording stopped shortly after the machine finished executing the program. The energy analyzer determined the amount of energy consumed by the machine during operation. For every joint, three tests were run (j = 1, 2, 3). A sample chart showing the total energy consumed obtained from the analyzer is presented in Figure 4.

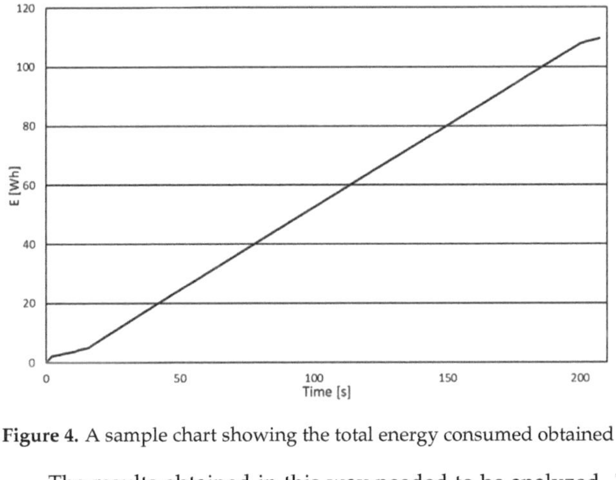

Figure 4. A sample chart showing the total energy consumed obtained from the analyzer.

The results obtained in this way needed to be analyzed. The energy consumption curve can be divided into three stages depending on the machine's operation and the power quality analyzer's activity. The first stage was the time between starting the recording on the analyzer and starting the program on the machine. The second stage involved the actual execution of the machining program, while the last stage was the time between completing the program execution and stopping the recording in the energy analyzer. The ranges of each stage are presented in Figure 5

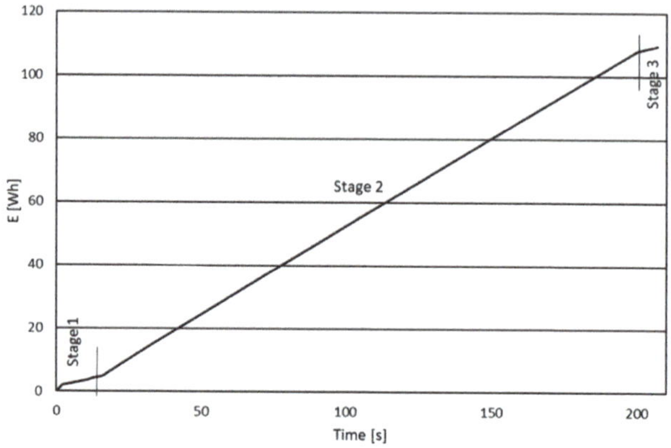

Figure 5. Range of each step of the energy consumption curve.

Key to this analysis is stage 2, during which the machining of the workpiece was carried out. Based on the results obtained from the power quality analyzer, charts were created only for stage 2. The results from the analyzer were converted by creating charts for stage 2, in which the energy determined by the analyzer was reduced by the value occurring at the transition from stage 1 to stage 2. For each connection, stage 2 for both the zero tests and the actual tests had to last the same amount of time, namely 184 s for the dowel, 220 s for the confirmat screws, and 280 s for the eccentric joints. Sample results for the dowel connection after excluding stage 2 are presented in Figure 6.

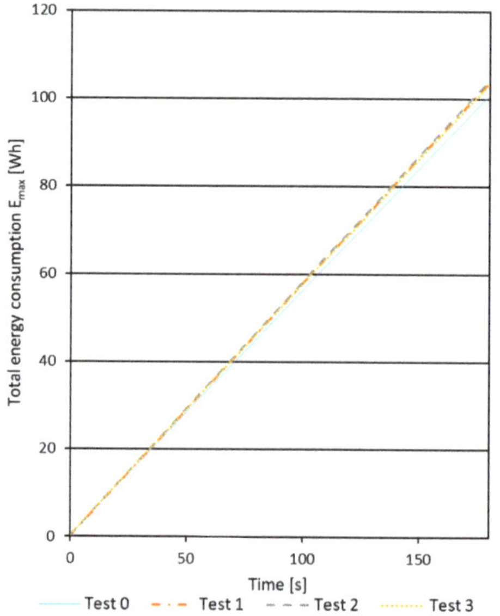

Figure 6. Sample results for the dowel joint after excluding stage 2.

The results obtained in this way could be used to calculate the actual cutting energy during the connection and the total energy required to make the joint. The total energy consumption to make the joint can be calculated from Equation (2):

$$E_p = \frac{E_{max}}{30} \quad (2)$$

E_p—total energy consumption to make the joint [Wh].

E_{max}—total energy consumption for executing the machining program for all holes to make a joint.

The specific cutting work (SCW) was calculated based on Equation (3):

$$SCW = \frac{E_{jmax} - E_{0max}}{\sum_{n=1}^{i} V_n} \quad (3)$$

SCW—specifying cutting work, J/mm^3.

E_{jmax}—total energy consumption during joint execution.

E_{0max}—total energy consumption during the zero test.

V_n—the volume of holes needed to make the connection.

i—number of holes depending on the connection.

j—test number.

Because three tests were performed on each joint, arithmetic averages for these three test results of the SCW were calculated. A Student's T-test was performed, and the standard deviation was determined. The data were collected in Tables 1–6.

Table 1. The energy consumption for dowels.

	Test 0	Test 1 (j = 1)	Test 2 (j = 2)	Test 3 (j = 3)	Average of Tests	Standard Deviation
		Wh				
Total energy consumption E_{max}	102.950	106.150	106.659	105.551	106.120	0.55
The energy consumption needed to drill holes for 30 furniture connections E_{jmax}–E_{0max}	-	3.199	3.707	2.6	3.169	0.55
The energy consumption needed to drill holes for one furniture connection E_p	-	3.538	3.555	3.518	3.537	0.02

Table 2. The energy consumption for eccentric joints.

	Test 0	Test 1 (j = 1)	Test 2 (j = 2)	Test 3 (j = 3)	Average of Tests	Standard Deviation
		Wh				
Total energy consumption E_{max}	160.964	166.779	165.063	166.682	166.174	0.96
The energy consumption needed to drill holes for 30 furniture connections E_{jmax}–E_{0max}	-	5.815	4.099	5.718	5.210	0.96
The energy consumption to drill holes for one furniture joint E_p	-	5.558	5.501	5.555	5.538	0.03

Table 3. The energy consumption confirmat screws.

	Test 0	Test 1 (j = 1)	Test 2 (j = 2)	Test 3 (j = 3)	Average of Tests	Standard Deviation
		Wh				
Total energy consumption E_{max}	93.305	97.598	96.424	97.370	97.131	0.62
The energy consumption needed to drill holes for 30 furniture connections E_{1max}–E_{0max}	-	4.292	3.119	4.064	3.825	0.62
The energy consumption needed to drill holes for one furniture connection E_p	-	3.253	3.214	3.246	3.238	0.02

Table 4. Summary of energy consumption for one selected furniture connection E_p.

	Test 1	Test 2	Test 3	Average of Tests	Standard Deviation
		Wh			
Dowels	3.538	3.555	3.518	3.537	0.02
Eccentric joints	5.558	5.501	5.555	5.538	0.03
Confirmat screws	3.253	3.214	3.246	3.238	0.02

Table 5. Volume for each furniture joint.

	V^1	V^2	V^3	Total V
		mm^3		
Dowels	377.00	1281.77	-	1658.77
Eccentric joints	225.80	829.37	2244.26	3299.43
Confirmat screws	1079.91	662.68	-	1742.59

Table 6. Specifying cutting work.

	Test 1	Test 2	Test 3	Average
		J/mm^3		
Dowels	230.05	266.46	186.89	227.89
Eccentric joints	210.63	148.51	207.31	190.63
Confirmat screws	295.54	212.90	278.90	261.68

3. Results and Discussion

3.1. Energy Consumption

The experimental data of energy measurements and calculated results of energy consumption were collected in Tables 1–4.

3.2. Discussion

The average energy consumption needed to drill holes for 30 dowels (E_{1max}–E_{0max}) is 3.169 Wh, and the energy consumption needed to drill holes for one furniture connection

(E_p) is 3.537 Wh, as shown in Table 1. The average energy consumption needed to drill holes for 30 eccentric joints (E_{1max}–E_{0max}) is 5.210 Wh, as shown in Table 2. The energy consumption needed to drill holes for one eccentric joint (E_p) was, on average, 5.538 Wh, which is 0.328 Wh more than the energy consumed for drilling one dowel. The energy consumption needed to drill holes for 30 furniture connections (E_{1max}–E_{0max}) is 3.825 Wh, while it was 3.238 Wh for drilling for one connection, as shown in Table 3. Energy data for all drilling tests were calculated for one furniture joint of each type, and their averages were also summarized. Comparing the average energy consumption needed for drilling holes for one furniture joint (E_p), the least energy is consumed while drilling for a confirmat screw (3.238 Wh), and the most energy was consumed while drilling for an eccentric joint (5.538 Wh). The difference between average energy consumption for dowels and confirmat screws and between eccentric joints and other selected furniture joints is smaller because only two holes are needed for dowels and confirmat screws and three are required for eccentric joints. A summary of energy consumption for one selected furniture connection (E_p) is shown in Table 4. A Student's t-test was performed, and the result was statistically significant. If there were no true difference between the groups, there would be very little chance of producing such a significant difference. As a result, we reject the null hypothesis and support the alternative hypothesis, which implies that the mean values of dowels and confirmat screws differ from one another. Because of the statistical significance of this difference, we may be certain that the results represent genuine group differences rather than just random fluctuations in the data. The standard deviation for one selected furniture connection for dowels and confirmat screws is 0.02; for eccentric joints, this figure is 0.03.

The statistical analysis conducted through t-tests offered valuable insights into the energy consumption disparities among the examined furniture joints. Firstly, the comparison between eccentric joints and confirmat screws revealed a statistically significant difference (p-value = 0.003958847), underscoring the distinct energy utilization patterns of these joint types. Additionally, the t-test conducted between dowels and eccentric joints also demonstrated a statistically significant disparity (p-value = 0.01477), highlighting differing energy consumption profiles between these two types of joints. Conversely, the t-test between dowels and confirmat screws yielded a non-significant result (p-value = 0.6490413), indicating a lack of compelling evidence to suggest a substantial discrepancy in energy consumption between these joint variants. These statistical findings complement the experimental measurements, shedding light on the relative energy efficiency of different furniture connectors and supporting informed decision-making in eco-design and sustainable manufacturing endeavors.

3.3. Specific Cutting Work

The data for calculation and calculated results of specific cutting work were collected in Tables 5 and 6.

The volume of the selected material for drilling holes for one furniture joint was 1658.77 mm^3 for the dowel, 3299.43 mm^3 for the eccentric joints, and 261.68 mm^3 for the confirmat screws, as shown in Table 5. Specific cutting work for one selected piece of furniture was 227.89 J/mm^3 for dowels, 190.63 J/mm^3 for eccentric joints, and 26.68 J/mm^3 for confirmat screws, as shown in Table 6. According to the physics of cutting, the confirmat screws have the smallest hole diameter and a large depth, which makes it most difficult for chips to escape during cutting, compared to a dowel or eccentric joint. It also has the highest specification cutting work rate.

The conducted research unveils a pioneering method for precisely determining specific cutting work. Leveraging the PQ-box 150 energy analyzer, a direct measurement of energy consumption during wood material cutting was conducted. By integrating the outcomes of these measurements with the volume of the resulting hole, the actual cutting work was calculated. In contrast to previous approaches reliant on measuring cutting forces to calculate cutting work, this methodology represents a significant advancement. Previous studies primarily employed rudimentary knives featuring a single cutting surface [21–25].

However, this novel approach enables the determination of appropriate cutting work across various machining tools.

4. Conclusions

The above data show that connecting furniture with dowels consumes the least energy but is the least user-friendly as it does not allow for disassembly and reassembly or easy transport. Despite higher energy consumption than dowels, eccentric joints allow furniture to be transported more conveniently and easily than dowels and self-assembly. Unfortunately, each time the furniture is unfolded again, it loses its strength due to the tearing of the material caused by the removal of the confirmat screw. The biggest energy consumption of drilling holes for one furniture connection, and at the same time the most durable and best in terms of technology and use, is for the eccentric joint, which allows the user to assemble the furniture independently and easily disassemble and transport it, without compromising the durability and strength of the furniture. Unfortunately, preparing the best furniture requires more energy due to the number of holes needed and their size. It can be concluded that increasing the usability of a piece of furniture increases the use of technology and, consequently, energy consumption. This study provides insightful information on the energy usage of various furniture joints, which is helpful for eco-design and sustainable manufacturing techniques. Our understanding will grow because of increased research and the constant observation of technical advancements, which will also direct the furniture sector towards more environmentally friendly production practices.

Author Contributions: Conceptualization, W.P. and T.R.; methodology, W.P. and Z.P.; software, Z.P.; validation, T.R. and K.W.; formal analysis, W.P. and B.P.; investigation, W.P. and B.P.; resources, B.P.; data curation, W.P. and Z.P.; writing—original draft preparation, B.P.; writing—review and editing, T.R.; visualization, W.P.; supervision, T.R.; project administration, T.R. All authors have read and agreed to the published version of the manuscript.

Funding: This research received no external funding.

Institutional Review Board Statement: Not applicable.

Data Availability Statement: Data is contained within the article.

Conflicts of Interest: The authors declare no conflicts of interest.

References

1. Burchart, D. Ekoprojektowanie—Holistyczne podejście do projektowania. *Probl. Ekol.* **2010**, *14*, 116–120.
2. Zhang, Q.M.; Zhang, W.M. Material Election and Ecological Design for Furniture Products Based on the Principles of Green Manufacturing. *Adv. Mater. Res.* **2013**, *610–613*, 502–506. [CrossRef]
3. Jayal, A.D.; Balaji, A.K. On a Process Modeling Framework for Sustainable Manufacturing: A Machining Perspective. In *Volume 15: Sustainable Products and Processes*; ASMEDC: Seattle, WA, USA, 2007; pp. 301–307. [CrossRef]
4. Wang, Q.; Zhang, D.; Chen, B.; Zhang, Y.; Wu, B. Energy Consumption Model for Drilling Processes Based on Cutting Force. *Appl. Sci.* **2019**, *9*, 4801. [CrossRef]
5. Gutowski, T.G.; Branham, M.S.; Dahmus, J.B.; Jones, A.J.; Thiriez, A.; Sekulic, D.P. Thermodynamic Analysis of Resources Used in Manufacturing Processes. *Environ. Sci. Technol.* **2009**, *43*, 1584–1590. [CrossRef] [PubMed]
6. Teimouri, R.; Amini, S.; Lotfi, M.; Alinaghian, M. Sustainable drilling process of 1045 steel plates regarding minimum energy consumption and desired work quality. *Int. J. Lightweight Mater. Manuf.* **2019**, *2*, 397–406. [CrossRef]
7. Rahim, E.A.; Rahim, A.A.; Ibrahim, M.R.; Mohid, Z. Experimental Investigation of Supercritical Carbon Dioxide (SCCO2) Performance as a Sustainable Cooling Technique. *Procedia CIRP* **2016**, *40*, 637–641. [CrossRef]
8. Rajemi, M.F.; Mativenga, P.T.; Aramcharoen, A. Sustainable machining: Selection of optimum turning conditions based on minimum energy considerations. *J. Clean. Prod.* **2010**, *18*, 1059–1065. [CrossRef]
9. Boothroyd, G. *Fundamentals of Metal Machining and Machine Tools*, 3rd ed.; CRC Press: Boca Raton, FL, USA, 1988.
10. Sydor, M.; Rogoziński, T.; Stuper-Szablewska, K.; Starczewski, K. The accuracy of holes drilled in the side surface of plywood. *BioResources* **2019**, *15*, 117–129. [CrossRef]
11. Sydor, M.; Majka, J.; Langová, N. Effective Diameters of Drilled Holes in Pinewood in Response to Changes in Relative Humidity. *BioResources* **2021**, *16*, 5407. [CrossRef]
12. Sydor, M.; Majka, J.; Rychlik, M.; Turbański, W. Application of 3D Scanning Method to Assess Mounting Holes' Shape Instability of Pinewood. *Materials* **2023**, *16*, 2053. [CrossRef] [PubMed]

13. Ji, M.; Xu, J.; Chen, M.; Mansori, M.E. Effects of Different Cooling Methods on the Specific Energy Consumption when Drilling CFRP/Ti6Al4V Stacks. *Procedia Manuf.* **2020**, *43*, 95–102. [CrossRef]
14. Jia, S.; Cai, W.; Liu, C.; Zhang, Z.; Bai, S.; Wang, Q.; Li, S.; Hu, L. Energy modeling and visualization analysis method of drilling processes in the manufacturing industry. *Energy* **2021**, *228*, 120567. [CrossRef]
15. Wilkowski, J.A.; Kozub, W.O.; Borysiuk, P.I.; Rousek, M.I.; Czarniak, P.A.; Gorski, J.A.; Podziewski, P.I.; Szymanowski, K.A. Machinability of particleboards bonded with SBR gum granulate. *Ann. Wars. Univ. Life Sci. SGGW For. Wood Technol.* **2014**, *85*, 191–195.
16. Wilkowski, J.; Grześkiewicz, M.; Czarniak, P.; Wojtoń, M. Cutting forces during drilling of thermally modified ash wood. *Ann. Wars. Univ. Life Sci. SGGW For. Wood Technol.* **2011**, *76*, 199–202.
17. Wilkowski, J.; Czarniak, P.; Górski, J.; Jabłoński, M.; Podziewski, P.; Szymanowski, K. Taguchi Method Optimization of the Total Power Consumption During Chipboard Cutting with the Panel Saw Machine. *Rieskové Beztrieskové Obrábanie Dreva* **2014**, *9*, 171–174.
18. Braun, J. Bezpieczeństwo energetyczne jako dobro publiczne—Miary i czynniki wpływające na jego poziom. *Stud. Ekon.* **2018**, *358*, 23–32.
19. Saloni, D.E.; Lemaster, R.L.; Jakson, S.D. Abrasive Machining Process Characterization on Material Removal Rate, Final Surface Texture, and Power Consumption for Wood—ProQuest. 2005. Available online: https://www.proquest.com/openview/a99e8550a54ea7cb1554f55652a2cc6c/1?pq-origsite=gscholar&cbl=25222 (accessed on 13 December 2023).
20. *PN–EN 310*; Wood-Based Panels–Determination of Modulus of Elasticity in Bending and of Bending Strength. Polish Committee for Standardization: Warszawa, Poland, 1994.
21. Atkins, A.G.; Vincent, J.F.V. An instrumented microtone for improved histological sections and the measurement of fracture toughness. *J. Mater. Sci. Lett.* **1984**, *3*, 310–312. [CrossRef]
22. Huang, X.; Jeronimidis, G.; Vincent, J.F.V. *The Instrumented Micro-Tome Cutting Tests on Wood from Transgenic Plants with Modified Lignification*; Spatz, H.C.H., Speck, T., Eds.; Badenweiler: Georg Thieme Verlag, 2000; pp. 475–482.
23. Kowaluk, G.; Dziurka, D.; Beer, P.; Sinn, G.; Tschegg, S. Influence of particleboards production parameters on work of fracture and work of chips formation during cutting. *Electron. J. Pol. Agric. Univ. Wood Technol.* **2004**, *7*, 1.
24. Beer, P.; Kowaluk, G.; Sinn, G.; Dziurka, D. Mechanical properties of particleboards induce cutting quality. In Proceedings of the 17th International Wood Machining Seminar, Rosenheim, Germany, 26–28 September 2005; pp. 26–28.
25. Kowaluk, G. Analyzing of Cutting Work due to Quality Machining off Chosen Laminated Chipboards. Ph.D. Thesis, University of Life Sciences in Poznań, Poznań, Poland, 2005.

MDPI AG
Grosspeteranlage 5
4052 Basel
Switzerland
Tel.: +41 61 683 77 34

Polymers Editorial Office
E-mail: polymers@mdpi.com
www.mdpi.com/journal/polymers

www.ingramcontent.com/pod-product-compliance
Lightning Source LLC
Chambersburg PA
CBHW041359210926
51850CB00013B/3125
* 9 7 8 3 7 2 5 8 4 0 1 2 0 *